Lecture Notes in Computer Science 15621

Founding Editors

Gerhard Goos
Juris Hartmanis

AF167584

The series Lecture Notes in Computer Science (LNCS), including its subseries Lecture Notes in Artificial Intelligence (LNAI) and Lecture Notes in Bioinformatics (LNBI), has established itself as a medium for the publication of new developments in computer science and information technology research, teaching, and education.

LNCS enjoys close cooperation with the computer science R & D community, the series counts many renowned academics among its volume editors and paper authors, and collaborates with prestigious societies. Its mission is to serve this international community by providing an invaluable service, mainly focused on the publication of conference and workshop proceedings and postproceedings. LNCS commenced publication in 1973.

Modesto Castrillón-Santana ·
Carlos M. Travieso-González ·
Oscar Deniz Suarez · David Freire-Obregón ·
Daniel Hernández-Sosa ·
Javier Lorenzo-Navarro · Oliverio J. Santana
Editors

Computer Analysis of Images and Patterns

21st International Conference, CAIP 2025
Las Palmas de Gran Canaria, Spain, September 22–25, 2025
Proceedings, Part I

Springer

Editors

Modesto Castrillón-Santana [ID]
University of Las Palmas de Gran Canaria
Las Palmas de Gran Canaria, Spain

Carlos M. Travieso-González [ID]
University of Las Palmas de Gran Canaria
Las Palmas de Gran Canaria, Spain

Oscar Deniz Suarez [ID]
University of Castilla-La Mancha
Ciudad Real, Spain

David Freire-Obregón [ID]
University of Las Palmas de Gran Canaria
Las Palmas de Gran Canaria, Spain

Daniel Hernández-Sosa [ID]
University of Las Palmas de Gran Canaria
Las Palmas de Gran Canaria, Spain

Javier Lorenzo-Navarro [ID]
University of Las Palmas de Gran Canaria
Las Palmas de Gran Canaria, Spain

Oliverio J. Santana [ID]
University of Las Palmas de Gran Canaria
Las Palmas de Gran Canaria, Spain

ISSN 0302-9743   ISSN 1611-3349 (electronic)
Lecture Notes in Computer Science
ISBN 978-3-032-04967-4   ISBN 978-3-032-04968-1 (eBook)
https://doi.org/10.1007/978-3-032-04968-1

Preface

CAIP 2025 marked the 21st edition of the International Conference on Computer Analysis of Images and Patterns—a well-established biennial series dedicated to cutting-edge research in computer vision, image processing, pattern recognition, and closely related fields. Over the years, CAIP has built a strong legacy, with past conferences hosted in locations such as Limassol, Salerno, Ystad, Valletta, York, Seville, Münster, Vienna, and Paris.

This year's scientific program was designed as a single-track event, fostering a cohesive and engaging atmosphere for discussion and exchange. From 109 submitted papers, each reviewed in a rigorous double-blind process by at least two experts, 65 were selected for oral presentation. These were grouped into the following nine thematic sessions:

SESSION 1: Facial and Video Recognition
SESSION 2: Image Segmentation
SESSION 3: Object Detection and Applications
SESSION 4: 3D Vision and Reconstruction
SESSION 5: Biomedical Imaging and Diagnostics
SESSION 6: Model Robustness and Generalization
SESSION 7: Multimodal and Vision-Language Models
SESSION 8: Robotics, Interaction and Intelligent Systems
SESSION 9: Emerging Methods and Vision Applications

In addition to the paper sessions, CAIP 2025 featured the Pedestrian Attributes Recognition (PAR) Contest, focusing on Multi-Task Learning—a timely challenge organized by Antonio Greco (University of Salerno) and Bruno Vento (University of Naples).

We were honored to welcome three distinguished keynote speakers, whose insights bridged academic excellence and societal impact: Anil K. Jain, Jacques Bulchand-Gidumal, and Nadia Bianchi-Berthouze.

We extend our heartfelt thanks to the organizing and technical program committees, and the many reviewers whose careful work ensured the quality and success of this conference. Our sincere appreciation goes to all authors who entrusted us with their research, and to all participants for enriching CAIP 2025 with your presence, ideas, and engagement.

We would like to express our gratitude for the sponsorship of the University of Las Palmas de Gran Canaria, the City Council of Las Palmas de Gran Canaria, the Elder Museum of Science and Technology, and the Gran Canaria Convention Bureau.

A very special thank you goes to AVANTE Canarias, and in particular to Marta Ortega and Raquel Granados, for their exceptional support, attention to detail, and tireless efforts behind the scenes to ensure everything ran smoothly.

September 2025

Modesto Castrillón-Santana
Carlos M. Travieso-González
Oscar Deniz Suarez
David Freire-Obregón
Daniel Hernández-Sosa
Javier Lorenzo-Navarro
Oliverio J. Santana

Organization

Honorary Chair

Nicolai Petkov····University of Groningen, Netherlands

General Chairs

Modesto Castrillón-Santana····University of Las Palmas de Gran Canaria, Spain
Carlos M. Travieso-González····University of Las Palmas de Gran Canaria, Spain

Program Chairs

Oscar Deniz Suarez····University of Castilla-La Mancha, Spain
David Freire-Obregón····University of Las Palmas de Gran Canaria, Spain
Daniel Hernández-Sosa····University of Las Palmas de Gran Canaria, Spain
Javier Lorenzo-Navarro····University of Las Palmas de Gran Canaria, Spain
Oliverio J. Santana····University of Las Palmas de Gran Canaria, Spain

Publicity Committee

Aythami Morales Moreno····Autonomous University of Madrid, Spain
Juan Tapia Farias····Hochschule Darmstadt University of Applied Sciences, Germany
Hazım Kemal Ekenel····Istanbul Technical University, Turkey
Enrique Alegre····University of León, Spain
Dorota Kamińska····Łódź University of Technology, Poland
Bruno Vento····University of Napoli, Italy
Sondos Mohamed····University of Cagliari, Italy

Steering Committee

Andreas Lanitis (Co-chair CAIP 2023)
Constantinos S. Pattichis (Co-chair CAIP 2021)
Mario Vento (Chair CAIP 2019)

Michael Felsberg (Chair CAIP 2017)
Nicolai Petkov (Permanent Member)

Program Committee

Ahmad Alsahaf	University Medical Center Groningen, Netherlands
Alain Tremeau	Jean Monnet University, France
Albert Ali Salah	Utrecht University, Netherlands
Alberto Marchisio	New York University Abu Dhabi, UAE
Alessia Saggese	University of Salerno, Italy
André P. Kelm	University of Hamburg, Germany
Andreas Lanitis	Cyprus University of Technology, Cyprus
Antonio Greco	University of Salerno, Italy
Aythami Morales	Universidad Autónoma de Madrid, Spain
Bastian Leibe	RWTH Aachen University, Germany
Bhavin Jawade	University at Buffalo, USA
Carmen Bisogni	Università degli Studi di Salerno, Italy
Christian Rathgeb	Hochschule Darmstadt, Germany
Cosimo Distante	National Research Council of Italy - Institute of Applied Sciences & Intelligent Systems, Italy
Cristina Carmona-Duarte	University of Las Palmas de Gran Canaria, Spain
Cristobal Curio	Reutlingen University, Germany
Daniel Cores	University of Santiago de Compostela, Spain
Daniel Riccio	University of Naples Federico II, Italy
Daniel Santana-Cedrés	University of Las Palmas de Gran Canaria, Spain
Delia A. Mitrea	Technical University of Cluj-Napoca, Romania
Di Huang	Beihang University, China
Dibio Leandro Borges	University of Brasília, Brazil
Elena Lazkano	University of the Basque Country, Spain
Eleni A. Dimitriadou	Cyprus University of Technology & CYENS, Cyprus
Fabio Narducci	University of Salerno, Italy
Fernando Alonso-Fernandez	Halmstad University, Sweden
Fernando C. Monteiro	Polytechnic Institute of Bragança, Portugal
Francesco Longobardi	University of Naples Federico II, Italy
Francisco M. Castro	University of Málaga, Spain
Gennaro Percannella	University of Salerno, Italy
Georg Stemmer	Intel Labs, Germany
George Azzopardi	University of Groningen, Netherlands
Giorgio Fumera	University of Cagliari, Italy

Contents – Part I

Object Detection and Applications

3D Vision and Reconstruction

Contents – Part II

Model Robustness and Generalization

Multimodal and Vision-Language Models

Robotics, Interaction and Intelligent Systems

Emerging Methods and Vision Applications

PAR Contest 2025

An Extended Dataset and a Baseline for Pedestrian Attribute Recognition with Advanced Neural Networks

Antonio Greco[1] and Bruno Vento[2(✉)]

[1] Department of Information and Electrical Engineering and Applied Mathematics (DIEM),University of Salerno, Via Giovanni Paolo II, 132, Fisciano 84084, SA, Italy
agreco@unisa.it
[2] Department of Electrical Engineering and Information Technology (DIETI), University of Naples, Via Claudio, 21, Naples 80135, NA, Italy
bruno.vento@unina.it

Abstract. Recent works and challenges on pedestrian attribute recognition demonstrated the necessity to collect extensive and representative datasets and to propose effective and efficient methods based on advanced neural networks. Following on the success of the first edition, the Pedestrian Attribute Recognition (PAR) 2025 Contest, organized within CAIP 2025, is an international competition designed to evaluate advanced neural networks for recognizing pedestrian attributes. Participants are provided with the new Mivia PAR KD Dataset, which includes 106,743 newly annotated images featuring a combination of labels and pseudo-labels obtained through a knowledge distillation method for attributes such as clothing color gender, and the presence or absence of accessories like bags and hats. Competing approaches have been assessed based on the mean accuracy metric, using a separate private test set comprising over 20,000 images, distinct from the training data and not provided to any participants, in order to ensure fairness in the evaluation of results. The goal was to push the participants to effectively leverage the latest advancements in neural network technologies and enhance the scalability and real-world applicability of PAR solutions. The contest teams advanced the state of the art by proposing approaches improving computational efficiency, reducing training time, better addressing class imbalance, and incorporating effective learning procedures and data augmentation strategies. The impressive 95.4% mean accuracy obtained by the winner confirms the achievement of the goal.

Keywords: Contest · Pedestrian Attribute Recognition · PAR

1 Introduction

The pedestrian attribute recognition task from cropped person images [36] has emerged as an increasingly important research topic, with applications spanning various real-world domains such as personalized recommendations [19], social

M. Castrillón-Santana et al. (Eds.): CAIP 2025, LNCS 15621, pp. 3–15, 2026.
https://doi.org/10.1007/978-3-032-04968-1_1

robotics [16] and multi-camera person re-identification [15]. In this context, considerable interest has been directed toward developing systems capable of simultaneously identifying multiple attributes of a pedestrian including clothing color [25], gender classification [13] and the presence or absence of accessories like bags and hats [24]. Accurate recognition of these diverse attributes not only enhances downstream tasks such as behavior analysis [18], people tracking [5] and surveillance [20], but also contributes to improve the robustness and adaptability of artificial intelligence systems operating in dynamic environments. Following the success of the first edition hosted at CAIP 2023 [21], the Pedestrian Attribute Recognition (PAR) 2025 Contest returns as a leading international competition dedicated to advancing the state of the art in pedestrian attribute recognition from images. In this second edition, participants will work with the updated MIVIA PAR KD Dataset, which features a newly enriched collection of images annotated with a combination of labels and pseudo-labels derived through knowledge distillation. These annotations cover attributes such as clothing color, gender and the presence or absence of accessories like bags and hats, supporting the training, validation and benchmarking of the proposed methods. The goal is to provide the participants with one of the most extensive, innovative and diverse datasets for pedestrian attribute recognition. In fact, despite the lack of a standard training set, recently a wide range of methods have been proposed to address the challenges of PAR, balancing both accuracy and computational efficiency. Notably, in the 2023 edition, the winning method successfully integrated Visual Question Answering (VQA) techniques with Large Language Models (LLMs), achieving an impressive 92% accuracy on the contest's private test set. This remarkable result highlighted the significant potential of Vision-Language Models (VLMs) in tackling these complex PAR tasks. Given the rapid advancements in VLM technology over the past two years, it is expected that many participants in this second edition will leverage these cutting-edge solutions. Nevertheless, the competition remains open to all innovative approaches and every original contribution is welcomed and highly valued, further advancing progress in this dynamic research field. To ensure fairness and realistic performance evaluation, the competing methods are assessed based on mean accuracy, using a private challenging test set composed of images different from those available in the training data and not provided to any participant. This test set will be exactly the same as the one adopted for PAR 2023, since the goal of the competition is to enable a direct comparison with the results of the previous edition and providing a benchmark for measuring progress against the baseline established two years ago.

2   Related Works

A wide range of publicly available datasets have played a foundational role in the development of PAR methods, each offering different sets of attributes, annotation strategies, and environmental conditions. The number of pedestrian attributes varies significantly across publicly available datasets, ranging from 5

Fig. 1. Example of the results produced by a pedestrian attribute recognition system for classifying upper and lower clothing color, gender, bag and hat.

to 69 binary and multi-class attributes. Several key datasets have contributed to the development of pedestrian attribute recognition (PAR) methods, including BAP [3], HAT [32], CAD [7], APiS [38], CRP [22], PARSE-27K [34], PETA [10], PA-100K [30], Market-1501 [29], DukeMTMC [29] and RAP v2.0 [28], as shown in Table 1. However, these datasets often present several limitations, including small sample sizes, acquisition restricted to either indoor or outdoor environments (but rarely both) and the presence of unreliable or ambiguous annotations. For example, BAP [3] was created by cropping samples from H3D [4] and PASCAL VOC 2010 [11], resulting in 2,003 images annotated with 9 binary attributes. HAT [32] includes 9,344 Flickr images labeled with 27 attributes related to pose, age and clothing. CAD [7] contains 1,856 Flickr images annotated with clothing attributes based on consensus among multiple annotators. APiS [38] aggregates pedestrian samples from various sources, annotated with both binary and multi-class attributes. Although CRP [22], PARSE-27K [34], PETA [10] and PA-100K [30] offer larger collections, they continue to face challenges related to annotation consistency and environmental diversity. Datasets such as Market-1501 [29] and DukeMTMC [29] are primarily designed for person re-identification tasks, providing attribute annotations only at the subject level. RAP v2.0 [28], instead, delivers a more comprehensive set of annotations collected across varied environments. The MIVIA PAR dataset was introduced during the PAR 2023 contest [21], marking a significant advancement in the field in terms of amount and variability of the samples. This dataset is larger than most existing collections, acquired in both indoor and outdoor environments and features reliable annotations verified by multiple expert annotators. Its variability and extensibility make it a valuable resource for investigating advanced learning strategies, attention mechanisms and methods for handling missing labels and unbalanced data distributions. However, in the dataset only

26,076 samples are fully annotated. The MIVIA PAR KD dataset, released for the PAR 2025 contest, further extends this resource by providing additional samples (106,743) and full annotations (through knowledge distillation) for all the images. This dataset allows the participants to train advanced neural network with a significant amount of representative samples for all the attributes of interest.

Table 1. Existing datasets for PAR. Datasets indicated with * are partially or ambiguously annotated. The proposed is the largest dataset with fully annotated binary and multiclass attributes.

Dataset	Pedestrians	Attributes
BAP [3]	8,035	9 binary
HAT [32]	9,344	11 binary - 2 multiclass
CAD [7]	1,856	23 binary - 3 multiclass
APiS [38]	3,661	11 binary - 2 multiclass
CRP* [22]	19,000	61 binary - 4 multiclass
PARSE-27K* [34]	27,000	8 binary - 2 multiclass
PETA* [10]	19,000	61 binary - 4 multiclass
PA-100K* [30]	100,000	26 binary
Market-1501 [29]	32,668	14 binary
DukeMTMC [29]	34,183	23 binary
RAP v2.0 [28]	84,298	69 binary - 3 multiclass
MIVIA PAR* [21]	105,244	3 binary - 2 multiclass
MIVIA PAR KD	106,743	3 binary - 2 multiclass

Despite the introduction of these new datasets, pedestrian attribute recognition remains a highly challenging task. The complexity of this task is further compounded in real-world video surveillance applications, where embedded systems typically operate under limited computational resources [14]. Early global-based methods, such as ACN [35], DeepSAR [27] and MTCNN [1], extracted features from entire images, prioritizing simplicity and computational efficiency. However, these approaches struggled with attributes requiring localized feature extraction, such as hats or bags. To overcome these limitations, part-based methods were proposed, focusing on extracting features from specific body regions [33]. More recently, attention-based mechanisms have been adopted to effectively combine the advantages of global and part-based approaches. Models such as MTANET [23] dynamically focus on the most relevant image regions without requiring explicit part localization. Additionally, transformer-based architectures like PARFormer [12] have further enhanced this paradigm by jointly addressing viewpoint and attribute classification tasks. To support broader human-centered perception tasks, including pose estimation, skeleton detection and attribute recognition, general-purpose frameworks have also emerged. Notable examples include UNIHCP [8] and HULK [37], both of which employ transformers to query human attributes from images. While these models demonstrate very impressive performance, their significant computational demands currently limit their deployment in real-time, resource-constrained embedded applications.

Recognizing the growing importance of pedestrian attribute recognition, the PAR 2025 contest has been organized to evaluate and promote the development of both effective and efficient approaches. Unlike other challenges such as UPAR [9], which primarily address binary attribute tasks, PAR 2025 introduces a more complex scenario involving three binary tasks (gender, bag presence and hat presence) alongside two multi-class tasks (upper and lower clothing color). As in PAR 2023 [17], the training dataset for PAR 2025 combines both public and unpublished samples, while the private test set is composed exclusively of unseen images, ensuring that the proposed solutions are evaluated on entirely novel data.

3   Contest Dataset and Task

For this contest, we provide participants with the MIVIA PAR KD Dataset, which includes annotations for five pedestrian attributes. The dataset consists of 106,743 images, each labeled with the following attributes:

– Clothing Color: The possible values for clothing colors are black, blue, brown, gray, green, orange, pink, purple, red, white and yellow, represented by the labels [1, 2, 3, 4, 5, 6, 7, 8, 9, 10, 11]. We provide separate labels for the upper and lower body clothing colors.
– Gender: The gender values are male and female, represented by [0, 1].
– Bag: Indicates the presence or absence of a bag, represented by [0, 1].
– Hat: Indicates the presence or absence of a hat, represented by [0, 1].

MIVIA PAR KD has been compiled from multiple sources, including existing datasets and private images, where pedestrians were manually extracted and labeled. Due to varying collection conditions, the dataset is heterogeneous in terms of image size, lighting pose and distance from the camera. Each image contains a single cropped person, as shown in Fig. 2. Participants will receive a folder containing the images and a CSV file with the labels for the training and validation samples. Unlike the previous edition, all the training samples in this dataset are now fully annotated, eliminating any missing labels through the use of knowledge distillation. The method employed to generate the missing labels is based on the winner of the PAR 2023 contest [6]. Additionally, new samples have been added to the training set to further assist participants in building robust models.

It is important to note that the dataset is unbalanced, so participants should consider implementing learning procedures that can effectively address this imbalance. Specifically, the training set includes the following distribution for upper body color: black (34,223), blue (14,222), white (12,002), gray (9,854), red (7,814), green (4,667) and brown (2,872). For the lower body, there is a clear bias,

Fig. 2. Examples of images from the MIVIA PAR KD dataset, showcasing the available attributes: upper color, lower color, gender, bag and hat.

with a predominance of black (52,861), blue (22,431) and gray (12,007) samples. This imbalance could also reflect a real-world distribution bias, as light-colored pants are less commonly worn. Additionally, the training set contains a majority of male samples (67,436), samples with no hat (74,352) and samples with no bag (62,455).

Participants wishing to enter the competition can request access to the training and validation samples of the MIVIA PAR KD dataset, along with their corresponding annotations, by following the procedure outlined on the contest website[1]. Comprehensive guidelines are also provided to help participants implement the necessary code for generating predictions across all specified pedestrian attributes. Competitors are encouraged to innovate by developing novel neural network architectures, proposing original training procedures and experimenting with custom loss functions (for example, to address data imbalance issues). Once submitted, the developed methods will be evaluated on a separate private test set. Performance will be assessed according to predefined criteria and a final ranking will be determined based on the rules outlined in the following section.

4   Evaluation Metrics

The proposed methods will be evaluated based on the average accuracy achieved across all tasks. The accuracy A is defined as the ratio between the number of correct predictions (i.e., when the prediction p_i matches the ground truth g_i) and the total number of samples K:

$$A = \frac{\sum_{i=1}^{K} (p_i == g_i)}{K} \tag{1}$$

Accuracy is computed separately for each of the five attributes as follows:

---

[1] https://mivia.unisa.it/par2025/.

- A_u: accuracy in recognizing the color of the upper-body clothing.
- A_l: accuracy in recognizing the color of the lower-body clothing.
- A_g: accuracy in recognizing the gender.
- A_b: accuracy in recognizing the presence of a bag.
- A_h: accuracy in recognizing the presence of a hat.

The higher the accuracy achieved by a method for a given task, the more effective it is in recognizing that specific pedestrian attribute. The final contest ranking is determined by the Mean Accuracy (mA), calculated as the average of the individual accuracies listed above:

$$mA = \frac{A_u + A_l + A_g + A_b + A_h}{5} \tag{2}$$

The method achieving the highest mA will be declared the winner of the PAR 2025 Contest, as it demonstrates the best overall performance across all tasks.

5 Contest Teams

This section provides a comprehensive overview of the methods submitted to the PAR 2025 Contest, emphasizing their core architectures, training approaches and key innovations. The teams and their methods, described in the following, are ordered alphabetically.

ARRAY [26] proposed a comprehensive multi-task classification pipeline for pedestrian images, exploiting a modern ConvNeXt Large backbone with 224×224 pixels input size and five classification heads, each responsible for predicting one of the visual attributes. The learning procedure, performed for each head, includes data augmentation (MixUp and CutMix), label smoothing applied to the CrossEntropy loss, an AdamW optimizer with weight decay and a Cosine Annealing learning rate scheduler.

CVPD adopts a pre-trained SigLIP VLM extended with five task-specific classification heads. The VLMPAR system was trained using the MIVIA PAR KD 2025 dataset. Images were preprocessed with the proprietary SigLIP2 image processor to ensure consistent resizing to 224×224 pixels and normalization. The training strategy uses mixed precision for efficiency, an AdamW optimizer with a linear learning rate scheduler and a high number of epochs with gradient accumulation. The inference benefits from GPU optimization and no-gradient mode, emphasizing resource efficiency and faster prediction.

FAST-NUCES developed a multi-task CNN based on ResNet50, where the original fully connected layer was replaced with an identity mapping to produce a 2048-dimensional feature vector. For each attribute, a dedicated linear classifier is added. Input images are loaded, resized to 256×128 pixels, converted into tensors and normalized. During training, batches of images and their labels are fed through the model, which generates five separate sets of logits per sample, one for each attribute. Each attribute's prediction error is computed using an appropriate loss function and the total loss is obtained by summing these individual losses.

The IMSLAB team [2] integrated a pre-trained CLIP ViT-B/32 with Graph Neural Networks (GNNs) to address both visual complexity and semantic dependencies among attributes. They extract features from two views of each pedestrian image: a full-body view that preserves global context for detecting attributes such as gender or the presence of accessories and a lower-body crop that focuses on the bottom half to more precisely capture features like trouser color. Simultaneously, attribute-specific textual prompts–such as "A pedestrian wearing a red shirt" or "A female pedestrian with a hat" are encoded by the CLIP text encoder to generate semantic text embeddings. Classification is achieved by computing cosine similarity scores between visual and textual embeddings for each attribute across both views. These similarity scores are then concatenated and fed into the GNN to explicitly model inter-attribute relationships through message passing, enhancing contextual representation. The training is conducted over 20 epochs with early stopping after 10 epochs without improvement and the AdamW optimizer with a batch size of 32.

The INVESTIGAI team employed a task-specific Visual Question Answering (VQA) framework by exploiting the pre-trained BLIP model. Each pedestrian image is paired with a fixed set of predefined natural language questions targeting specific semantic attributes. To facilitate consistent evaluation and integration, textual outputs are mapped to numeric codes representing attribute classes. A robust fallback mechanism addresses out-of-vocabulary or unexpected answers by defaulting to majority-class labels.

The iROC-ULPGC team [31] propose a new method, inspired by the approach that won the PAR 2023 contest. The algorithm has been improved in terms of the number and type of queries done to the vision language models (VLMs), the adoption of different VLMs (BLIP-2, Paligemma1 and Paligemma 2) and the introduction of an ensemble of these VLM-based classifiers. Unlike conventional ensemble techniques that combine model outputs indiscriminately, the method adopts a decision-driven selective fusion strategy, where each model is independently queried with carefully engineered, attribute-specific natural language prompts that align with the model's respective strengths. For every image-prompt pair, the models perform conditional text generation, producing answers that are then normalized to standardize synonyms and spelling variations, thus maintaining consistent terminology across predictions.

The SCLAB method is a custom adaptation of the pre-trained ViT-B16 architecture, called PARViTModel, with five classification heads and an input size equal to 224×224. Final class predictions for each attribute are determined by selecting the highest-scoring class as the predicted output.

The SIG2PAR team adopts the Siglip2VisionModel, whose final five layers of the visual encoder are fine-tuned with the five classification heads, enhancing its ability to capture cues relevant to pedestrian attributes while preserving the stability of earlier layers. The heads are fully connected neural networks whose architecture is tailored to the complexity of the classification task. For multi-class tasks such as upper and lower clothing color recognition, deeper heads incorporating SEBlock modules (Squeeze-and-Excitation) are employed to recalibrate channel-wise feature responses and improve the model's focus on attribute-relevant information. Simpler heads are used for binary classification tasks. Each task head typically consists of a sequence of linear layers, interleaved with GELU activations, dropout layers for regularization and layer normalization, optimized with an asymmetric loss function.

6   Contest Results

The comparative evaluation highlights significant differences in accuracy and consistency across teams, reflecting the architectural and training choices of their models. IROC-ULPGC's leading performance with a mean accuracy of 0.954 and the lowest variance 0.028 indicates a robust multi-task learning framework that likely benefits from both a powerful vision backbone and well-optimized task-specific heads. Their ability to maintain excellent accuracy across all attributes, including the notoriously challenging bag detection, suggests an architecture that effectively balances shared feature extraction with task-specific specialization. SIG2PAR, closely trailing with a mean accuracy of 0.940, benefits from the use of deeper heads with SE blocks for color attributes and shallower heads for binary attributes likely contributes to its strong performance on clothing colors and gender classification. However, the relative drop in bag detection accuracy 0.889 compared to IROC-ULPGC reveals a limitation intrinsic to its backbone strategy: while freezing the encoder stabilizes the feature representation, it may constrain adaptability for subtle accessory-related cues. SCLab and INVESTI-GAI demonstrate moderate overall accuracy around 0.910 with similar variance profiles, indicating architectures potentially less specialized or with less capacity dedicated to accessory attributes. Their lower bag and hat detection accuracies may stem from limited architectural attention to these smaller and less salient objects or less effective integration of spatial and semantic cues necessary to disambiguate such features. FAST-NUCES achieves solid accuracy on clothing colors and gender but lags behind ROC-ULPGC, SIG2PAR, and SCLab, which all show higher mean accuracy and better accessory detection. Its accessory performance, especially on bags, notably trails these top methods. CVPD team shows a mean accuracy of 0.756 with a high variance, reflecting uneven performance. The method demonstrates strong results on gender and clothing color attributes (Au=0.786, Al=0.853, Ag=0.834) and a pronounced drop in bag presence detection accuracy 0.466 signals. The high variance underscores that VLMPAR's multi-task learning setup might benefit from improved loss balancing to mitigate underfitting on complex attributes. IMSLAB and ARRAY's lower accuracies and higher variances reveal that their solutions either lack sufficiently expressive visual backbones or effective multi-task integration mechanisms.

Table 2. Final results of the PAR 2025 contest. The methods are ranked based on their mA, with the best-performing approach listed first. The highest values in each column are highlighted in bold.

Team	A_u	A_l	A_g	A_b	A_h	mA	$Variance$
IROC-ULPGC	**0.927**	**0.913**	**0.996**	**0.961**	0.975	**0.954**	**0.028**
SIG2PAR	0.925	0.911	**0.996**	0.889	**0.976**	0.940	0.037
SCLab	0.906	0.904	0.948	0.856	0.934	0.910	0.029
INVESTIGAI	0.910	0.905	0.947	0.826	0.937	0.905	0.039
FAST_NUCES	0.779	0.866	0.939	0.708	0.892	0.837	0.076
CVPD	0.786	0.853	0.834	0.466	0.842	0.756	0.134
IMSLAB	0.805	0.813	0.792	0.422	0.661	0.698	0.136
ARRAY	0.386	0.483	0.815	0.871	0.807	0.672	0.181

7   Conclusion

The PAR 2025 contest aims to foster innovation in pedestrian attribute recognition by providing participants with the MIVIA PAR KD dataset, the largest fully annotated dataset with the attributes of interest, featuring both labels and pseudo-labels. This comprehensive resource offers a robust foundation for developing advanced neural network solutions. By promoting fair evaluation through a private test set, the competition is positioned to drive meaningful progress in the field and contribute significantly to the broader research community. Moreover, while the recent literature showcased a variety of innovative approaches, PAR 2025 opened new opportunities for improvement. The enhancements demonstrated by the proposed methods in terms of computational efficiency, training time and learning procedures advance the scalability and real-world applicability of PAR systems, enabling the deployment of models that are both effective and resource-efficient across diverse real scenarios.

Acknowledgements. The research activities behind the organization of the contest have been partially supported by A.I. Tech srl (SA), Italy.

References

1. Abdulnabi, A.H., Wang, G., Lu, J., Jia, K.: Multi-task CNN model for attribute prediction. IEEE Trans. Multimedia **17**(11), 1949–1959 (2015)
2. Bhat, M.H.: A region-aware multi-modal framework for pedestrian attribute recognition via clip and graph neural networks. In: International Conference on Computer Analysis of Images and Patterns (CAIP) (2025)

3. Bourdev, L., Maji, S., Malik, J.: Describing people: A poselet-based approach to attribute classification. In: International Conference on Computer Vision, pp. 1543–1550 (2011)
4. Bourdev, L., Malik, J.: Poselets: Body part detectors trained using 3d human pose annotations. In: IEEE International Conference on Computer Vision, pp. 1365–1372 (2009)
5. Carletti, V., Greco, A., Saggese, A., Vento, M.: Multi-object tracking by flying cameras based on a forward-backward interaction. IEEE Access **6**, 43905–43919 (2018)
6. Castrillón-Santana, M., Sánchez-Nielsen, E., Freire-Obregón, D., Santana, O.J., Hernández-Sosa, D., Lorenzo-Navarro, J.: Evaluation of a visual question answering architecture for pedestrian attribute recognition. In: Computer Analysis of Images and Patterns, pp. 13–22 (2023)
7. Chen, H., Gallagher, A., Girod, B.: Describing clothing by semantic attributes. In: European Conference on Computer Vision (ECCV), pp. 609–623. Springer (2012)
8. Ci, Y., et al.: Unihcp: A unified model for human-centric perceptions. In: IEEE/CVF Conference on Computer Vision and Pattern Recognition, pp. 17840–17852 (2023)
9. Cormier, M., et al.: Upar challenge 2024: Pedestrian attribute recognition and attribute-based person retrieval-dataset, design, and results. In: IEEE/CVF WACV, pp. 359–367 (2024)
10. Deng, Y., Luo, P., Loy, C.C., Tang, X.: Pedestrian attribute recognition at far distance. In: ACM International Conference on Multimedia, pp. 789–792 (2014)
11. Everingham, M., Van Gool, L., Williams, C., Winn, J., Zisserman, A.: The pascal visual object classes challenge (2010)
12. Fan, X., Zhang, Y., Lu, Y., Wang, H.: Parformer: Transformer-based multi-task network for pedestrian attribute recognition. IEEE Trans. Circuits Syst. Video Technol. **34**(1), 411–423 (2024)
13. Foggia, P., Greco, A., Percannella, G., Vento, M., Vigilante, V.: A system for gender recognition on mobile robots. In: International Conference on Applications of Intelligent Systems, pp. 1–6 (2019)
14. Foggia, P., Greco, A., Saggese, A., Vento, M.: Multi-task learning on the edge for effective gender, age, ethnicity and emotion recognition. Eng. Appl. Artif. Intell. **118**, 105651 (2023)
15. Gou, M., Karanam, S., Liu, W., Camps, O., Radke, R.J.: Dukemtmc4reid: A large-scale multi-camera person re-identification dataset. In: IEEE Conference on Computer Vision and Pattern Recognition Workshops, pp. 10–19 (2017)
16. Greco, A., Roberto, A., Saggese, A., Vento, M., Vigilante, V.: Emotion analysis from faces for social robotics. In: International Conference on Systems, Man and Cybernetics (SMC), pp. 358–364. IEEE (2019)
17. Greco, A., Saggese, A., Sansone, C., Vento, B.: An experimental evaluation of smart sensors for pedestrian attribute recognition using multi-task learning and vision language models. Sensors **25**(6), 1736 (2025)
18. Greco, A., Saggese, A., Vento, B.: A robust and efficient overhead people counting system for retail applications. In: International Conference on Image Analysis and Processing, pp. 139–150. Springer (2022)
19. Greco, A., Saggese, A., Vento, M.: Digital signage by real-time gender recognition from face images. In: IEEE International Workshop on Metrology for Industry 4.0 & IoT, pp. 309–313. IEEE (2020)

20. Greco, A., Saldutti, S., Vento, B.: Fast and effective detection of personal protective equipment on smart cameras. In: International Conference on Pattern Recognition, pp. 95–108. Springer (2022)
21. Greco, A., Vento, B.: Par contest 2023: pedestrian attributes recognition with multi-task learning. In: International Conference on Computer Analysis of Images and Patterns, pp. 3–12. Springer (2023)
22. Hall, D., Perona, P.: Fine-grained classification of pedestrians in video: Benchmark and state of the art. In: IEEE Conference on Computer Vision and Pattern Recognition, pp. 5482–5491 (2015)
23. Ji, Z., Hu, Z., He, E., Han, J., Pang, Y.: Pedestrian attribute recognition based on multiple time steps attention. Pattern Recogn. Lett. **138**, 170–176 (2020)
24. Jia, J., Huang, H., Chen, X., Huang, K.: Rethinking of pedestrian attribute recognition: A reliable evaluation under zero-shot pedestrian identity setting. arXiv preprint arXiv:2107.03576 (2021)
25. Jia, J., Huang, H., Yang, W., Chen, X., Huang, K.: Rethinking of pedestrian attribute recognition: realistic datasets with efficient method. arXiv preprint arXiv:2005.11909 (2020)
26. Kairanbay, M., Salman, A.: Multi-task pedestrian attribute classification using convnext with advanced data augmentation. In: International Conference on Computer Analysis of Images and Patterns (CAIP) (2025)
27. Li, D., Chen, X., Huang, K.: Multi-attribute learning for pedestrian attribute recognition in surveillance scenarios. In: IEEE Asian Conference on Pattern Recognition (ACPR), pp. 111–115 (2015)
28. Li, D., Zhang, Z., Chen, X., Huang, K.: A richly annotated pedestrian dataset for person retrieval in real surveillance scenarios. IEEE Trans. Image Process. **28**(4), 1575–1590 (2018)
29. Lin, Y., Zheng, L., Zheng, Z., Wu, Y., Hu, Z., Yan, C., Yang, Y.: Improving person re-identification by attribute and identity learning. Pattern Recogn. **95**, 151–161 (2019)
30. Liu, X., Zhao, H., Tian, M., Sheng, L., Shao, J., Yi, S., Yan, J., Wang, X.: Hydraplus-net: Attentive deep features for pedestrian analysis. In: IEEE International Conference on Computer Vision (ICCV), pp. 350–359 (2017)
31. Salas-Cáceres, J.: Leveraging generalist vqa models to improve zero-shot pedestrian attribute recognition. In: International Conference on Computer Analysis of Images and Patterns (CAIP) (2025)
32. Sharma, G., Jurie, F.: Learning discriminative spatial representation for image classification. In: British Machine Vision Conference (BMVC), pp. 1–11 (2011)
33. Sooksatra, S., Rujikietgumjorn, S.: Skeleton-based attention mask for pedestrian attribute recognition network. J. Imaging **7**(12) (2021)
34. Sudowe, P., Leibe, B.: Patchit: Self-supervised network weight initialization for fine-grained recognition. In: British Machine Vision Conference (BMVC). vol. 1, pp. 24–25 (2016)
35. Sudowe, P., Spitzer, H., Leibe, B.: Person attribute recognition with a jointly-trained holistic cnn model. In: IEEE International Conference on Computer Vision Workshops, pp. 87–95 (2015)
36. Wang, X., Zheng, S., Yang, R., Zheng, A., Chen, Z., Tang, J., Luo, B.: Pedestrian attribute recognition: a survey. Pattern Recogn. **121**, 108220 (2022)
37. Wang, Y., et al.: Hulk: A universal knowledge translator for human-centric tasks. arXiv preprint arXiv:2312.01697 (2023)

38. Zhu, J., Liao, S., Lei, Z., Yi, D., Li, S.: Pedestrian attribute classification in surveillance: database and evaluation. In: IEEE International Conference on Computer Vision Workshops, pp. 331–338 (2013)

Leveraging Generalist VQA Models to Improve Zero-Shot Pedestrian Attribute Recognition

José Salas-Cáceres(✉)

Universidad de Las Palmas de Gran Canaria, Instituto Universitario de Sistemas Inteligentes y Aplicaciones Numéricas en Ingeniería (SIANI), Las Palmas de Gran Canaria, Spain
jose.salas@ulpgc.es

Abstract. Pedestrian Attribute Recognition (PAR) plays a key role in surveillance scenarios where classical biometric traits, such as facial features, are often unavailable due to low image quality, occlusions, or variable conditions. By extracting soft biometric attributes, such as gender, clothing type, and carried objects, PAR provides essential contextual information that can support tasks like person re-identification and behavior analysis. In this work, a novel approach is proposed based on Visual Question Answering (VQA) models, which avoids the limitations of supervised learning methods by leveraging general-purpose models without the need for additional training. This extends the PAR2023-winning strategy by introducing two state-of-the-art models, PaliGemma 1 and PaliGemma 2, along with a refined set of attribute-specific questions and an innovative fusion mechanism that combines both models' strengths. Experimental results on the PAR2025 dataset demonstrate that the proposed system surpasses previous methods, achieving a mean accuracy of 95.4% on the private set, outranking previous approaches on this task.

Keywords: Pedestrian Attribute Recognition · Visual Question Answering · Vision Language Model · Contest

1 Introduction

Soft biometrics refer to human characteristics that are not uniquely identifiable, such as hair and clothing style, gender, or carried objects. These traits can provide valuable contextual information to complement traditional biometric systems [11]. In this context, Pedestrian Attribute Recognition (PAR) aims to predict a set of attributes from a predefined list that describe the characteristics of a pedestrian [18]. This task is important in applications where the quality of the image makes it difficult to extract classical biometric features, such as facial details [4]. Surveillance scenarios are a typical example, where reliably inferring pedestrian attributes is crucial for person re-identification or behavior analysis.

M. Castrillón-Santana et al. (Eds.): CAIP 2025, LNCS 15621, pp. 16–26, 2026.
https://doi.org/10.1007/978-3-032-04968-1_2

PAR has traditionally been tackled using supervised learning models, that is, models trained on specific datasets to perform well on designated tasks [18]. Although these models have exhibited high performance across a variety of datasets, they possess limitations that can be particularly impactful in the domain of attribute recognition. These limitations primarily stem from the nature of the data encountered in the field. Datasets in this area are often highly imbalanced, as each individual may exhibit a different number of attributes, and some attributes may be significantly more common than others [18]. This issue is exemplified by the "presence-of-hat" or "presence-of-bag" attributes challenge included in the PAR2025 contest [8], where, as detailed in Sect. 2, it is far more common for a pedestrian not to wear a hat or bag than to wear one.

In addition, images often correspond to outdoor scenes, leading to varying illumination conditions, as well as heterogeneous viewing angles and pedestrian poses. These challenges substantially increase the complexity of training supervised models, as they exacerbate difficulties during model optimization and tend to introduce bias toward more prevalent classes.

With this in mind, zero-shot models have proven to be a promising alternative, as they do not require training on the target data and therefore avoid the associated limitations [14,15]. This is exemplified by [3], which, by utilizing a Visual Question Answering (VQA) model originally not specifically trained for PAR, but for general image captioning, was able to outperform competing approaches and win the 2023 edition of the PAR challenge [7].

Building upon the mentioned methodology proposed in [3], this work presents the approach developed by the IROC-ULPGC team for the PAR2025 contest [8]. The method is based on the use of various general-purpose VQA models, including BLIP-2 [13], PaliGemma 1 [2], and PaliGemma 2 [16].

2   Dataset

The dataset provided to all participants of the competition is the MIVIA PAR KD Dataset 2025, which contains over 105,000 pedestrian images of varying quality, resolution, poses, and distances. Each image is annotated with five attributes: upper-body clothing color, lower-body clothing color, gender, presence of a bag, and presence of a hat [8].

The color-related attributes can take one of 11 possible values, corresponding to the following classes: black, blue, brown, gray, green, orange, pink, purple, red, white, and yellow. The remaining attributes are binary: gender is annotated as either male (0) or female (1), while the presence of a bag or hat is indicated by 0 (absent) or 1 (present).

This dataset aggregates images from multiple sources, including established benchmarks such as PETA [5] and RAP [12], along with additional images manually cropped and annotated by the contest organizers. This variety of sources, as previously discussed, introduces a high degree of variability in terms of scenes, illumination conditions, and camera viewpoints, making the dataset particularly challenging.

For evaluation purposes, the dataset is divided into three subsets: training, validation, and test. However, at the time of this study, only the training and validation sets were publicly available. All experiments presented in this work were conducted exclusively on the validation set, which comprises 12,449 samples. Since all methods follow a zero-shot learning strategy, no training phase was required. Although the test set is private, Sect. 4 provides the results of the final proposal on it.

It is important to note that, unlike the previous edition of the competition [7], the current dataset contains no images with missing labels. This improvement was made possible by applying the winning approach from the prior edition [3], which employed the BLIP-2 architecture to infer missing attribute annotations. Several representative samples from the dataset are shown in Fig. 1.

Fig. 1. Samples of the MIVIA PAR KD Dataset 2025 dataset. Extracted from [10].

Figure 2 illustrates the class distribution for each of the five attributes within the validation set. A significant class imbalance is observed across all attributes, particularly in the color classes of both upper- and lower-body clothing.

3   Methodology

This section presents the methodology proposed for the PAR2025 contest. Given its similarity to the winning solution of PAR2023, special attention is given to highlighting the differences and improvements introduced in this approach compared to the 2023 version.

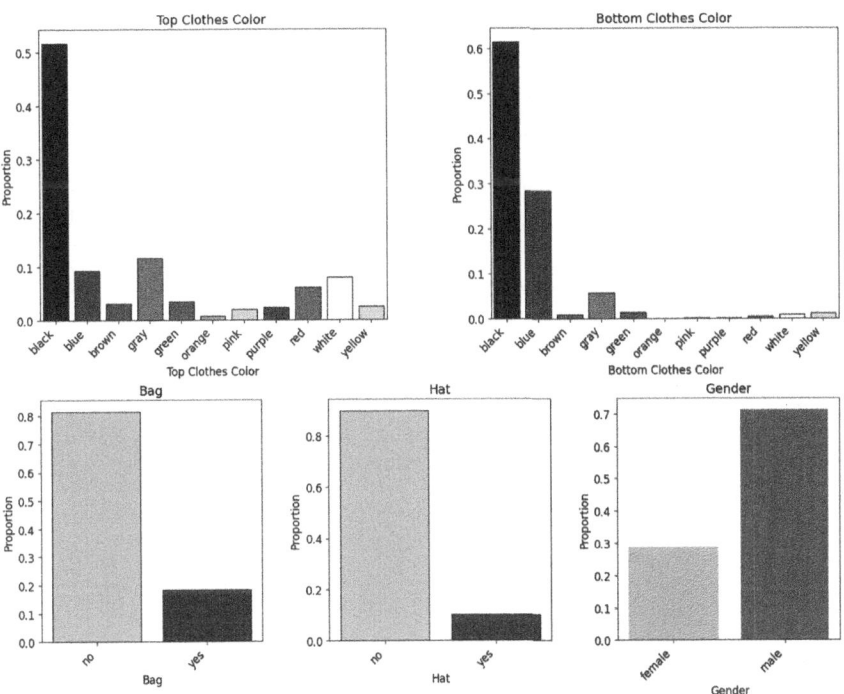

Fig. 2. Class distribution of the five attributes in the validation set.

3.1   Visual Question Answering

Recent advancements in VQA's offer a promising alternative to the traditional machine learning models. VQA models, which integrate computer vision and natural language processing, can interpret images and answer questions about their content [1,19]. This capability enables zero-shot learning, allowing models to generalize to new tasks without explicit retraining. In surveillance, VQA models can be employed to infer pedestrian attributes by posing specific, contextual questions about an image [4]. For instance, queries like "What color is the person's shirt?" or "Is the person carrying a bag?" can elicit informative responses that aid in attribute recognition. Other works have demonstrated the effectiveness of this approach, with VQA models achieving high accuracy rates in surveillance tasks by leveraging large-scale pre-trained language models and vision-language representations [4,14].

Incorporating contextual queries is essential when utilizing VQA models for PAR. Contextual information, such as the presence of objects or environmental cues, enhances the model's understanding of the scene, leading to more accurate attribute inference. By formulating queries that consider the broader context, like asking about the presence of a jacket before determining its color, VQA models can better handle the variability and complexity inherent in real-world surveillance images. This approach improves the robustness and comprehensiveness of pedestrian attribute recognition, facilitating more effective surveillance and public safety measures.

3.2 Proposed Method

Taking these aspects into account, the proposed solution, building upon the approach introduced in [3], relies on the use of contextual queries about the image content. The formulation of these queries is a critical component, as variations in prompt design can significantly influence the model's overall accuracy [17]. The approach adopted in this work expands on the original question set used in [3], consisting of:

1. Is the person male or female?
2. What color is the person's shirt?
3. What color is the person's trousers?
4. Does the person wear a bag?
5. Does the person wear a hat?
6. Does the person wear a cap?
7. Does the person wear a jacket?
8. What color is the person's jacket?

The original formulation did not account for garments such as skirts or long dresses. To address this limitation, a new set of questions was refined through an iterative process. Additional queries targeting previously overlooked clothing items, such as coats, dresses, and skirts, were progressively introduced, and those yielding improved accuracy on the validation set were empirically selected.

Another significant improvement to the original question set focused on the extraction of color information. It was observed that the model frequently returned responses containing multiple color terms, introducing ambiguity. To address this, two additional strategies were explored: querying for a one-word color summary and querying for the predominant color. After multiple rounds of validation and refinement, the final question set was as follows:

1. Is the person male or female?
2. What color is the person's shirt?
3. What color are the person's trousers?
4. Does the person wear a bag?
5. Does the person wear a hat?
6. Does the person wear a cap?
7. Does the person wear a jacket?
8. What color is the person's jacket?
9. Does the person wear a dress?
10. What color is the person's dress?
11. Is the person wearing a long dress?
12. What color is the person's shirt? (one word)
13. What color are the person's trousers? (one word)
14. What color is the person's dress? (one word)

15. What color is the person's jacket? (one word)

These queries were subsequently organized into a decision rule set, where attribute classification could depend on a single model response, as in the case of the gender attribute, or on a combination of responses, such as in hat detection, which aggregates answers from both "hat" and "cap" queries.

For example, the logic used for determining upper and lower body colors followed these rules expressed in pseudo-code notation:

```
if person wears a jacket then
    color of upper body clothes = jacket color
else if person wears a dress
    color of upper body clothes = dress color
else
    color of upper body clothes = shirt color
endif

if person wears a long dress
    color of bottom body clothes = dress color
else
    color of bottom body clothes = trousers color

if person wears hat or cap
    hat presence = yes
else
    hat presence = no
```

Additionally, to handle multi-word color outputs, the following fallback rule was implemented:

```
if length of color > 1 word
    color = response from the one-word color question
else
    color = original color response
```

It is worth noting that if the color generated by the models does not appear in the labels, it is converted into one of the valid ones by a map strategy. If despite this, the answer is still invalid, a random response is selected among the possible ones.

To further enhance the performance of the system, an ensemble strategy was explored by fusing the outputs of multiple VQA models. Specifically, three distinct models were used: PaliGemma 1, PaliGemma 2, and BLIP-2. Three fusion approaches were evaluated.

The first approach applied of a majority voting scheme, where the final answer was determined by the most frequently predicted response among all three models. In cases of a tie, the final answer was chosen at random from the tied predictions.

The second approach employed a weighted selection mechanism based on the individual strengths of each model, as measured on the validation set. For each attribute, the response was taken from the model that demonstrated the highest accuracy for that specific attribute. This evaluation resulted in a final configuration that used only the two versions of PaliGemma: PaliGemma 2 for color-related questions, due to its superior color recognition performance, and PaliGemma 1 for clothing presence detection, where it performed better.

Finally, a hybrid method was proposed, which combined these two strategies by selecting the majority-voted answer when there was a clear consensus, and defaulting to the response from the best-performing model in the event of a tie.

3.3   Evaluation Metrics

As specified in the competition guidelines, the primary evaluation metric used to assess the performance of the proposed approaches is classification accuracy. For each attribute, accuracy is computed individually as shown in Eq. 1, where K denotes the total number of samples, p_i is the predicted label for the i-th sample, and g_i is the corresponding ground truth label.

The mean accuracy (mA) is then calculated as the average of the individual accuracies, according to Eq. 2, where u, l, g, b, and h refer to upper-body clothing, lower-body clothing, gender, presence of a bag, and presence of a hat, respectively.

$$A = \frac{\sum_{i=1}^{K}(p_i = g_i)}{K} \tag{1}$$

$$mA = \frac{A_u + A_l + A_g + A_b + A_h}{5} \tag{2}$$

4   Results and Discussion

Table 1 presents the results of the non-fusion techniques. Two different sets of questions are evaluated: 'IROC-2023', which replicates the original questions used in [3], and 'IROC-2025', which extends the original with additional clothing-related questions and more concise descriptions of colors, as described in Sect. 3.1. The first row, corresponding to the BLIP-2 model with the original question set, reflects the expected performance of the PAR2023 winning approach on the updated dataset.

The new set of questions consistently improves mean accuracy by approximately 0.3-0.7%. While the improvement is minimal, it builds directly upon the previous version and consistently outperforms it, making it a strict enhancement. Both PaliGemma models outperform BLIP-2, with PaliGemma 1 achieving the highest mean accuracy of 88.30%. This improvement, nearly 20% over BLIP-2, is primarily due to a 40-50% increase in the accuracy of detecting bag and hat attributes. Additionally, PaliGemma 2 performs better than PaliGemma 1 on color-related questions, while PaliGemma 1 excels in detecting clothing items.

Table 1. Table of model performances across different parameters.

Model	Questions Set	A_u (%)	A_l (%)	A_g (%)	A_b (%)	A_h (%)	mA (%)
BLIP-2	IROC-2023	80.34	83.50	91.05	50.28	52.29	71.49
	IROC-2025	81.60	83.73	91.05	50.28	52.29	71.79
Paligemma 1	IROC-2023	80.52	85.15	**95.41**	**90.27**	**87.44**	87.76
	IROC-2025	82.92	85.45	**95.41**	**90.27**	**87.44**	**88.30**
Paligemma 2	IROC-2023	80.77	85.74	95.19	84.27	84.64	86.12
	IROC-2025	**83.55**	**86.57**	95.19	84.31	84.64	86.85

Table 2 shows the performance of the three fusion strategies. All experiments were conducted using the 'IROC-2025' of questions. The voting strategy reduced overall accuracy, dropping from 88.30%, achieved by PaliGemma 1, to 86.73%. This decrease is likely caused by the inclusion of BLIP-2 in the voting process, which has a much lower accuracy in the bag and hat attributes and therefore introduces noise. In contrast, the weighted selection strategy achieved the best results. By combining the strengths of both PaliGemma models, using PaliGemma 2 for color-related questions and PaliGemma 1 for the rest, it reached a mean accuracy of 88.66%. Based on this performance, the weighted approach was selected for the final submission to the PAR2025 contest.

Figure 3 displays the confusion matrices for the proposed method on the validation set. It is worth noting that, in the lower body color attribute, although the model reaches over 50% accuracy in only four color classes, the overall accuracy remains above 85%. This discrepancy is explained by the significant class imbalance in this attribute, where the colors black and blue add up close to 90% of the whole set, as is shown in Fig. 2. The matrix also reveals a frequent confusion between several colors and brown.

Table 2. Table of fusion method performances across different parameters.

Fusion Strategy	A_u (%)	A_l (%)	A_g (%)	A_b (%)	A_h (%)	mA (%)
Voting	83.53	86.33	95.22	84.88	83.66	86.73
Weighted Selection	83.36	**86.61**	**95.48**	**90.43**	**87.42**	**88.66**
Mix	**83.65**	**86.61**	95.24	84.88	83.66	86.76

Before the final rankings of the competition were published, the PAR2025 organizers provided the results of the proposed method on the private test set. Table 3 presents its performance in comparison with top-rank approaches from the previous edition. The proposed method clearly outperforms all previous solutions, with notable improvements in non-color attributes. Particularly remarkable is the performance on the gender attribute, where the model achieved an accuracy of 99.6%.

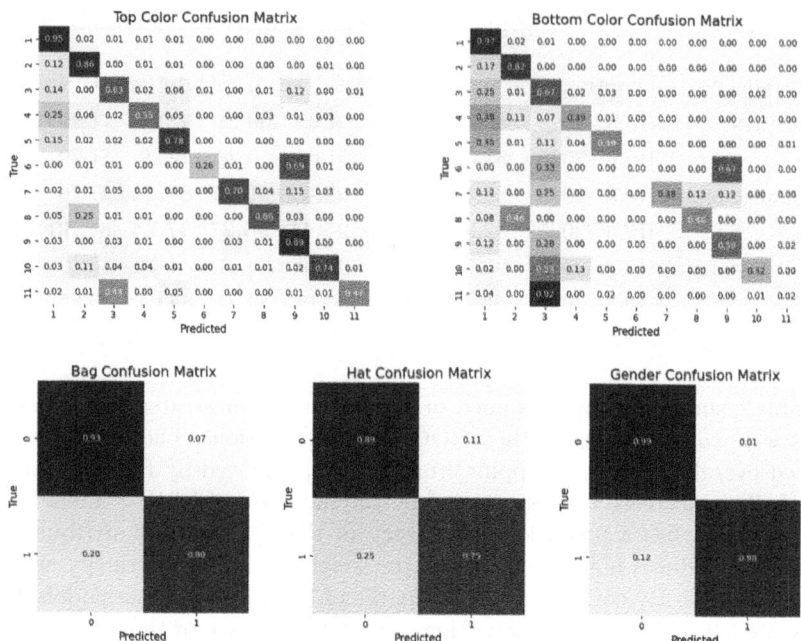

Fig. 3. Confusion matrices of the proposed method in the validation set.

Table 3. Comparison of the proposal method with scores of various teams in the PAR2023 contest. Scores extracted from [6,9].

Team	A_u (%)	A_l (%)	A_g (%)	A_b (%)	A_h (%)	mA (%)
IROC-ULPGC (2025)	**92.70**	**91.30**	**99.60**	**96.10**	**97.50**	**95.40**
IROC-ULPGC (2023) [3]	92.07	90.81	92.72	92.15	92.79	92.11
HUSTNB	72.84	74.78	84.48	63.08	86.78	76.39
SPARKY	73.87	90.41	88.33	66.35	57.72	75.33
PAR23-Baseline [7]	62.80	72.40	72.10	64.10	83.20	70.90

5   Conclusion

This work presents a VQA-based approach for the PAR2025 contest, building upon the methodology introduced in the winning solution of PAR2023. The proposed method leverages general-purpose VQA models to extract soft biometric attributes from pedestrian images without requiring additional training or fine-tuning. Among the evaluated models, PaliGemma 1 and PaliGemma 2 demonstrated a significant improvement over previous solutions, increasing overall accuracy by nearly 20% points compared to BLIP-2.

The introduction of an extended and more refined set of questions, further contributed to performance gains. Additionally, a weighted fusion strategy was designed to combine the strengths of both PaliGemma models, assigning PaliGemma 2 to color-related questions and PaliGemma 1 to all others, achieving the highest mean accuracy of 88.66%, generating an increase of almost 17% on the previous winning approach in the validation set.

These results confirm that VQA models, even when not specifically trained for attribute recognition, are well-suited for this task. Their ability to generalize and reason about visual and semantic content enables them to handle complex attribute queries with high reliability, making them a strong alternative to traditional supervised approaches in highly variable and unbalanced scenarios.

The current methodology requires significant computational resources and GPU time. Therefore, a potential direction for future work is the distillation of the model into a smaller and more efficient version, capable of running on embedded devices and supporting real-time inference.

Acknowledgments. This publication is part of the project PID2021-122402OB-C22, funded by MCIN/ AEI/10.13039/501100011033/FEDER, EU, the ACIISI-Gobierno de Canarias and FEDER under project ULPGC Facilities Net and Grant EIS 2021 04, and by the Consejería de Universidades, Ciencia e Innovación y Cultura (Gobierno de Canarias) and the European Social Fund Plus (FSE+) under the funding framework for doctoral research.

Finally, the author would like to thank to the IROC-ULPGC 2023 team members for their guidance, support, and valuable feedback throughout the development of this work.

References

1. Barra, S., Bisogni, C., De Marsico, M., Ricciardi, S.: Visual question answering: which investigated applications? Pattern Recogn. Lett. **151**, 325–331 (2021). https://doi.org/10.1016/j.patrec.2021.09.008. https://www.sciencedirect.com/science/article/pii/S0167865521003147
2. Beyer*, L., et al.: PaliGemma: a versatile 3B VLM for transfer. arXiv preprint arXiv:2407.07726 (2024)
3. Castrillón-Santana, M., Sánchez-Nielsen, E., Freire-Obregón, D., Santana, O.J., Hernández-Sosa, D., Lorenzo-Navarro, J.: Evaluation of a visual question answering architecture for pedestrian attribute recognition. In: Tsapatsoulis, N., et al. (eds.) Computer Analysis of Images and Patterns, pp. 13–22. Springer, Cham (2023). https://doi.org/10.1007/978-3-031-44237-7_2
4. Castrillón-Santana, M., Sánchez-Nielsen, E., Freire-Obregón, D., Santana, O.J., Hernández-Sosa, D., Lorenzo-Navarro, J.: Visual question answering models for zero-shot pedestrian attribute recognition: a comparative study. SN Comput. Sci. **5**(6) (2024). https://doi.org/10.1007/s42979-024-02985-0
5. Deng, Y., Luo, P., Loy, C.C., Tang, X.: Pedestrian attribute recognition at far distance. In: Proceedings of the 22nd ACM International Conference on Multimedia, p. 789–792. Association for Computing Machinery, New York (2014). https://doi.org/10.1145/2647868.2654966
6. Greco, A., Saggese, A., Sansone, C., Vento, B.: An experimental evaluation of smart sensors for pedestrian attribute recognition using multi-task learning and

vision language models. Sensors **25**(6) (2025). https://doi.org/10.3390/s25061736. https://www.mdpi.com/1424-8220/25/6/1736

7. Greco, A., Vento, B.: PAR contest 2023: pedestrian attributes recognition with multi-task learning. In: 20th International Conference on Computer Analysis of Images and Patterns: CAIP 2023. Springer, Heidelberg (2023). https://doi.org/10.1007/978-3-031-44237-7_1

8. Greco, A., Vento, B.: Par contest 2025: pedestrian attributes recognition with advanced neural networks. In: 21st International Conference on Computer Analysis of Images and Patterns: CAIP 2025. Springer, Heidelberg (2025)

9. Greco, A., Vento, B.: Pedestrian attribute recognition (PAR) contest 2023 (2025). https://mivia.unisa.it/par2023/. Accessed 10 June 2025

10. Greco, A., Vento, B.: Pedestrian attribute recognition (PAR) contest 2025 (2025). https://mivia.unisa.it/par2025/. Accessed 04 June 2025

11. Jain, A.K., Dass, S.C., Nandakumar, K.: Soft biometric traits for personal recognition systems. In: Zhang, D., Jain, A.K. (eds.) ICBA 2004. LNCS, vol. 3072, pp. 731–738. Springer, Heidelberg (2004). https://doi.org/10.1007/978-3-540-25948-0_99

12. Li, D., Chen, X., Huang, K.: Multi-attribute learning for pedestrian attribute recognition in surveillance scenarios. In: ACPR, pp. 111–115 (2015)

13. Li, J., Li, D., Xiong, C., Hoi, S.: Blip: bootstrapping language-image pre-training for unified vision-language understanding and generation. In: 40 International Conference on Machine Learning (ICML) (2022)

14. Munir, F., Azam, S., Mihaylova, T., Kyrki, V., Kucner, T.P.: Pedestrian vision language model for intentions prediction. IEEE Open J. Intell. Transport. Syst. **6**, 393–406 (2025). https://doi.org/10.1109/OJITS.2025.3554387

15. Ngo, B.H., Ngo, S.T., Le, P.D., Phan, Q.M., Tran, M.T., Le, T.N.: Crosspar: enhancing pedestrian attribute recognition with vision-language fusion and human-centric pre-training. In: Proceedings of the Asian Conference on Computer Vision (ACCV), pp. 1301–1315 (2024)

16. Steiner, A., et al.: Paligemma 2: a family of versatile vlms for transfer. arXiv preprint arXiv:2412.03555 (2024)

17. Wang, L., et al.: Prompt engineering in consistency and reliability with the evidence-based guideline for LLMs. NPJ Digit. Med. **7**(1), 41 (2024)

18. Wang, X., et al.: Pedestrian attribute recognition: a survey. Pattern Recogn. **121**, 108220 (2022). https://doi.org/10.1016/j.patcog.2021.108220. https://www.sciencedirect.com/science/article/pii/S0031320321004015

19. Zhang, J., Huang, J., Jin, S., Lu, S.: Vision-language models for vision tasks: a survey. IEEE Trans. Pattern Anal. Mach. Intell. **46**(8), 5625–5644 (2024). https://doi.org/10.1109/TPAMI.2024.3369699

Multi-task Pedestrian Attribute Classification Using ConvNeXt with Advanced Data Augmentation

Magzhan Kairanbay$^{(\boxtimes)}$ 🆔 and Ali Salman 🆔

Array Innovation, Al Baraka Tower A, Building 372, Road 4611, Block 346, Manama, Bahrain

magzhan.kairanbay@array.world
https://www.array.world

Abstract. We propose a multi-task deep learning framework for pedestrian attribute classification, jointly predicting five key attributes: upper and lower clothing colors, gender, and the presence of a bag and hat. Our approach employs a ConvNeXt-Large backbone pretrained on ImageNet-22k, enhanced with advanced data augmentation techniques including MixUp and CutMix. To address class imbalance, we introduce a conditional augmentation strategy targeting challenging color classes (e.g., black and gray). Experiments on the MIVIA PAR KD Dataset 2025, comprising over 105,000 annotated pedestrian images, demonstrate the effectiveness of our method, achieving a peak mean accuracy of 93.14% across all attributes. These results highlight the advantages of combining modern convolutional architectures, task-aware augmentation, and multi-task learning for robust pedestrian analysis in real-world settings.

Keywords: Pedestrian attribute recognition · Multi-task learning · ConvNeXt · Data Augmentation · Deep Learning

1 Introduction

Pedestrian attribute recognition is crucial in computer vision, supporting applications like surveillance, autonomous driving, and smart cities. It involves identifying semantic attributes—such as gender, clothing color, and accessories—from a single image, providing valuable context for tasks like tracking, behavior analysis, and person search.

Despite its importance, pedestrian attribute recognition remains challenging due to real-world factors such as low resolution, occlusion, lighting variations, diverse poses, and changing viewpoints. These conditions cause high intra-class variation and inter-class similarity, which hinder the generalization capabilities of models.

Traditional approaches typically treat each attribute independently, overlooking the correlations among them. This often results in suboptimal performance. To overcome this, multi-task learning (MTL) frameworks have been

M. Castrillón-Santana et al. (Eds.): CAIP 2025, LNCS 15621, pp. 27–38, 2026.
https://doi.org/10.1007/978-3-032-04968-1_3

introduced, leveraging shared representations to jointly learn related attributes. This approach captures attribute dependencies, enhances generalization, and reduces overfitting [2, 8, 10].

Building on this, recent deep MTL architectures further improve pedestrian attribute recognition by sharing features and tasks, leading to more robust and semantically consistent representations [14, 21].

Parallel to these advances, new deep learning architectures have emerged that bridge convolutional neural networks (CNNs) and vision transformers. ConvNeXt [15] is a notable example, integrating transformer-inspired design elements into a pure convolutional model. This hybrid design achieves competitive or superior performance with improved computational efficiency and scalability.

Another critical factor in boosting model robustness is data augmentation. Techniques like MixUp [24], which interpolates images and labels, and CutMix [23], which replaces image patches with adjusted labels, have been shown to improve calibration and regularization. These augmentations are especially valuable in scenarios with limited or imbalanced pedestrian attribute data, expanding training diversity and enhancing generalization (see Fig. 1).

Fig. 1. Visualization of data augmentation techniques: original images (top two images), MixUp (linear interpolation), and CutMix (top-bottom patch composition). Both enhance model generalization by blending image and label information.

Motivated by these developments, we propose a comprehensive multi-task learning framework for pedestrian attribute recognition that addresses challenges such as class imbalance, visual ambiguity, and attribute diversity. Our approach integrates state-of-the-art architectures and targeted training strategies to deliver robust, generalizable performance. The key components and contributions are:

- **ConvNeXt-Large backbone:** Utilizes ConvNeXt-Large for powerful, fine-grained feature extraction combining convolutional and transformer benefits.
- **Multi-task architecture:** Shared backbone with attribute-specific heads enables joint learning of common and task-specific features, improving accuracy and robustness.
- **Data augmentation:** Applies MixUp and CutMix to enhance generalization and reduce overfitting, especially in imbalanced or limited data scenarios.
- **Conditional augmentation:** Targets visually ambiguous attributes (e.g., white, black, gray) with selective transformations to improve discrimination without affecting clearer classes.
- **Empirical evaluation:** Extensive experiments on benchmark datasets establish strong baselines and confirm each component's contribution via ablation studies.

Together, these innovations form a more resilient and generalizable framework for pedestrian attribute recognition, paving the way for improved accuracy and efficiency in real-world applications.

2 Related Works

PAR is a key computer vision task with applications in surveillance, reidentification, and behavior analysis. Early methods used hand-crafted features (e.g., color histograms, edge orientations) with classifiers like SVMs [12,25]. Although somewhat effective, they required expert feature design and struggled in complex, real-world settings.

The paradigm shifted with the advent of deep learning, which offered automatic feature extraction and end-to-end learning capabilities. CNNs, in particular, became the backbone of modern PAR systems due to their hierarchical structure and strong visual representation power. Li et al. [13] introduced a pioneering deep learning model for joint attribute prediction, showcasing the advantage of learning shared features for multiple attributes in a unified framework. Building on this foundation, subsequent research explored deeper CNN architectures, attention mechanisms, and localized part-based models to address the challenges posed by occlusions, viewpoint variations, and attribute co-occurrence [20,21].

A significant portion of recent work in PAR has focused on integrating MTL principles to leverage the inherent interdependencies among pedestrian attributes. MTL allows a single model to simultaneously learn multiple attribute classifiers, promoting shared feature learning and regularization through task-level inductive biases [2,19]. For instance, the correlation between attributes such as gender and clothing type (e.g., skirts typically associated with females) can be exploited to improve prediction consistency and robustness. Studies have demonstrated that MTL architectures, which typically consist of a shared backbone followed by task-specific heads, not only enhance accuracy but also reduce model complexity and training data requirements [13,21]. Nonetheless, MTL introduces its own set of challenges, such as gradient conflict and task imbalance. Solutions like task uncertainty weighting [11], gradient surgery [22], and dynamic task routing [18] have been proposed to mitigate these issues and stabilize training.

In terms of backbone architecture, recent advances have introduced more powerful and efficient CNN designs tailored to the needs of vision tasks. ConvNeXt [15] exemplifies this trend by modernizing the classical ResNet structure using insights drawn from Vision Transformers (ViTs) [5]. By incorporating elements such as larger convolution kernels, inverted bottleneck designs, and layer normalization [1], ConvNeXt achieves state-of-the-art performance on benchmarks like ImageNet, while preserving the efficiency and inductive bias advantages of convolutions. This architecture has become increasingly relevant in tasks like PAR, where both spatial precision and computational efficiency are critical.

Another crucial component in the development of robust PAR systems is *data augmentation*, especially given the limited diversity and imbalance often found in pedestrian datasets. Beyond conventional transformations, modern augmentation techniques such as MixUp [24] and CutMix [23] generate new training samples by interpolating or patching images and labels. These methods improve model generalization, enhance robustness to label noise, and contribute to better calibration, which are particularly beneficial in multi-task and multi-label settings like PAR [4,9].

In summary, the field of pedestrian attribute recognition has evolved significantly, transitioning from manual feature engineering to sophisticated deep multi-task architectures. The integration of modern CNN designs like ConvNeXt, together with advanced augmentation strategies, continues to push the boundaries of performance. However, challenges such as attribute imbalance, occlusion, and real-world variability persist, making PAR an active and impactful area of research in computer vision.

3   Dataset

We utilize the **MIVIA PAR KD Dataset 2025** [6,7], a large-scale and diverse dataset designed for pedestrian attribute recognition. It consists of over **105,000** annotated images (refer to Fig. 2), each depicting a single pre-cropped pedestrian and labeled with five semantic attributes:

- **Clothing Colors:** Both upper and lower clothing are categorized into 11 predefined color classes—black, blue, brown, gray, green, orange, pink, purple, red, white, and yellow—represented by integer labels [1–11].
- **Gender:** A binary classification indicating male (0) or female (1).
- **Accessories:** Binary labels indicating the presence (1) or absence (0) of a *bag* and a *hat*.

The dataset aggregates samples from several well-known public pedestrian attribute datasets, including **PETA**, **RAP**, and **Colorful**, as well as manually labeled private images. As a result, the dataset exhibits significant variation in terms of *camera viewpoints*, *illumination conditions*, *pedestrian poses*, and *image resolutions*, effectively simulating real-world scenarios.

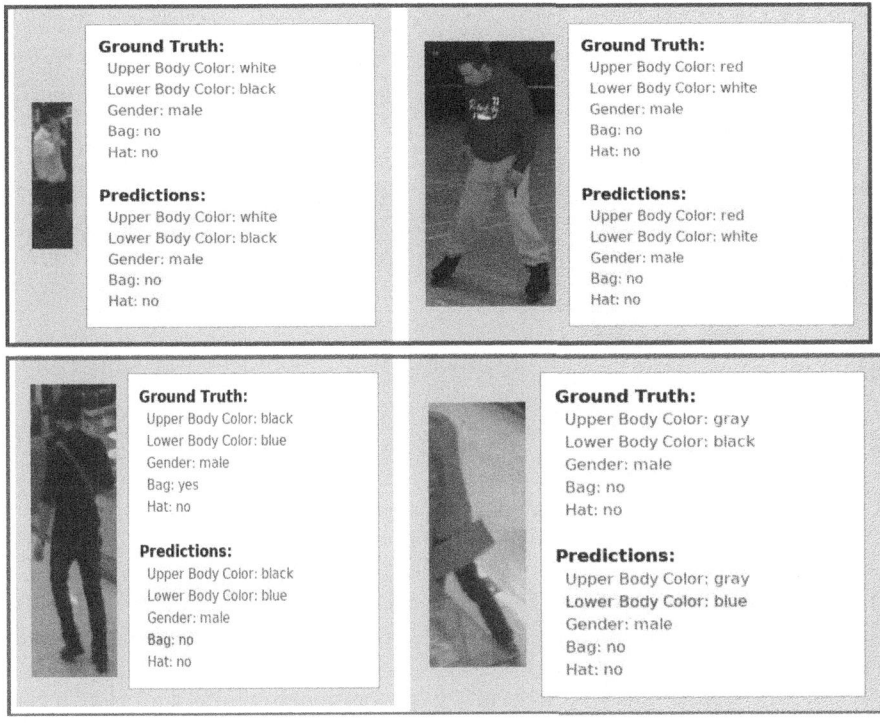

Fig. 2. Samples from MIVIA PAR KD 2025 Dataset.

All images are annotated completely—a major improvement over previous iterations—by leveraging knowledge distillation methods. These methods were derived from the winning solution of the PAR 2025 competition and were used to infer and supplement missing labels. In addition, newly collected samples have been added to the training set to further enhance robustness.

To ensure a fair and unbiased evaluation, model performance will be assessed on a **private test set** that is not publicly released and has no overlap with the training data.

4   Proposed Architecture

Our proposed multi-task architecture is designed to jointly learn multiple pedestrian attributes by leveraging shared representations extracted from a high-capacity convolutional backbone, followed by dedicated task-specific heads. The overall structure is illustrated in Fig. 3.

We adopt the ConvNeXt-Large model [15] as a shared backbone for feature extraction. The model is initialized with weights pretrained on ImageNet-22k and fine-tuned on ImageNet-1k at a resolution of 224×224. We remove the

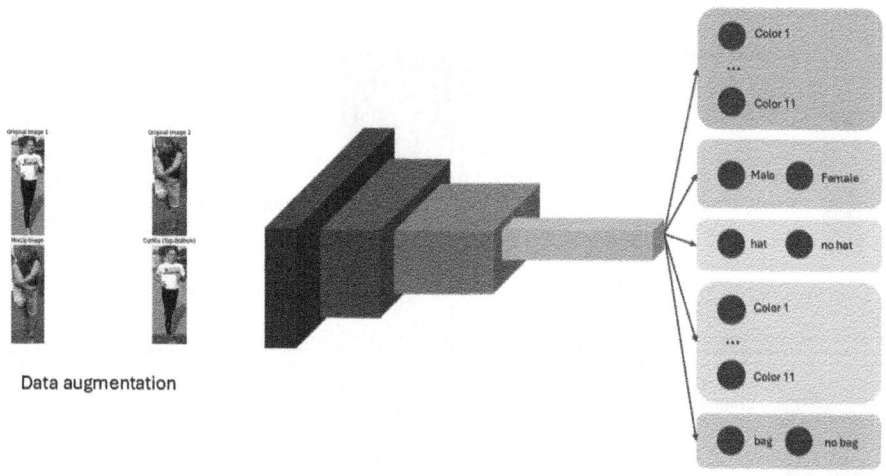

Fig. 3. The main architecture of proposed system

final classification layer to repurpose the network as a generic feature extractor, outputting a high-dimensional feature representation $\mathbf{f} \in \mathbb{R}^d$ for each input image.

ConvNeXt-Large offers a modernized convolutional architecture that incorporates several design principles inspired by Vision Transformers, including:

- Larger depthwise convolutional kernels
- Layer normalization instead of batch normalization
- GELU activations and inverted bottleneck structures

These enhancements result in a backbone that delivers transformer-level performance with convolutional efficiency, making it well-suited for attribute recognition tasks that benefit from local spatial details.

To enable multi-task learning, we attach five independent fully connected (FC) layers to the shared backbone, each responsible for predicting a specific pedestrian attribute:

$$\mathbf{y}_i = \text{FC}_i(\mathbf{f}), \quad i \in \{1, 2, 3, 4, 5\} \tag{1}$$

where $\mathbf{f}$ is the shared feature vector produced by the ConvNeXt backbone, and $\mathbf{y}_i$ is the output prediction for the i-th attribute. Each classification head is designed with output dimensions corresponding to the number of classes for its respective task (e.g., 11 for clothing colors, 2 for gender, etc.).

This design allows the model to jointly optimize for multiple correlated tasks, benefiting from shared spatial and semantic cues while maintaining specialization via the attribute-specific heads.

4.1   Data Augmentation Strategy

To enhance the generalization capability of our model and reduce overfitting, we employ a diverse set of data augmentation techniques during training. These strategies are particularly important for handling class imbalance and limited intra-class variation in pedestrian attribute datasets.

We apply conventional image augmentations such as random cropping, horizontal flipping, rotation, and color jittering. These transformations increase the variability of training data, helping the model become robust to common real-world variations including viewpoint changes, lighting conditions, and minor deformations.

Certain attribute classes, such as upper body clothing colors "black" and "gray", tend to dominate the dataset, often exhibiting lower visual diversity due to their low saturation and luminance. To mitigate this issue, we introduce conditional augmentation strategies specifically targeted at these challenging classes. For images belonging to these categories, we apply intensified transformations including increased color jitter, affine distortions, and geometric perturbations. This encourages the model to learn more discriminative features despite the low color variance and helps counteract class imbalance.

We incorporate MixUp, a data interpolation technique that linearly combines pairs of training images and their corresponding labels. Given two samples $(\mathbf{x}_i, \mathbf{y}_i)$ and $(\mathbf{x}_j, \mathbf{y}_j)$, a synthetic training example is generated as:

$$\tilde{\mathbf{x}} = \lambda \mathbf{x}_i + (1 - \lambda)\mathbf{x}_j \tag{2}$$

$$\tilde{\mathbf{y}} = \lambda \mathbf{y}_i + (1 - \lambda)\mathbf{y}_j \tag{3}$$

where $\lambda \sim \text{Beta}(\alpha, \alpha)$ for a hyperparameter $\alpha \in (0, 1)$. MixUp encourages the model to behave linearly in-between training examples, which has been shown to improve robustness and calibration.

We also employ CutMix, an augmentation technique that replaces a random patch of one image with a patch from another while adjusting the labels accordingly. Formally:

$$\tilde{\mathbf{x}} = \mathbf{M} \odot \mathbf{x}_i + (1 - \mathbf{M}) \odot \mathbf{x}_j \tag{4}$$

$$\tilde{\mathbf{y}} = \lambda \mathbf{y}_i + (1 - \lambda)\mathbf{y}_j \tag{5}$$

where $\mathbf{M}$ is a binary mask defining the region to be replaced, and λ represents the proportion of the original image preserved. CutMix is particularly effective in improving localization and enhancing robustness to occlusion and partial views.

All augmentation strategies are applied online during training, with MixUp and CutMix used probabilistically in combination with standard and conditional augmentations. This comprehensive augmentation pipeline ensures the model is exposed to a broad distribution of appearance variations, which is crucial for real-world pedestrian attribute recognition tasks.

4.2   Training Strategy

Loss Function. We formulate the training objective as a sum of cross-entropy losses over the five pedestrian attributes: upper clothing color, lower clothing color, gender, bag, and hat. To improve generalization and prevent the model from becoming overconfident, we incorporate label smoothing into the cross-entropy loss formulation. The smoothed target distribution helps mitigate the impact of noisy labels and regularizes the classifier:

$$\mathcal{L} = \sum_{i=1}^{5} \mathcal{L}_{\text{CE}}^{(i)}(\mathbf{y}_i, \mathbf{t}_i) \tag{6}$$

where $\mathcal{L}_{\text{CE}}^{(i)}$ denotes the cross-entropy loss for the i-th attribute, $\mathbf{y}_i$ is the predicted logit vector, and $\mathbf{t}_i$ is the ground-truth label distribution with a label smoothing factor $\epsilon = 0.05$. Label smoothing distributes a portion of the target probability mass across all classes, making the loss less sensitive to incorrect predictions on uncertain or ambiguous samples.

We adopt the AdamW optimizer [17], which decouples weight decay from the gradient-based updates and has been shown to improve performance in large-scale vision tasks. The initial learning rate is set to 1×10^{-4}, with a weight decay factor of 1×10^{-2}. A cosine annealing learning rate schedule [16] is applied over 9 training epochs, gradually related workducing the learning rate in a smooth, non-monotonic fashion to avoid premature convergence and promote better generalization.

Training is conducted using a batch size of 32. During each training iteration, MixUp and CutMix augmentations are each applied with a probability of 1/3. For both augmentation strategies, we set the Beta distribution hyperparameter $\alpha = 0.4$, which controls the degree of interpolation or patch replacement. These stochastic augmentations are integrated with conventional data augmentation techniques to introduce additional diversity into the training data, further regularizing the learning process.

Overall, this training pipeline is designed to promote robust multi-task learning across heterogeneous attribute categories by combining principled regularization, advanced optimization, and diverse augmentation strategies.

4.3   Experimental Setup

All experiments are conducted using the PyTorch framework on NVIDIA GPUs. Each model is trained for 9 epochs with a batch size of 32. For evaluation, we report both attribute-wise accuracy and the overall mean accuracy (mA), defined as the average of individual attribute accuracies. The mA metric provides a holistic measure of multi-task performance, accounting for the relative difficulty of each attribute classification task.

5   Results

Table 1 illustrates the training progression across epochs, including the total loss, individual attribute accuracies, and mean accuracy. Our model consistently improves throughout training, achieving a peak mean accuracy of **93.14%** at epoch 8. Notably, the model performs strongest on the `hat` and `gender` attributes, while slightly lower performance is observed for `upper` clothing color, reflecting higher visual ambiguity in those classes.

Table 1. Training progression and attribute-wise accuracies. Mean accuracy (mA) is computed as the average of all attribute accuracies.

Epoch	Loss	Upper	Lower	Gender	Bag	Hat	mA
1	2.869	87.68	89.63	95.04	92.79	98.65	92.76
2	2.589	88.11	89.57	95.35	92.54	98.80	92.87
3	2.467	87.36	89.46	96.19	93.07	98.67	92.95
4	2.322	88.18	89.63	96.47	93.23	98.69	93.24
5	2.190	87.76	89.54	96.51	93.21	98.64	93.13
6	2.076	87.33	89.24	96.39	93.41	98.71	93.02
7	1.961	87.40	89.13	96.47	93.33	98.84	93.03
8	1.925	87.76	89.05	96.76	93.33	98.80	**93.14**
9	1.918	87.77	89.12	96.65	93.35	98.77	93.13

If we compare our results with the winner of the PAR 2023 contest ([3], mA = 0.921), we observe a minor yet notable improvement. The following table (refer to Table 2) summarizes the results of our proposed solution and those of other participants for each task, along with the overall mean accuracy (mA), standard deviation, and inference speed (FPS).

Table 2. Comparison with PAR 2023 participants on individual attribute accuracies (A_u: upper, A_l: lower, A_g: gender, A_b: bag, A_h: hat), mean accuracy (**mA**), standard deviation, and Frames Per Second (FPS).

Team	A_u	A_l	A_g	A_b	A_h	mA	St. Dev.	FPS
iROC-ULPGC [3]	0.921	0.908	0.927	0.921	0.928	0.921	0.007	<0.5
HUSTNB	0.728	0.748	0.845	0.631	0.868	0.764	0.086	25
SPARKY	0.739	0.904	0.883	0.663	0.577	0.753	0.126	<0.5
Baseline [7]	0.628	0.724	0.721	0.641	0.832	0.709	0.073	150
CODY	0.597	0.743	0.848	0.446	0.549	0.636	0.143	10
AWESOMEPAR	0.182	0.138	0.936	0.771	0.756	0.557	0.330	11
Our proposed solution	0.876	0.890	0.967	0.933	0.988	**0.931**	–	–

5.1   Ablation Studies

To better understand the contribution of various components in our pipeline, we perform ablation studies focusing on augmentation techniques, optimization strategies, and targeted class handling.

Impact of Data Augmentation: We evaluate the effect of MixUp and Cut-Mix augmentations by disabling them during training. The absence of these techniques leads to an average decrease of approximately 1% in validation accuracy across all attributes, indicating their critical role in improving generalization and robustness to overfitting.

Optimization Strategy: Replacing the AdamW optimizer with standard Adam, and removing the cosine annealing learning rate schedule, results in a consistent drop of 0.1–0.2% in mean accuracy. This suggests that AdamW with scheduled decay facilitates better convergence and performance stability.

Conditional Augmentation: We specifically target underrepresented classes—particularly black and gray upper clothing—by applying additional augmentations. This targeted strategy yields noticeable improvements in color classification accuracy, highlighting its effectiveness in mitigating class imbalance and enhancing feature diversity for visually similar categories.

5.2   Analysis and Discussion

The experimental results yield several noteworthy observations:

Architecture Choice: The adoption of ConvNeXt-Large as the shared backbone strikes a strong balance between computational efficiency and expressive capacity. Its hierarchical design and modernized convolutional structure effectively capture discriminative features necessary for fine-grained pedestrian attribute recognition.

Multi-task Learning Benefits: The shared representation learned across tasks enables the model to leverage inter-attribute correlations (e.g., between clothing color and accessory presence), yielding better generalization than training independent models for each attribute. This joint learning strategy enhances both performance and parameter efficiency.

Augmentation Effectiveness: The synergy between MixUp, CutMix, and conditional augmentations fosters improved generalization, especially for visually ambiguous or imbalanced attribute categories. Notably, the targeted augmentation strategy boosts classification accuracy for underrepresented color classes, validating its impact.

Training Stability: Employing label smoothing and cosine annealing contributes to more stable convergence and mitigates overconfidence in predictions. These regularization techniques help maintain training stability while subtly improving overall model calibration.

6 Conclusion

We proposed a robust multi-task learning framework for pedestrian attribute classification that leverages a ConvNeXt-Large backbone in conjunction with advanced augmentation strategies. Our method achieves strong performance by combining the representational power of modern convolutional architectures with regularization techniques such as MixUp, CutMix, and conditional augmentations tailored to address class imbalance.

The use of a shared backbone and task-specific heads enables efficient representation learning across multiple attributes, enhancing both accuracy and computational efficiency. The proposed conditional augmentation strategy further demonstrates its utility in handling real-world dataset challenges involving imbalanced and ambiguous classes.

For future work, we plan to investigate attention-based mechanisms for improved spatial localization, ensemble learning for increased robustness, and the extension of this framework to larger and more diverse attribute datasets.

References

1. Ba, J.L., Kiros, J.R., Hinton, G.E.: Layer normalization. arXiv preprint arXiv:1607.06450 (2016)
2. Caruana, R.: Multitask learning. In: Machine Learning, vol. 28, pp. 41–75. Springer, Heidelberg (1997)
3. Castrillón-Santana, M., et al.: Evaluation of a visual question answering architecture for pedestrian attribute recognition. In: International Conference on Computer Analysis of Images and Patterns, pp. 13–22. Springer, Heidelberg (2023)
4. Cubuk, E.D., Zoph, B., Mane, D., Vasudevan, V., Le, Q.V.: Autoaugment: learning augmentation policies from data. In: Proceedings of the IEEE Conference on Computer Vision and Pattern Recognition (CVPR), pp. 113–123 (2019)
5. Dosovitskiy, A., et al.: An image is worth 16x16 words: transformers for image recognition at scale. In: International Conference on Learning Representations (ICLR) (2021)
6. Greco, A., Vento, B.: An extended dataset and a baseline for pedestrian attribute recognition with advanced neural networks. In: 21st International Conference on Computer Analysis of Images and Patterns (CAIP) (2025)
7. Greco, A., Vento, B.: Par contest 2023: pedestrian attributes recognition with multi-task learning. In: International Conference on Computer Analysis of Images and Patterns, pp. 3–12. Springer, Heidelberg (2023)
8. Han, X., Luo, H., Zhu, Y., Wang, X.: Attribute recognition by joint recurrent learning of context and correlation. In: Proceedings of the IEEE International Conference on Computer Vision (ICCV), pp. 4291–4299 (2017)

9. Hendrycks, D., Mu, N., Cubuk, E.D., Gilmer, J., Lakshminarayanan, B.: Augmix: a simple data processing method to improve robustness and uncertainty. In: International Conference on Learning Representations (ICLR) (2020)

10. Jing, Y., Liu, J.: Deep learning for pedestrian attribute recognition: a survey. In: Proceedings of the International Conference on Image Processing (ICIP). IEEE (2015)

11. Kendall, A., Gal, Y., Cipolla, R.: Multi-task learning using uncertainty to weigh losses for scene geometry and semantics. In: Proceedings of the IEEE Conference on Computer Vision and Pattern Recognition (CVPR), pp. 7482–7491 (2018)

12. Layne, R., Hospedales, T.M., Gong, S.: Person re-identification by attributes. In: British Machine Vision Conference (BMVC) (2012)

13. Li, Y., Liu, C., Yang, J.: Deep learning for pedestrian attribute recognition: a multi-task learning approach. In: Proceedings of the 22nd ACM International Conference on Multimedia (ACM MM), pp. 843–846 (2015)

14. Li, Y., Zhu, C., Gong, S.: Deepmar: deep multi-attribute recognition for pedestrian attribute recognition. In: Proceedings of the British Machine Vision Conference (BMVC) (2015)

15. Liu, Z., et al.: A convnet for the 2020s. In: Proceedings of the IEEE/CVF Conference on Computer Vision and Pattern Recognition (CVPR), pp. 11976–11986 (2022)

16. Loshchilov, I., Hutter, F.: SGDR: stochastic gradient descent with warm restarts. arXiv preprint arXiv:1608.03983 (2016)

17. Loshchilov, I., Hutter, F.: Decoupled weight decay regularization. In: International Conference on Learning Representations (ICLR) (2019). arXiv:1711.05101

18. Rosenbaum, C., Klinger, T., Riemer, M.: Routing networks: adaptive selection of non-linear functions for multi-task learning. In: International Conference on Learning Representations (ICLR) (2018)

19. Ruder, S.: An overview of multi-task learning in deep neural networks. arXiv preprint arXiv:1706.05098 (2017)

20. Sarfraz, M.H., Schumann, A., Stiefelhagen, R.: Deep view-sensitive pedestrian attribute inference in an end-to-end model. In: British Machine Vision Conference (BMVC) (2017)

21. Wang, Z., Xu, Z., Gong, S., Zhu, C., Wang, C.: Attribute recognition by adaptive learning. In: Proceedings of the IEEE Conference on Computer Vision and Pattern Recognition (CVPR), pp. 1234–1242 (2017)

22. Tianhe, Yu., Kumar, S., Gupta, A., Levine, S., Hausman, K., Finn, C.: Gradient surgery for multi-task learning. Adv. Neural Inf. Process. Syst. (NeurIPS) **33**, 5824–5836 (2020)

23. Yun, S., Han, D., Oh, S.J., Chun, S., Choe, J., Yoo, Y.: Cutmix: regularization strategy to train strong classifiers with localizable features. In: Proceedings of the IEEE/CVF International Conference on Computer Vision (ICCV), pp. 6023–6032 (2019)

24. Zhang, H., Cisse, M., Dauphin, Y.N., Lopez-Paz, D.: mixup: beyond empirical risk minimization. In: International Conference on Learning Representations (ICLR) (2018)

25. Zhao, W.S., Ouyang, W., Wang, X., Wang, X.: Person re-identification based on color histograms of clothing with forward-backward ranking. In: Chinese Conference on Pattern Recognition, pp. 1–5. IEEE (2011)

A Region-Aware Multi-modal Framework for Pedestrian Attribute Recognition via CLIP and Graph Neural Networks

Mudasir Hussain Bhat[ID] and Daw-tung Lin[(✉)][ID]

Department of Computer Science and Information Engineering, National Taipei
University, New Taipei City, Taiwan
{s711183401,dalton}@gm.ntpu.edu.tw

Abstract. Pedestrian Attribute Recognition (PAR) is a fundamental
task in surveillance and intelligent vision systems, aiming to identify
attributes such as clothing color, gender, and accessories carried from
pedestrian images. In this paper, we propose a region-aware, prompt-
guided, and graph-based multi-task framework developed as a solution
to the PAR 2025 Contest. Our method integrates a fully fine-tuned
CLIP ViT-B/32 vision-language encoder with a Graph Attention Network (GAT)-based classifier that models inter-attribute dependencies
through attention-driven message passing. The system extracts visual
features from both full-body and lower-body views, computes similarity
scores with a curated set of 160 handcrafted textual prompts, and feeds
these semantically aligned representations into a graph-based classifier.
Evaluated on the private test set, our framework achieves a mean accuracy of 69.8%, demonstrating strong generalization and robustness under
real-world conditions.

Keywords: Pedestrian Attribute Recognition · Multimodal Learning ·
Graph Neural Networks

1 Introduction

Understanding and recognizing human attributes from visual data is a fundamental problem in computer vision. In this context, pedestrian attribute recognition (PAR) focuses on identifying semantic characteristics, such as gender, clothing type, and accessories carried, from pedestrian images [1]. As a fine-grained
recognition task, PAR contributes directly to high-level downstream applications
including person re-identification [2], surveillance, and behavioral understanding
[3].

Despite its practical importance, PAR remains challenging due to complex
visual conditions, such as occlusions, pose variations, and background clutter. In

---

D.-T. Lin—This work was supported in part by the National Science and Technology
Council (NSTC), Taiwan, under Grant No. NSTC 113-2221-E-305-011.

M. Castrillón-Santana et al. (Eds.): CAIP 2025, LNCS 15621, pp. 39–52, 2026.
https://doi.org/10.1007/978-3-032-04968-1_4

addition, attributes often exhibit contextual dependencies, and their distribution is typically long-tailed, making many categories underrepresented. Although convolutional neural networks (CNNs) have shown success in extracting local visual patterns, they struggle to model global semantic relationships and inter-attribute interactions [4]. These challenges have driven research toward more advanced architectures capable of leveraging external knowledge and multi-modal signals.

Inspired by the success of Transformer-based architectures in a wide range of natural language processing (NLP) tasks, vision-language models (VLMs) have emerged as a powerful paradigm for bridging visual and textual modalities [5,6]. These models leverage key capabilities of Transformers, such as contextual representation learning and multi-head attention mechanisms, to enhance semantic understanding in visual tasks [7]. A notable example is CLIP (Contrastive Language–Image Pretraining), which aligns images and textual descriptions within a shared embedding space using contrastive learning on large-scale image–text pairs. CLIP has demonstrated strong generalization and zero-shot learning capabilities, making it particularly suitable for tasks such as Pedestrian Attribute Recognition (PAR), where rich semantic reasoning and cross-modal alignment are essential [5]. However, directly applying CLIP to multi-attribute recognition tasks presents inherent limitations, particularly in modeling the relational structure among attributes and incorporating task-specific language priors. [8]

This paper presents our solution to the PAR 2025 contest. We propose a unified framework that integrates CLIP-based vision–language embeddings with graph neural networks and prompt-guided supervision. In the proposed approach, each pedestrian attribute is represented as a node in a fully connected graph, and inter-attribute dependencies are captured using a Graph Attention Network (GAT). CLIP serves as the backbone for extracting dense, semantically rich features from both image inputs and attribute-specific text prompts, thereby enabling explicit language-driven guidance. This integration of multi-modal representation learning and relational reasoning results in a robust, accurate, and interpretable system for fine-grained pedestrian attribute recognition.

Our primary contributions are summarized as follows:

1. **Region-Aware Visual Preprocessing:** To enhance attribute localization, particularly for lower-body attributes, we adopt a dual-view input strategy. Both the full-body image and a cropped lower-body region are independently encoded by the CLIP backbone. The resulting features are then selectively routed such that predictions for lower-body attributes are derived from the lower-body representation, while the full-body features support the remaining attributes. This region-aware decoding improves fine-grained recognition without compromising global context.

2. **Prompt-Guided Vision-Language Representation Learning**: We utilize a pre-trained CLIP model to encode pedestrian images by computing similarity scores with a set of manually constructed textual prompts that describe visual attributes. These prompts cover both individual attribute categories (e.g., clothing color, gender, accessories) and composite descriptions

combining multiple attributes. The resulting image–text similarity vectors serve as high-level, semantically aligned feature representations that guide downstream attribute classification.

3. **Multi-Task Architecture with Graph Reasoning:** To model the contextual dependencies among pedestrian attributes, we introduce a Graph Neural Network (GNN) architecture based on Graph Attention Networks (GAT). Each attribute is represented as a node in a fully-connected graph, enabling the network to perform message passing between attribute nodes. This design allows the model to capture inter-attribute correlations (e.g., between clothing style and accessory presence), thereby improving the consistency and accuracy of multi-attribute predictions.

4. **Class Imbalance Mitigation Strategies**: The long-tailed distribution of pedestrian attributes is addressed using two complementary strategies. First, we adopt a task-specific Focal Loss with class weights inversely proportional to label frequency, which emphasizes learning from hard and underrepresented examples. Second, we implement a WeightedRandomSampler that increases the sampling probability of rare attribute combinations during training, ensuring better exposure to minority classes and reducing prediction bias.

2   Literature Review

Pedestrian Attribute Recognition (PAR) has evolved beyond early convolutional approaches [9] that focused solely on visual appearance, advancing toward multimodal and graph-based frameworks. These modern paradigms emphasize semantic understanding, spatial context, and reduced reliance on supervision, drawing significant interest from the vision community. As a result, researchers have increasingly focused on cross-modal alignment, attribute reasoning, and prompt-driven learning.

Among notable advances in PAR, the top method [10] in the PAR Challenge 2023 [11] employed a zero-shot VQA approach using the BLIP-2 framework, which frames attributes as natural language queries with handcrafted prompts. Without task-specific fine-tuning, it achieved 92% mean accuracy, demonstrating the strong generalization capabilities of pretrained vision-language models. Extending large-scale vision-language models, the PromptPAR framework [5] leverages CLIP to align pedestrian images with semantic attribute descriptors via prompt-driven fusion. This approach bridges visual features and attribute semantics, shifting the focus from fixed label prediction to contextual semantic reasoning.

To further enhance semantic understanding, recent approaches [12] have introduced language modeling techniques to capture inter-attribute dependencies. These methods formulate multi-attribute recognition as a masked language modeling task, treating attributes as words in a descriptive sentence. This allows a transformer-based model to infer masked attributes using both visual cues and contextual semantics, enabling the learning of nuanced co-occurrence patterns.

ViTA-PAR [13] employs learnable visual and textual prompts to effectively integrate global and local attribute information, thus improving person representations with high-level descriptors, such as gender, and fine-grained details, including specific clothing and accessories. This vision-language prompting framework attains state-of-the-art performance on multiple pedestrian attribute recognition benchmarks, exemplifying the capacity of adaptive prompt learning combined with pre-trained language models to advance contextual understanding in pedestrian attribute analysis.

In parallel, graph-based frameworks [14] have been developed to model rich dependencies among pedestrian attributes in a multi-task setting. Earlier approaches captured attribute dependencies using graph-based structures to facilitate relational reasoning, allowing predictions for one attribute to inform others. A representative example is the Relation-Aware GCN model [15], which jointly learns attribute and contextual relations by constructing two complementary graphs: an attribute graph that captures semantic correlations among attributes (e.g., hat, bag, clothing types), and a context graph that encodes spatial relationships across image regions. Graph convolutional layers propagate information within and across these graphs, enabling the model to capture both co-occurrence patterns and spatial dependencies. This dual-graph integration enhances multi-attribute prediction by emphasizing meaningful interactions—such as associating 'wearing a hat' with head-region features—while attenuating irrelevant correlations.

Building on this direction, recent works [16] have employed inter-attribute attention and graph-based models (e.g., GCNs, GATs) to exploit attribute dependencies, enabling enhanced contextual reasoning and multi-task learning. These approaches, including vision-language fusion and graph-attentional modeling, collectively advance the robustness and accuracy of Pedestrian Attribute Recognition (PAR). They illustrate the potential of vision-language models (VLMs) and text-semantic learning in PAR, leveraging large pretrained language priors and adaptive prompts to significantly improve recognition of clothing and accessory attributes in context.

3   PAR Contest Overview

The PAR 2025 Contest evaluates methods for multi-attribute pedestrian prediction from cropped images. Each method must predict five attributes: upper-body clothing color, lower-body clothing color (each with 11 categories), gender, and the presence of a bag and a hat. Color categories include black, blue, brown, gray, green, orange, pink, purple, red, white, and yellow, represented by labels [1–10]. On the contrary, the remaining three attributes (gender, bag, and hat) are treated as binary classification problems. These are encoded as follows: gender (0 for male, 1 for female), bag (0 for absence, 1 for presence), and hat (0 for absence, 1 for presence).

Participants are provided with the MIVIA PAR KD Dataset 2025, containing 94,281 training and 12,449 validation images. Each image depicts a single, fully annotated pedestrian and reflects diverse visual conditions in pose, resolution, and lighting.

Final evaluation is conducted on a private test set and submissions are ranked according to the *mean accuracy (mA)* computed across all five attributes. Further details and evaluation protocols can be found in the official contest publication [17] and website[1].

4 Methodology

Pedestrian Attribute Recognition (PAR) has increasingly benefited from advances in multimodal learning, where the integration of visual and textual modalities enables more robust and semantically grounded predictions. Our proposed method comprises two principal components: (1) a CLIP-based visual-language encoder for multimodal representation learning, and (2) a Graph Neural Network (GNN) multi-task classifier for inter-attribute reasoning.

Figure 1 provides a complete overview of the proposed system, the subsequent subsections describe its core components and the overall design methodology in detail.

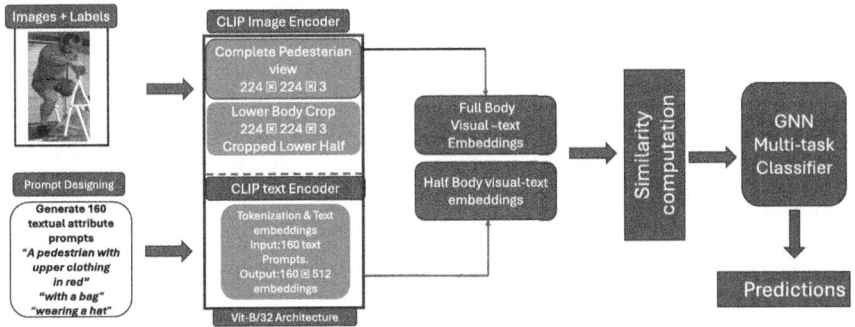

Fig. 1. Overview of the proposed system. The CLIP vision encoder (left) extracts features from full-body and lower-body pedestrian views, while the CLIP text encoder (right) embeds attribute-specific prompts. Cosine similarities between visual and textual embeddings are computed and passed to a Graph Neural Network (GNN) for inter-attribute reasoning and final prediction.

4.1 Class Imbalance Mitigation

The training data exhibits severe class imbalance across all attributes. For instance, "black" appears 52,861 times in the *lower_color* attribute, whereas

---

[1] https://mivia.unisa.it/par2025/.

"orange" occurs only 208 times. Similarly, the "no hat" class dominates "hat" with 74,352 versus 19,942 samples.

To address this, we apply class weights inversely proportional to label frequency:

$$w_{t,c} = \frac{1}{n_{t,c} + \epsilon} \tag{1}$$

$$\tilde{w}_{t,c} = \frac{w_{t,c}}{\sum_{j=1}^{K_t} w_{t,j}} \cdot K_t \tag{2}$$

These weights are integrated into a focal loss formulation:

$$\mathcal{L}_{\text{focal}} = -\tilde{w}_{t,c} \cdot (1 - p_{t,c})^{\gamma} \cdot \log(p_{t,c}) \tag{3}$$

where $\gamma = 2.0$, placing more emphasis on hard and underrepresented examples. In addition, we employ a Weighted Random Sampler to ensure rare classes are sampled more frequently during training without altering the original dataset distribution.

4.2   Region-Aware Visual Encoding with CLIP

In pedestrian attribute recognition, different attributes exhibit spatial locality. Global attributes such as gender and accessory presence require holistic visual context, whereas fine-grained attributes like lower-body clothing color are confined to specific spatial regions. Relying solely on global features can suppress critical localized cues. To address this limitation, we adopt a region-aware encoding strategy using CLIP ViT-B/32. For each pedestrian image I, we extract two spatially distinct views:

– I_{full}: the full-body image, capturing contextual and global cues.
– I_{low}: a cropped lower-body image, focusing on pants and leg-related features.

Each view is passed through the CLIP vision encoder f_{vision}:

$$\phi_{\text{full}} = f_{\text{vision}}(I_{\text{full}}), \quad \phi_{\text{low}} = f_{\text{vision}}(I_{\text{low}}) \tag{4}$$

These embeddings $\phi \in \mathbb{R}^{512}$ maintain alignment with the text embedding space for downstream semantic matching.

4.3   Prompt Design and Semantic Supervision

To support semantic supervision within our multimodal framework for Pedestrian Attribute Recognition (PAR), we construct a curated set of 160 textual prompts. These are categorized into two main groups :attribute-specific prompts, which provide focused supervision for individual traits, and context-aware prompts, which encode co-occurrence patterns by combining multiple attributes as outlined in Table 1.

Attribute-Specific Prompts. These prompts are designed to independently capture individual pedestrian characteristics, thereby enhancing the model's capacity to isolate and learn localized visual features for each attribute. Specifically, we define:

- 11 prompts describing upper-body clothing colors, e.g., "A pedestrian with upper body clothing in red."
- 11 prompts for lower-body clothing colors, e.g., "A pedestrian with lower body clothing in blue."
- 2 prompts for gender, e.g., "A pedestrian who is female."
- 4 prompts targeting accessory presence, such as bags and hats, e.g., "A pedestrian wearing a hat."

These 28 prompts provide isolated supervision cues, facilitating explicit learning of attribute-specific representations.

Context Aware Prompts. To enrich contextual learning and capture co-occurrence dependencies among attributes, we compose 132 compound prompts by combining color descriptors with gender or accessory-related semantics. For example:

- "A male pedestrian with upper body clothing in green."
- "A female pedestrian with lower body clothing in black."
- "A pedestrian with upper body clothing in orange wearing a hat."
- "A pedestrian with lower body clothing in gray with a bag."

These prompts model inter-attribute dependencies and provide semantically entangled cues that reinforce contextual attribute reasoning.

The integration of both prompt types offers a balanced supervision signal, bridging visual perception with high-level semantics, and fostering a robust alignment between image regions and natural language descriptors.

Table 1. Distribution of Generated Textual Prompts for PAR

Prompt Category	Number of Prompts
Upper-Body Color (Attribute specific)	11
Lower-Body Color (Attribute specific)	11
Gender (Attribute specific)	2
Accessories (Attribute specific: Hat and Bag)	4
Gender + Upper-Body Color (Context aware)	22
Gender + Lower-Body Color (Context aware)	22
Accessories + Upper-Body Color (Context aware)	44
Accessories + Lower-Body Color (Context aware)	44
Total	**160**

Each prompt $p_m \in P = \{p_1, \ldots, p_{160}\}$ is encoded via the CLIP text encoder f_{text}, producing a language embedding:

$$t_m = f_{\text{text}}(p_m) \tag{5}$$

Given a visual embedding ϕ extracted from the image encoder, we compute its cosine similarity with each textual embedding t_m to derive an attribute-aligned semantic projection:

$$s_{\phi,m} = \frac{\phi \cdot t_m}{\|\phi\| \, \|t_m\|}, \quad S = [s_{\phi,1}, \ldots, s_{\phi,M}] \tag{6}$$

In practice, we compute two such vectors: S_{full} and S_{low}, corresponding to the global (full-body) and local (lower-body) image views. These similarity vectors serve as semantically enriched inputs to our downstream multi-task classification model.

To be specific, for each prompt t_m where $m = 1, \ldots, M$, the cosine similarities are calculated as follows:

$$s_{\text{full},m} = \frac{\phi_{\text{full}} \cdot t_m}{\|\phi_{\text{full}}\| \, \|t_m\|} \tag{7}$$

$$s_{\text{low},m} = \frac{\phi_{\text{low}} \cdot t_m}{\|\phi_{\text{low}}\| \, \|t_m\|} \tag{8}$$

This dual-pass decoding strategy ensures spatially aware classification without introducing additional trainable parameters for fusion. By leveraging attention-based relational modeling and integrating region-specific representations, the GNN module supports robust, semantically grounded predictions across multiple attributes.

In contrast to linear multi-task heads, the proposed GNN classifier explicitly learns inter-task correlations, enabling the model to make contextually consistent decisions—such as resolving ambiguity in *hat* detection based on gender or upper clothing. This final integration step completes the pipeline by unifying semantic similarity with relational reasoning.

4.4   Graph Neural Network for Inter-Attribute Reasoning

To effectively capture contextual dependencies among pedestrian attributes—such as the co-occurrence of *bag* and *hat*, or the occlusion of *upper body clothing* by carried items—we model attribute prediction as a graph-based multi-task learning problem. Each of the five pedestrian attributes $\mathcal{A} = \{\text{upper\_color}, \text{lower\_color}, \text{gender}, \text{bag}, \text{hat}\}$ is represented as a node in a fully connected graph $\mathcal{G} = (\mathcal{V}, \mathcal{E})$, allowing information to flow between tasks.

The input to the graph model is the visual-textual similarity vector $S \in \mathbb{R}^M$, where $M = 160$, computed from CLIP-based alignment between image embeddings and handcrafted prompts. This vector encodes rich semantic cues across both single-attribute and compound prompts.

We project S into a latent space via a learnable transformation:

$$h_{\text{proj}} = \text{ReLU}(W_{\text{proj}}S + b_{\text{proj}}) \tag{9}$$

This projected vector $h_{\text{proj}} \in \mathbb{R}^d$ initializes all $K = 5$ attribute nodes:

$$h_k^{(0)} = h_{\text{proj}}, \quad \forall k \in \{1, \ldots, K\} \tag{10}$$

A two-layer Graph Attention Network (GAT) is employed for inter-attribute reasoning. At each layer l, node features are linearly transformed:

$$z_k^{(l)} = W^{(l)} h_k^{(l-1)} \tag{11}$$

Then, pairwise attention coefficients are computed as:

$$e_{kj}^{(l)} = \text{LeakyReLU}(a^{(l)\top}[z_k^{(l)} \| z_j^{(l)}]) \tag{12}$$

These are normalized using softmax:

$$\alpha_{kj}^{(l)} = \frac{\exp(e_{kj}^{(l)})}{\sum_{j'} \exp(e_{kj'}^{(l)})} \tag{13}$$

Finally, node features are updated through weighted message aggregation:

$$h_k^{(l)} = \sigma\left(\sum_j \alpha_{kj}^{(l)} z_j^{(l)}\right) \tag{14}$$

This attention-based message passing mechanism enables each node to integrate cues from semantically or spatially correlated attributes. For instance, *gender* and *lower color* may jointly influence predictions for *bag presence* in certain contexts.

To enforce spatial locality, we apply the GNN classifier to two distinct similarity vectors: S_{full} (from the full-body view) and S_{low} (from the cropped lower-body view). Each yields predictions:

$$Y_{\text{full}} = f_{\text{GNN}}(S_{\text{full}}), \quad Y_{\text{low}} = f_{\text{GNN}}(S_{\text{low}}) \tag{15}$$

Final task-specific outputs are fused deterministically:

$$y_k^{\text{final}} = \begin{cases} y_{k,\text{low}}, & \text{if } k = \text{lower\_color} \\ y_{k,\text{full}}, & \text{otherwise} \end{cases} \tag{16}$$

This dual-pass decoding strategy eliminates additional fusion parameters while preserving spatial sensitivity. The GNN's architecture thus serves a dual purpose: capturing inter-attribute correlations via attention-based message passing and respecting region-specific visual cues via selective input routing.

Compared to conventional linear classifiers that treat tasks independently, our GNN-based multi-task head promotes joint learning of semantically entangled attributes, enhancing prediction consistency and interpretability in pedestrian attribute recognition.

5   Experimental Results

We adopt CLIP-ViT-B/32[2] as the vision–language encoder, paired with a GNN-based multi-task classifier. Following [18], we fine-tune the entire CLIP model with a learning rate of 2×10^{-5} for 20 epochs, using AdamW and early stopping (patience = 5). A batch size of 32 and mixed-precision training are used. We apply focal loss with inverse-frequency class weights and a weighted sampler to address label imbalance. Each image is processed into full- and lower-body crops, augmented, and encoded using 160 attribute prompts. Cosine similarities from both views are input to the GNN for attribute prediction. For a comprehensive overview of the dataset and evaluation metrics, please refer to Sect. 3.

5.1   Validation Set Evaluation

The proposed model was evaluated on the official validation set released by the contest organizers, comprising 12,449 images. While the primary evaluation metric for the contest is mean accuracy (mA), additional metrics, including weighted precision, recall, and F_1-score, were computed to better assess performance under class imbalance and multi-class conditions. Table 2 presents the results, where the model achieved a mean accuracy of 0.8718. Notably, F_1-scores were high for *hat* (0.955), *gender* (0.846), and *lower-body color* (0.873). The lowest performance was observed for the *bag* attribute ($F_1 = 0.823$), potentially due to frequent occlusion and substantial intra-class variability.

Table 2. Validation performance on the PAR validation set.

Attribute	Acc.	Prec. (w)	Rec. (w)	F1 (w)
Upper body color (Au)	0.819	0.831	0.819	0.819
Lower body color (Al)	0.877	0.876	0.877	0.873
Gender (Ag)	0.859	0.870	0.859	0.846
Bag (Ab)	0.845	0.827	0.845	0.823
Hat (Ah)	0.959	0.959	0.959	0.955
Mean Accuracy (mA): 0.8718				

5.2   Private Test Set Evaluation

Final performance is assessed on a private test set, with evaluation conducted solely by the contest organizers. Table 3 reports mean accuracy (mA) per attribute. The model sustains competitive results across upper-body color (0.805), lower-body color (0.813), and gender (0.792), indicating strong generalization. However, the bag attribute remains a challenge, achieving an accuracy

---

[2] https://huggingface.co/openai/clip-vit-base-patch32.

of only 0.422. The overall mean accuracy across all five attributes on the private test set is 69.8%.

Table 3. Accuracy obtained on the Private test set for each attribute.

Attribute	Acc.
Upper color (Au)	0.805
Lower color (Al)	0.813
Gender (Ag)	0.792
Bag (Ab)	0.422
Hat (Ah)	0.661
Mean Accuracy (mA): 0.698	

Comparison with Other Teams: As the official ranking for the PAR 2025 Contest was not released at the time of submission, we conducted a comparative analysis against the top-performing approaches reported in the official PAR 2023 Contest [19]. All methods were evaluated on a common private test set[3], ensuring a fair and consistent basis for comparison. Our method achieved a mean accuracy of 69.8%, corresponding to the fifth-highest result reported. A detailed breakdown is provided in Table 4.

Table 4. Performance comparison across all five attributes on the private test set among top-performing methods.

Team	A_u	A_l	A_g	A_b	A_h	mA
IROC-ULPGC	**0.921**	**0.908**	**0.927**	**0.921**	**0.928**	**0.921**
HUSTNB	0.728	0.748	0.845	0.631	0.868	0.764
SPARKY	0.739	0.904	0.883	0.663	0.577	0.753
BASELINE	0.628	0.724	0.721	0.641	0.832	0.709
IMSLAB (Ours)	0.805	0.813	0.792	0.422	0.661	0.698
CODY	0.597	0.743	0.848	0.446	0.549	0.636
AWESOMEPAR	0.182	0.138	0.936	0.771	0.756	0.557

---

[3] The private test set used in the PAR 2025 Contest was identical to that employed in PAR 2023.

5.3 Qualitative Results

To visually examine the performance of the model in various real-world scenarios, we present selected qualitative results from the validation set, as illustrated in Fig. 2. These examples highlight the model's strengths and limitations of the model under challenging visual conditions such as pose variation, partial occlusion, ambiguous visual cues, and annotation inconsistencies. Each example displays a cropped pedestrian image alongside its predicted attributes, with correct predictions indicated in green and incorrect ones marked in red.

In the first example, the model successfully identified all attributes despite significant pose variation, partial occlusion, and reduced spatial resolution, highlighting the effectiveness of the model in handling challenging pedestrian appearances. The second case presents a clean frontal view image where all attributes are confidently and correctly predicted, reflecting high performance under ideal conditions.

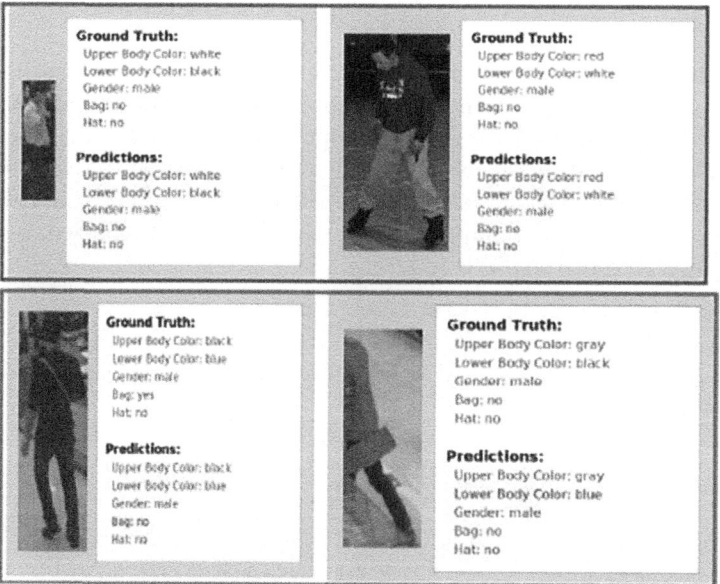

Fig. 2. Visualization results on validation samples, with correct predictions highlighted in green and incorrect ones in red. The examples illustrate the model's performance across diverse scenarios. (Color figure online)

In the bottom row, the third image illustrates a rear-view scenario in which only a bag strap is partially visible, while the actual bag remains occluded. The model fails to recognize the presence of a bag, underscoring the inherent limitations of reasoning from incomplete visual evidence. Although prompt-based guidance can be leveraged to suggest that visible straps indicate the presence of a bag, such an approach remains prone to false positives, as similar visual cues may also originate from clothing such as jeans or tops. This underscores a fundamental challenge in multimodal learning tasks, how to integrate textual guidance effectively when visual evidence is incomplete, without allowing semantic assumptions to override the true visual context. The final example presents a partially visible pedestrian, where the model predicts the lower-body clothing color as blue, while the ground truth annotation specifies black. Upon visual inspection, the prediction appears more consistent with the visual content, suggesting that the discrepancy is likely due to annotation inconsistency rather than an error in the model's attribute prediction. We observed multiple such cases during evaluation, indicating the presence of minor label noise within the dataset.

6   Conclusion

In this work, we presented a robust vision-language framework for pedestrian attribute recognition (PAR), integrating a fully fine-tuned CLIP ViT-B/32 encoder with a GNN-based multi-task classifier. By leveraging handcrafted textual prompts and computing dual-view similarity scores from full-body and lower-body regions, our method effectively captures both global and localized semantic cues. Experimental results on the contest-provided validation set demonstrate strong performance across multiple attributes, particularly in well-represented classes. The final evaluation on the private test set confirms the model's ability to generalize under held-out conditions, achieving a mean accuracy of 69.8%.

Our findings underscore the benefits of full model fine-tuning, multi-view semantic reasoning, and inter-attribute relational modeling via GNNs. Future work could explore dynamic prompt generation, vision-language pretraining specific to surveillance domains, or integrating temporal context for video-based PAR. Overall, our approach demonstrates competitive performance in a challenging real-world setting and contributes a scalable architecture for fine-grained attribute recognition in pedestrian analysis.

References

1. Vandenhende, S., Georgoulis, S., Van Gansbeke, W., Proesmans, M., Dai, D., Van Gool, L.: Multi-task learning for dense prediction tasks: a survey. IEEE Trans. Pattern Anal. Mach. Intell. **44**(7), 3614–3633 (2021)
2. Gou, M., Karanam, S., Liu, W., Camps, O., Radke, R.J.: DukeMTMC4ReID: a large-scale multi-camera person re-identification dataset. In: IEEE Conference on Computer Vision and Pattern Recognition Workshops, pp. 10–19 (2017)
3. Vishwakarma, S., Agrawal, A.: A survey on activity recognition and behavior understanding in video surveillance. Vis. Comput. **29**(10), 983–1009 (2012)
4. Mou, L., Hua, Y., Zhu, X.X.: A relation-augmented fully convolutional network for semantic segmentation in aerial scenes. In: Proceedings of the IEEE/CVF Conference on Computer Vision and Pattern Recognition, pp. 12416–12425 (2019)

5. Wang, X., et al.: Pedestrian attribute recognition via clip based prompt vision-language fusion. IEEE Trans. Circ. Syst. Video Technol. (2024)

6. Kumar, P.: Large language models (LLMs): survey, technical frameworks, and future challenges. Artif. Intell. Rev. **57**(10), 260 (2024)

7. Islam, S., et al.: A comprehensive survey on applications of transformers for deep learning tasks. Expert Syst. Appl. **241**, 122666 (2024)

8. Jing, D., et al.: Fineclip: self-distilled region-based clip for better fine-grained understanding. Adv. Neural. Inf. Process. Syst. **37**, 27896–27918 (2024)

9. Junejo, I.N., Ahmed, N.: Depthwise separable convolutional neural networks for pedestrian attribute recognition. SN Comput. Sci. **2**(2), 100 (2021)

10. Greco, A., Vento, B.: PAR Contest 2023: pedestrian attributes recognition with multi-task learning. In: 20th International Conference on Computer Analysis of Images and Patterns: CAIP 2023. Springer, Cham (2023)

11. Castrillón-Santana, M., et al.: Evaluation of a visual question answering architecture for pedestrian attribute recognition. In: International Conference on Computer Analysis of Images and Patterns. Springer, Cham (2023)

12. Li, W., et al.: Label2label: a language modeling framework for multi-attribute learning. In: European Conference on Computer Vision. Springer, Cham (2022)

13. Park, M., Park, H., Kim, J.: ViTA-PAR: visual and textual attribute alignment with attribute prompting for pedestrian attribute recognition. arXiv preprint arXiv:2506.01411 (2025)

14. Chen, Z.M., et al.: Multi-label image recognition with graph convolutional networks. In: Proceedings of the IEEE/CVF Conference on Computer Vision and Pattern Recognition (2019)

15. Tan, Z., et al.: Relation-aware pedestrian attribute recognition with graph convolutional networks. In: Proceedings of the AAAI Conference on Artificial Intelligence, vol. 34, no. 07 (2020)

16. Bica, I., et al.: Learning "what-if" explanations for sequential decision-making. arXiv preprint arXiv:2007.13531 (2020)

17. Greco, A., Vento, B.: An extended dataset and a baseline for pedestrian attribute recognition with advanced neural networks. In: 21st International Conference on Computer Analysis of Images and Patterns, CAIP 2025 (2025)

18. Liu, M., et al.: Fully fine-tuned CLIP models are efficient few-shot learners. Knowl.-Based Syst. 113819 (2025)

19. Greco, A., Saggese, A., Sansone, C., Vento, B.: An experimental evaluation of smart sensors for pedestrian attribute recognition using multi-task learning and vision language models. Sensors **25**(6), 1736 (2025). https://doi.org/10.3390/s25061736

Adapting to the Wild: From Human Face to Animal Face Recognition

Maria De Marsico[1]([⊠]) [iD], Anil K. Jain[2] [iD], Michele Miranda[1] [iD],
and Alessio Orlando[1] [iD]

[1] Sapienza University of Rome, Rome, Italy
{demarsico,miranda}@di.uniroma1.it, orlando.1792394@studenti.uniroma1.it
[2] Michigan State University, East Lansing, MI, USA
jain@cse.msu.edu

Abstract. Animal recognition can boost, e.g., the search for lost pets and the recognition of individual animals from endangered species. These tasks currently rely on invasive methods like microchips, tattoos, or collars, which are neither reliable nor humane for pets. The lack of large datasets focused on animal face recognition hinders training a deep architecture from scratch to this aim. We investigated the use of pre-trained network models with some transfer learning to animal face recognition, also assessing the influence of possible different image quality. The experiments tested two models, FaceNet trained on large databases of human faces, and ViT trained on generic object categories, hypothesizing that the former should have achieved better performance, and further compared these with State of The Art (SOTA). The benchmarks were a dataset of dog images of good quality, and another with groups of endangered primates (lemurs, golden monkeys, and chimpanzees) photographed in more 'in the wild' conditions. We compared the achieved results for each dataset with ad-hoc trained deep networks representing SOTA for the two specific recognition problems. As expected, the best performance was achieved with dog faces due to better image quality. Less expected was that ViT, pre-trained on ImageNet object database, overcame both FaceNet and, most noticeably, the SOTA reaching a mean verification accuracy of 92.7% and a rank-1 identification rate of 69.6%, compared with 88.8% and 39.74% respectively for SOTA that uses an ad-hoc architecture. The results with endangered primates are encouraging too, but the performance varies with the animal class and the task (verification or identification) and does not always outperform SOTA.

Keywords: Transfer Learning · Animal Face Recognition · Dog Face Recognition · Primate Face Recognition

1 Introduction

About a third of U.S. domestic cats and dogs get lost, and 80% of these are never found[1]. The desirability of the possible automatic search for a lost pet in a

---

[1] https://peeva.co/blog/missing-pet-epidemic-facts-and-figures -Acc. Jul. 2025.

© The Author(s), under exclusive license to Springer Nature Switzerland AG 2026
M. Castrillón-Santana et al. (Eds.): CAIP 2025, LNCS 15621, pp. 53–64, 2026.
https://doi.org/10.1007/978-3-032-04968-1_5

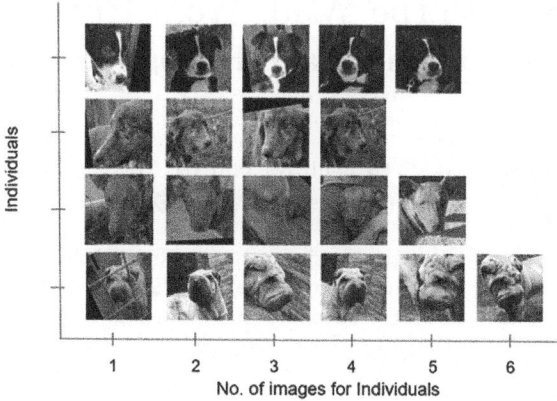

Fig. 1. Examples of inter-/intra-individual face variance in the dog dataset.

database is testified by articles in the media.[2] In 2021 the pet food company Iams launched NOSEiD, a "nose-centric" app to identify lost dogs.[3] Less affective but more critical issues arise with the protection of endangered species. The annual Red List of Threatened Species by the International Union for Conservation of Nature (IUCN) monitors the status of global diversity.[4] At present, 27% of mammal species are listed as 'critically endangered', 'endangered', or 'vulnerable', i.e., 'threatened by extinction'. These include more than 66% of species in the order of Primates, and the trend is toward an increment. Watching biodiversity and wildlife often requires recognizing and tracking individual animals. Current animal identification relies on microchips or tattoos, which can be painful and considered inhumane, or collars, which can be taken off rather easily. In the case of wild animals, capture for implantation of a tracking device can be difficult and potentially harmful. Animal biometric recognition seems a promising alternative. The face is a shared suitable trait that can be captured from a distance and without the subjects' cooperation. Human face recognition is an extensively investigated task [18]. Deep learning models have significantly improved [6] to the point where state-of-the-art systems achieve above 99% accuracy in controlled as well as semi-controlled conditions (see, for example, [12]). It is not the same for animal face recognition. The best systems specially trained for the recognition of dogs or of some primate endangered species stabilize around 90% accuracy [21] [4]. The main reason could be the lack of sufficient training data in contrast with the high variability in the individuals to recognize. The well-known

---

[2] For instance, the July 2018 "Find Your Lost Pet With Facial Recognition" at https://findbiometrics.com/find-your-lost-pet-facial-recognition-507106/ talks about FindingRover app.

[3] see "Petco Love Uses Pet Biometrics to Help Find Missing Cats and Dogs" at https://idtechwire.com/petco-love-uses-pet-biometrics-help-find-missing-cats-dogs-042701/ - April 27, 2021 - Acc. Jul. 2025.

[4] See updated statistics at https://www.iucnredlist.org/resources/summary-statistics.

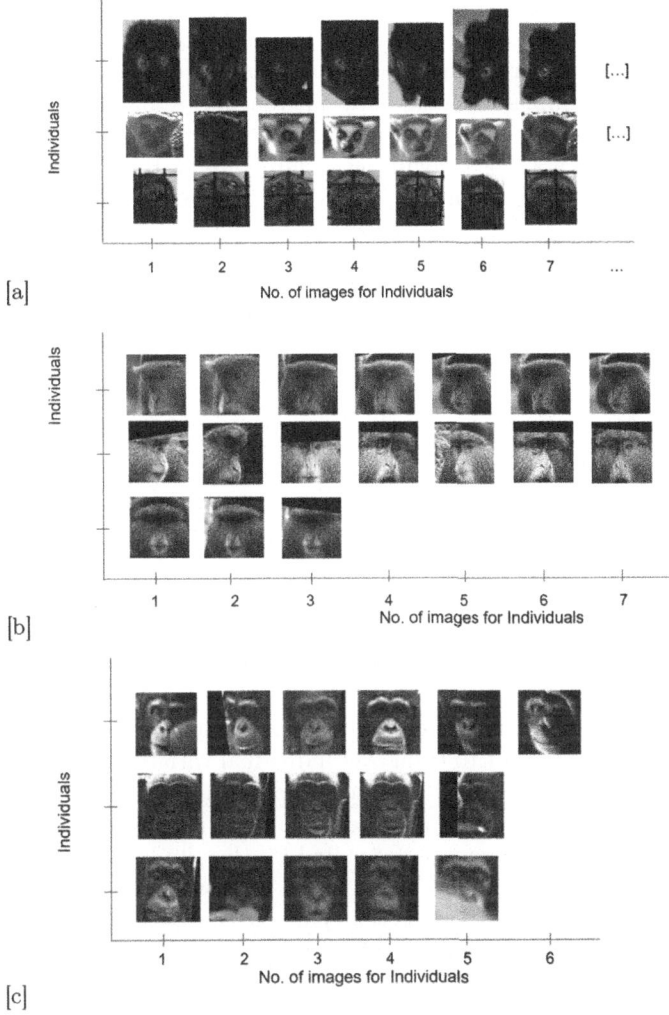

Fig. 2. Examples of inter-/intra-individual face variance for monkey groups in primate dataset:(a) lemurs; (b) golden monkeys; (c) chimpanzees.

"other race effect" [15] studied for humans, is even worse when trying to recognize animal individuals. The general structure of eyes-nose-mouth is shared by a lot of animals, but very large variations in head shapes, ear styles, muzzle sizes, etc., significantly differentiate breeds and species. Figure 1 and Figs. 2 (a), (b), and (c) exemplify intra- and inter-individual variations in the animal datasets used here (Sect. 3). The capture conditions affect image quality and raise further challenges. This paper investigates the possibility that transfer learning allows using the advances in human face or object recognition for animal face recognition using less data and without devising specific architectures. In particular, the experiments consider two of the most accurate present systems for human face or object recognition, trained with such a huge amount of data 'that would be hardly available for the problem at hand. The chosen architectures are fine-tuned and tested on two animal datasets with different image quality. The responses to the following research questions represent the paper's contributions. RQ1) Is transfer learning from a face recognition architecture better than transfer learning from an object recognition architecture for animal face recognition? RQ2) Does transfer learning achieve results comparable with SOTA ah-hoc architectures for the same problem? RQ3) Does image quality influence the results of transfer learning?

2   Related Work

Species identification is useful for animal population studies in natural habitats. The authors of [20] use transfer learning from object categories to species. We rather tackle the problem of animal re-identification, i.e., recognizing a single individual upon re-encounter. This is a core ability in the study of ecosystem evolution. Related works usually rely on features like stains, cuts, or scratches that may not apply to all species (see [10, 13, 19] for whales, and [2] for elephants). We instead test face as a more general trait. An exhaustive list of proposals and methods for individual animal recognition can be found in [16].

Most publicly available animal datasets are devoted to species or breed recognition and not individual animal recognition. They contain, at most, a few images per individual, not sufficient to train over individual variations. A large identity-annotated Holstein-Friesian cattle dataset is Cows2021 [8], which contains 10,402 RGB labeled images capturing the breed's individually distinctive black and white coat pattern. Therefore it is not suited for the aim of this paper that uses face trait. Some datasets deal with muzzle-based cow recognition, but the cow face is not always visible. A very recent paper [1] proposes a dataset of 2,893 images taken from 459 distinct cows, with a variable number of samples per individual (several animals have only from 1 to 3 images), but it was too recent to be included in this study and will be a possible object for future extension of the proposed experiments. A lemur dataset is presented in [9], but it was collected in controlled conditions, contrary to the primate dataset used here to assess the influence of 'in the wild' capture on the results of transfer learning.

Regarding the animals considered in the presented experiments, lemur face identification for species protection is tackled in [3]. The authors of [4] tackle the recognition of individuals in three different groups of primates and propose the dataset used here. The work proposing DogFaceNet [14] for dog face recognition presents the dog face dataset used here. In their experiments, dogs with less than 5 samples are discarded. A better performance is reported in [21] that uses the same dataset though releasing the constraint on the number of samples.

3   Datasets and Experimental Setup

The aim of the datasets used here is to allow comparison of the transfer learning approach on different animals and on data with different quality.

3.1   Dog Dataset

The exploited dataset of dog face images extends the one collected and used with DogFaceNet [14]. It contains 8,363 images of 1,393 dog individuals (from 2 to 41 images per dog with a peak at 5)[5]. Figure 1 shows examples of inter-individual (vertical) and intra-individual (horizontal) sample variations. The dog face images are scraped from the internet. They are taken in uncontrolled conditions, entailing PIE (Pose, Illumination, and Expression) variations, but present quite good overall quality regarding resolution and sharpness. Each RGB image is normalized to 224×224 size. The authors provide the train/test split used here: 6,128 training samples from 1,252 dogs; 1,538 validation samples from 869 dogs; 537 test samples from 93 dogs. It is worth underlining that both in the original work and in our experiments, the training and testing sets have no overlapping identities (identity-based partition). This increases the generalizability of the obtained models that is important for a real-world consumer application.

Original Preprocessing. The black portion of some borders testifies the counter-rotations to keep the eyes horizontally aligned in the upper part of the images and the muzzle a bit lower than the centers.

3.2   Endangered Primate Dataset

The Endangered Primate Dataset groups three collections with lemurs, chimpanzees, and golden monkeys. Each collection comes from the same camera in the same place, either a national park or a zoo. Images are smaller and more challenging than the dog dataset due to capture distance, resolution, and blur. Subsets present different unbalance as for the number of images per individual.

LemurFace Dataset. LemurFace contains 3,000 images of 129 lemurs. Pictures were taken by one of the authors of [4] at the Duke Lemur Center (North Carolina, USA) on two consecutive days, both indoors and outdoors. Most individuals appear in 15 to 28 images. Figure 2(a) shows examples of inter-individual (vertical) and intra-individual (horizontal) sample variations.

GoldenMonkeyFace Dataset. GoldenMonkeyFace contains 1450 images of 49 golden monkeys, extracted from 241 videos recorded in the Volcanoes National Park in Rwanda. The distribution of images is quite uneven, with most individuals appearing in 0 to 50 images each. Figure 2(b) shows examples of inter-individual (vertical) and intra-individual (horizontal) sample variations.

ChimpanzeeFace Dataset. ChimpanzeeFace combines two datasets, C-Tai and C-Zoo, presented by Loos and Ernst in [11] and then extended by Freytag

---

[5] https://github.com/GuillaumeMougeot/DogFaceNet/releases/ - Dataset 1.

et al. in [7]. C-Zoo was collected in the Leizpig Zoo in Germany and contains 2109 faces of 24 individuals. C-Tai is captured in the Tai National Park in Cote d'Ivoire and contains 5078 face images of 78 subjects. The final Chimpanzee dataset does not contain 'the animals with less than three samples, so the total is 5559 face images of 90 chimpanzees. The number of images per individual varies, but most primates appear in 1 to 80 images. Figure 2(c) shows examples of inter-individual (vertical) and intra-individual (horizontal) sample variations.

Original Preprocessing. Manually marked eyes and mouth landmarks allow aligning the images, which are cropped around the face and rotated: the horizontally aligned eyes are in the upper part, and the mouth is in the lower part.

3.3   Experimental Setup

Models. This work compares transfer learning from two pre-trained models for animal face recognition. 1) FaceNet[6] is designed for face recognition; it is based on InceptionResNetV1 [17] . 2) The Vision Transformer ViT [5] is a high-performance transformer model for robust object recognition.

Additional Preprocessing. The original datasets preprocessing was extended by model-dependent preprocessing for the presented experiments. Using FaceNet, images were further cropped (150×150) to reduce the amount of *context* fed to the network. Color-normalization was separately applied by mean and standard deviation to the RGB channels. The ViT preprocessing implemented the *vit-base-patch16-224*[7] model pre-training, with image resizing (224×224) and a color normalization with fixed values of 0.5 for both mean and standard deviation.

Training. The present study exploits the triplet loss with a Siamese network (either two identical FaceNet networks or two identical ViTs). To reduce the hard triplet mining requirements, semi-hard mining was used for the first epochs.

4   Evaluation Results

The tests with FaceNet were made with just the last ResNet block unfrozen, since the small size of the animal datasets allowed to retrain only a small part of the network from scratch. ViT is more flexible, and it was entirely unfrozen.

Fair comparison and understanding of whether the proposed approach can reach SOTA require using the same datasets, protocols, and metrics of [14] and [21] for dogs and of [4] for primates, which all use dedicated architectures and learning strategies. The investigated models are referred to as DogCNN (based on FaceNet) and DogViT when retrained on dogs and as PrimCNN (based on FaceNet) and PrimViT on primates. When appropriate, we report the results of changing the *patience* hyperparameter, i.e., the number of epochs to wait before the early stopping of training if no progress is achieved on the validation set.

---

[6]   https://github.com/timesler/facenet-pytorch.
[7]   https://huggingface.co/google/vit-base-patch16-224.

4.1 Verification (1:1 Identity Comparison)

DogFace Dataset. The tests followed the same protocol used in the compared works, using 100 draws, each randomly selecting 2,500 authentic pairs and 2,500 impostor pairs from all possible pairs. The results represent the average performance. Table 1 shows that the compared SOTA model reaches a mean accuracy of 88.8%. The best model tested in this work, using ViT as a backbone, achieved a mean accuracy of 92.7%, setting a new state of the art. The model based on FaceNet unexpectedly only achieved a significantly lower 80.8%.

Table 1. Mean accuracy of each method on dogs dataset

Learning Method	Accuracy
DogFaceNet [14]	76.9%
SOTA [21]	88.8%
DogCNN	80.8%
DogViT	**92.7%**

Table 2. Verification performance for the different primate groups expressed in terms of *mean ± s.d.* of TAR at FAR=1%.

Lemur Face			
Method	No re-training	Patience 1	Patience 10
A	63.30% ± 5.11%	67.35% ± 3.44%	67.22% ± 4.03%
B	72.60% ± 10.10%	73.7% ± 9.30%	81.70% ± 9.49%
C	83.11% ± 5.31%		

Golden Monkey Face			
Method	No re-training	Patience 1	Patience 10
A	67.70% ± 8.50%	69.04% ± 6.91%	70.54% ± 5.35%
B	52.60% ± 5.81%	52.89% ± 6.90%	53.90% ± 7.50%
C	78.72% ± 5.80%		

Chimpanzees Face			
Method	No re-training	Patience 1	Patience 10
A	30.54% ± 8.87%	48.86% ± 6.35%	48.43% ± 7.31%
B	47.40% ± 7.01%	60.81% ± 5.00%	62.65% ± 5.51%
C	59.87% ± 3.34%		

Primate Face Dataset. Still following the same protocol of the compared work, the verification performance is computed differently from DogFaceNet, by dividing each primate subset into 5 folds and partitioning the test samples into probe and gallery. The probe/gallery split is repeated until every image has been used as a probe. The highest similarity score is selected when more gallery templates belong to the same individual. Verification results are the *mean ± s.d.* of the True Acceptance Rate (TAR) when FAR= 1%. Table 2 shows the results: A stands for PrimCNN, B for PrimViT, and C for PrimNet [4].

Lemur Face Dataset. The proposed methods do not outperform SOTA. PrimViT gets very close (with longer patience=10), yet with a considerably higher standard deviation. The PrimViT advantage is the quite close performance to SOTA using transfer learning instead of a dedicated architecture.

Golden Monkey Face Dataset. It is interesting to notice from Table 2 that for this dataset, the best of the two proposed models is the simplest one, i.e., PrimCNN. Anyway, none of the proposed models could outperform SOTA.

Chimpanzee Face Dataset. For Chimpanzee Dataset the ViT-based model outperforms the state of the art, yet again with higher variance (Table 2).

4.2   Identification (1:N Comparison)

Dog Face Dataset. For the identification test with the dog dataset, the gallery contains one sample for each identity in the test set (it is worth reminding that test identities do not appear in the training set) and the remaining images go in the probe. The two compared works do not evaluate open set performance. Therefore, the results here are consistently limited to closed set and reported in terms of Cumulative Match Scores (CMS) at Rank 1 and Rank 5. The data split for probe and gallery composition and subsequent tests are repeated 1000 times. Table 3 presents the results. DogViT outperforms the other proposed method and also the current state of the art. In addition, the accuracy of 87.20% at Rank 5 makes it feasible to apply a shortlist-based search.

Table 3. Mean Cumulative Match Score at Rank 1 and 5 for each method on dogs.

Learning Method	Rank 1	Rank 5
DogFaceNet [14]	10.96%	32.58%
SOTA [21]	39.74%	68.80%
DogCNN	38.42%	57.31%
DogViT	**69.60%**	**87.20%**

Primate Face Dataset. In both closed and open set identification, evaluation is 5-fold, and there are 100 trials for the gallery/probe split to smooth out the randomness. A random sample for every identity is chosen as a probe, and the rest are stored in the gallery. The results of the identification closed set are reported in terms of *mean ± s.d.* of CMS at Rank 1. The results of the identification open set are reported in terms of *mean ± s.d.* of Detection and Identification Rate at Rank 1 when FAR= 1%. As above, in Table 4, A stands for PrimCNN, B for PrimViT, and C for PrimNet [4].

Lemur Face Dataset. The trend continues whereby the ViT-based proposed model outperforms SOTA, but with a (slightly) higher variance. For lemurs, this only applies in closed set, while in open set the results are quite bad.

Golden Monkey Face Dataset. The results confirm the observed trend. Results are higher than SOTA for identification closed set with ViT backbone, with a (slightly) higher variance. Differently from lemurs, in an open set, the results on golden monkeys are worse but closer to SOTA, which is low anyway.

Chimpanzee Face Dataset. The performances of PrimViT with Chimpanzees in closed set are slightly higher than SOTA but with a higher variance. Open set of PrimViT works quite badly, but also the compared work shows lower results.

Table 4. Performance of identification closed and open set for the different primates.

Lemur Face

	Closed set (*mean ± s.d.* of CMS at Rank 1)			Open set (*mean ± s.d.* of DIR at FAR = 1%)		
Method	No re-training	Patience 1	Patience 10	No re-training	Patience 1	Patience 10
A	79.40% ± 2.40%	79.47% ± 2.89%	76.56% ± 1.21%	11.90% ± 2.10%	11.49% ± 2.17%	10.76% ± 2.34%
B	94.40% ± 1.30%	93.80% ± 1.90%	94.00% ± 2.20%	17.30% ± 2.90%	16.90% ± 3.00%	17.60% ± 1.50%
C	93.76% ± 0.90%			81.73% ± 2.36%		

Golden Monkey Face

	Closed set (*mean ± s.d.* of CMS at Rank 1)			Open set (*mean ± s.d.* of DIR at FAR = 1%)		
Method	No re-training	Patience 1	Patience 10	No re-training	Patience 1	Patience 10
A	77.90% ± 3.30%	69.41% ± 8.56%	71.54% ± 6.01%	13.71% ± 5.22%	11.31% ± 6.05%	15.95% ± 7.22%
B	87.10% ± 5.06%	87.1% ± 7.02%	92.80% ± 2.90%	40.50% ± 11.20%	41.71% ± 11.59%	50.10% ± 11.89%
C	90.36% ± 0.92%			66.11% ± 7.99%		

Chimpanzees Face

	Closed set (*mean ± s.d.* of CMS at Rank 1)			Open set (*mean ± s.d.* of DIR at FAR = 1%)		
Method	No re-training	Patience 1	Patience 10	No re-training	Patience 1	Patience 10
A	53.44% ± 5.58%	52.95% ± 6.01%	48.04% ± 7.02%	1.25% ± 0.49%	7.05% ± 1.42%	7.26% ± 1.13%
B	64.40% ± 5.10%	73.95% ± 2.79%	75.95% ± 3.57%	8.50% ± 2.10%	14.99% ± 3.96%	15.93% ± 3.35%
C	75.82% ± 1.25%			37.08% ± 11.22%		

4.3 Discussion and Final Notes on Experiments

It is possible to draw some responses to the introduced research questions. RQ1) Surprisingly enough, transfer learning from ViT object recognition model works better than transfer learning from FaceNet for animal face recognition, despite the different image distributions: for instance, LemurFace contains 3000 images of 129 lemurs, while DogFace has more than 8300 images of 1393 dogs with a very different number of samples per individual. RQ2) ViT-based approach achieves outstanding results over the present SOTA for dogs, demonstrating that transfer learning from the transformer can work also with unbalanced data distributions. RQ3) The quality of target images definitely affects the effectiveness of transfer learning, causing results with endangered primates to be not as good as for dogs and quite rambling; however, they still encourage further investigations.

The misclassified pairs of dogs provide some further interesting cues. The qualitative analysis of false rejections seems to suggest that pose is critical, especially the amount of pitch that makes images diverge from human ones due to the tridimensional structure of the muzzle. On the contrary, considering false accept responses, it seems that similar colors get confused, causing an incorrect evaluation of overall similarity especially when the geometrical features are, on the contrary, quite similar. Wrong identification rankings enforce this observation. The same happens with primate images. Blurring is also a big issue, especially with images extracted from videos. Given the capture conditions, this is frequently observed with primates. The same issues can also be noticed in correct rankings, even when the first position is occupied by the correct animal.

5   Conclusions and Future Work

Methods proposed in the literature use complex dedicated networks trained from scratch, e.g., for dogs or primates. The present proposal investigates how to address the lack of data that negatively affects training by taking advantage of well-known and established networks designed for different tasks: human face recognition and object recognition. Tests are made on two datasets with different animals and image quality. The approach seems very promising, but its real strength is in the possibility to work with fewer data and with fewer weights to train, making learning lighter and faster. It is possible to observe that ViT is pre-trained on ImageNet, with more than 100 different dog races and also many monkey classes, including lemurs and chimpanzees. This may partially explain the better outcomes. A stronger evaluation protocol for dogs would select impostor pairs only from dog faces of different individuals but same-race. This is not possible, given the dataset labeling. On the other end, we still achieve better results than SOTA using the same setting. The system does not turn into a dog race classification system, because the used dataset includes more individual animals per race. Furthermore, training on ImageNet may allow distinguishing races more accurately, not individual animals, as for other animal datasets. The experiments also demonstrate that, when the image quality decreases, as in the case of the primate dataset, the transfer learning strategy becomes jeopardized and seldom reaches SOTA. In the future, it will be worth making a thorough analysis of the misclassifications for the two models to gain a deeper insight into the different covariates, e.g., color or quality, that can affect the performance of models obtained by transfer learning. Some complementary behaviors may suggest the design of a fusion strategy of different models to improve results, especially for lower-quality images as those extracted from videos.

As a final remark, it is worth underlining that this work is a proof of concept and the conclusions may not generalize due to the relatively small databases used in the experiments, and the peculiarities of the animals considered. However, a more thorough evaluation was not possible due to the lack of either larger datasets or datasets with different animals capturing the same features. In addition, more advanced architectures could achieve better performance, but our aim was just to verify whether transfer learning can overcome ad-hoc, hard-to-train architectures.

References

1. Ahmed, S.U., Frnda, J., Waqas, M., Khan, M.H.: Dataset of cattle biometrics through muzzle images. Data in Brief, p. 110125 (2024)
2. Ardovini, A., Cinque, L., Sangineto, E.: Identifying elephant photos by multi-curve matching. Pattern Recogn. **41**(6), 1867–1877 (2008)
3. Crouse, D., et al.: LemurFaceID: a face recognition system to facilitate individual identification of lemurs. BMC Zool. **2**(1), 1–14 (2017)
4. Deb, D., Wiper, S., Gong, S., Shi, Y., Tymoszek, C., Fletcher, A., Jain, A.: Face recognition: primates in the wild, pp. 1–10 (2018). https://doi.org/10.1109/BTAS.2018.8698538
5. Dosovitskiy, A., et al.: An image is worth 16x16 words: transformers for image recognition at scale (2020). arXiv:2010.11929
6. Du, H., Shi, H., Zeng, D., Zhang, X.P., Mei, T.: The elements of end-to-end deep face recognition: a survey of recent advances. ACM Comput. Surv. (CSUR) **54**(10s), 1–42 (2022)
7. Freytag, A., Rodner, E., Simon, M., Loos, A., Kühl, H.S., Denzler, J.: Chimpanzee faces in the wild: log-Euclidean CNNs for predicting identities and attributes of primates. In: Rosenhahn, B., Andres, B. (eds.) GCPR 2016. LNCS, vol. 9796, pp. 51–63. Springer, Cham (2016). https://doi.org/10.1007/978-3-319-45886-1_5
8. Gao, J., Burghardt, T., Andrew, W., Dowsey, A.W., Campbell, N.W.: Towards self-supervision for video identification of individual Holstein-Friesian cattle: the cows2021 dataset (2021). arXiv:2105.01938
9. Guan, Y., et al.: Face recognition of a Lorisidae species based on computer vision. Global Ecol. Conserv. **45**, e02511 (2023)
10. Hiby, L., Lovell, P.: Computer aided matching of natural markings: a prototype system for grey seals. Rep. Int. Whaling Comm. **12**, 57–61 (1990)
11. Loos, A., Ernst, A.: Detection and identification of chimpanzee faces in the wild. In: 2012 IEEE International Symposium on Multimedia, pp. 116–119 (2012). https://doi.org/10.1109/ISM.2012.30
12. Mishra, N.K., Dutta, M., Singh, S.K.: Multiscale parallel deep CNN (MPDCNN) architecture for the real low-resolution face recognition for surveillance. Image Vis. Comput. **115**, 104290 (2021)
13. Mizroch, S.A., Beard, J.A., Lynde, M.: Computer assisted photo-identification of humpback whales. Rep. Int. Whaling Comm. **12**, 63–70 (1990)
14. Mougeot, G., Li, D., Jia, S.: A deep learning approach for dog face verification and recognition. In: Nayak, A.C., Sharma, A. (eds.) PRICAI 2019. LNCS (LNAI), vol. 11672, pp. 418–430. Springer, Cham (2019). https://doi.org/10.1007/978-3-030-29894-4_34
15. O'toole, A.J., Deffenbacher, K.A., Valentin, D., Abdi, H.: Structural aspects of face recognition and the other-race effect. Memory Cogn. **22**, 208–224 (1994)
16. Schneider, S., Taylor, G.W., Linquist, S., Kremer, S.C.: Past, present and future approaches using computer vision for animal re-identification from camera trap data. Methods Ecol. Evol. **10**(4), 461–470 (2019)

17. Szegedy, C., et al.: Going deeper with convolutions. In: Proceedings of the IEEE Conference on Computer Vision and Pattern Recognition, pp. 1–9 (2015)
18. Wanyonyi, D., Celik, T.: Open-source face recognition frameworks: a review of the landscape. IEEE Access **10**, 50601–50623 (2022)
19. Whitehead, H.: Computer assisted individual identification of sperm whale flukes. Rep. Int. Whaling Comm. **12**, 71–77 (1990)
20. Wilber, M.J., et al.: Animal recognition in the Mojave desert: vision tools for field biologists. In: 2013 IEEE Workshop on Applications of Computer Vision (WACV), pp. 206–213. IEEE (2013)
21. Yoon, B., So, H., Rhee, J.: A methodology for utilizing vector space to improve the performance of a dog face identification model. Appl. Sci. **11**(5) (2021). https://doi.org/10.3390/app11052074

Facial and Video Recognition

Enhanced Deep Learning DeepFake Detection Integrating Handcrafted Features

Alejandro Hinke-Navarro[1]([✉]), Mario Nieto-Hidalgo[1][iD], Juan M. Espín[1][iD], and Juan E. Tapia[2][iD]

[1] Facephi Biometrics S.A., Alicante, Spain
{alejandrohinke,marionieto,jmespin}@facephi.com
[2] da/sec-Biometrics and Internet Security Research Group, Darmstadt, Darmstadt, Germany
juan.tapia-farias@h-da.de

Abstract. The rapid advancement of deepfake and face swap technologies has raised significant concerns in digital security, particularly in identity verification and onboarding processes. Conventional detection methods often struggle to generalize against sophisticated facial manipulations. This study proposes an enhanced deep-learning detection framework that combines handcrafted frequency-domain features with conventional RGB inputs. This hybrid approach exploits frequency and spatial domain artifacts introduced during image manipulation, providing richer and more discriminative information to the classifier. Several frequency handcrafted features were evaluated, including the Steganalysis Rich Model, Discrete Cosine Transform, Error Level Analysis, Singular Value Decomposition, and Discrete Fourier Transform.

Keywords: Face Manipulation · Handcrafted Features · Digital Forensics

1 Introduction

Image manipulation has become a widely discussed topic over the years, with its detection posing an increasingly complex challenge because of the rapid advancements in generative techniques. Among the various forms of digital face manipulation, key methods include face swap, identity swap, attribute manipulation, and entire face synthesis [15].

Face swapping involves replacing a target individual's face with another person's, effectively altering the subject's appearance while retaining their original context. In contrast, full-face synthesis refers to the complete generation of facial images from scratch using advanced generative models, such as Generative Adversarial Networks (GANs) or diffusion models. These techniques enable the creation of highly realistic facial representations, often indistinguishable from authentic images.

© The Author(s), under exclusive license to Springer Nature Switzerland AG 2026
M. Castrillón-Santana et al. (Eds.): CAIP 2025, LNCS 15621, pp. 67–78, 2026.
https://doi.org/10.1007/978-3-032-04968-1_6

Nowadays, most users perceive this technology as harmless entertainment; however, it is increasingly being misused for malicious purposes, such as spreading fake news, generating illicit content, and engaging in political manipulation, among others. These harmful applications have a significant impact on social media, undermining trust and contributing to a crisis of authenticity in digital content across the Internet.

Traditional methods, based on RGB pixel values and convolutional neural networks (CNNs), perform well on intra-datasets but struggle to generalize across unseen datasets [15]. This limitation arises because artifacts indicative of manipulation can be significantly diminished due to factors such as image compression or manual editing, making it challenging for these models to detect subtle inconsistencies across diverse sources.

To address this challenge, the following research questions are explored:

- How can generalization across datasets be improved?
- Which frequency-domain representations are most effective?
- How can frequency-domain features be integrated into deep learning models?

This work focuses specifically on identity swapping and full-face synthesis, two of the most prevalent manipulation techniques in digital media.

The main contributions of this work are:

- A study of frequency-domain features for face manipulation detection.
- An evaluation of several handcrafted features, identifying Discrete Cosine Transform as the most effective.
- A demonstration that minimum score-level fusion between intensity pixel values and frequency features yields improved performance over baseline models.

The remainder of this article is structured as follows: Sect. 2 reviews related work on deepfake and face manipulation detection. Section 3 describes the datasets and the proposed method. Section 4 reports experimental results. Finally, Section. 5 concludes the paper and outlines directions for future research.

2   Related Works

Traditional methods based on intensity values (RGB) images and convolutional neural networks (CNNs) have demonstrated high performance on intra-datasets but lower generalization capabilities to perform with high rates on cross-datasets. To overcome this challenge, new approaches like frequency domain have been explored based on the changes in frequencies (high and low) that are produced when the image is manipulated. A similar effect is dedicated to compression.

Conventional deepfake detection approaches predominantly leverage spatial domain features extracted from pixel values across RGB images using deep neural networks.

Studies such as Luo et al. [6] highlight that CNN-based models often overfit to method-specific color textures, which limits their generalization capabilities when tested against unseen manipulations.

Similarly, the work by Ibsen et al. [1] emphasizes that RGB-only models struggle to differentiate real from synthetic faces when exposed to novel generative models or post-processing operations. These limitations underline the need for more robust detection techniques that incorporate additional feature representations beyond RGB data.

Recent advances in deepfake detection have highlighted the effectiveness of frequency-domain analysis in identifying manipulated content [6, 16].

Wang et al. [16] introduced a Frequency Domain Filtered Residual Network, which enhances detection robustness by fusing wavelet-transformed frequency information with RGB data, particularly improving performance on compressed deepfake images.

Luo et al. [6] showed that multi-scale SRM filtering strengthens cross-dataset generalization by detecting high-frequency noise residuals.

More recently, Tan et al. [12] proposed FreqNet, a frequency-aware model that enhances deepfake detector generalization by learning high-frequency features independently of their source.

Li et al. [4] introduced FreqBlender, a method that synthesizes pseudo-fake faces by manipulating frequency information, improving the learning of generic forgery traces, and enhancing detection accuracy.

Tapia et al. [14] also demonstrate that frequency-based filters can be used to detect digital manipulation attacks, such as Morphing.

Rahaman et al. [10] introduced the concept of spectral bias, demonstrating through Fourier analysis that neural networks exhibit a learning preference for low-frequency functions. This spectral bias explains why neural networks often generalize well to natural data and highlights the robustness of low-frequency components to parameter perturbations.

Many of these findings suggest that the frequency domain contains valuable information that can be effectively leveraged to improve the detection of manipulated images.

3   Proposed Method

This work proposes a method for detecting face digital manipulation attacks, based on handcrafted frequency features and fusion with intensity values at the score level. Several datasets were employed to evaluate the generalization capabilities of the proposed approach. A diagram illustrating the method is presented in Fig. 1.

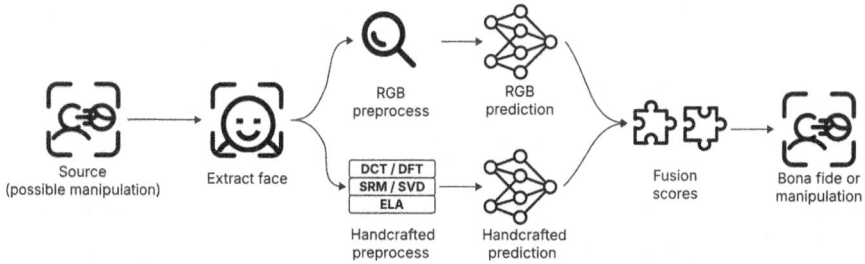

Fig. 1. Manipulation Attack Detection Framework.

3.1   Datasets

In this study, six different datasets of digitally manipulated face images were used:

- **FaceForensics++** [11]: A widely used dataset for deepfake detection, comprising 4,320 videos, including 720 original videos sourced from YouTube and 3,600 manipulated videos generated using FaceShifter, FaceSwap, Face2Face, Deepfakes, and NeuralTextures. The official dataset split was followed, with 720 videos for training, 140 for validation, and 140 for testing. Five random frames per video were used in this study.
- **Celeb-DF** [5]: A deepfake dataset specifically designed for identity-swapping manipulations, containing 5,639 deepfake videos generated from 590 original videos sourced from YouTube. Due to its real-world origin, the dataset is highly compressed, often exhibiting lower visual quality and compression artifacts, making detection more challenging. Only one frame per video was used in this study.
- **DeepfakeTIMIT** [2]: This dataset comprises videos where faces are swapped using a GAN-based approach developed from the original autoencoder-based Deepfake algorithm. It includes 620 videos with faces swapped, using the VidTIMIT database as the source. Two different qualities are provided: lower quality (LQ) with 64×64 input/output size models and higher quality (HQ) with 128×128 size models. One frame per video was used in this study.
- **DeePhy** [9]: This dataset employs sequential face swapping based on a phylogenetic approach. It contains 468 spoof videos sourced from YouTube, encoded in MPEG4 format with a resolution of 720p, using a single frame per video. One frame per video was used in this study.
- **Defacto** [7]: This dataset includes face-swapped images generated from MS-COCO images through automated forgery generation techniques, resulting in semantically meaningful and detailed manipulations. It contains 3,000 spoof images of variable sizes. One frame per video was used in this study.
- **SWAN-DF** [3]: The first high-fidelity publicly available dataset of realistic audio-visual deepfakes, where both faces and voices appear and sound like the target person. Based on the public SWAN database of real videos recorded

in HD on iPhone and iPad Pro, it includes 30 pairs of manually selected individuals. Faces and voices were swapped using several autoencoder-based face-swapping models and blending techniques from DeepFaceLab, along with voice conversion methods such as YourTTS, DiffVC, HiFiVC, and FreeVC. A random selection of 10% of the dataset was used in this study.

Table 1 shows a summary of all the datasets used in this research.

Table 1. Summary of datasets. DF, FS, NT, FSW, and F2F represent DeepFake, FaceShifter, NeuralTransfer, FaceSwap, and Face-to-Face, respectively.

Database	No. of Images	Manipulation algorithm
FF++	25,000 fake, 5000 real	DF, FS, NT, FSW, F2F
CelebDF	5,639 fake, 590 real	Improved DF
DeepfakeTIMIT	640 fake	GAN-based (face swap-GAN)
DeePhy	468 fake, 100 real	Phylogenetic sequential FS
Defacto	3,000 fake, 200 real	Automated semantic FS
SWAN-DF	11,940 fake	Autoencoder-based (DeepFaceLab)

3.2 Metrics

To evaluate the effectiveness of the proposed method, the ISO/IEC 30107-3 was followed[1]. Detection Equal Error Rate (D-EER) metric was employed, which represents the point at which the Attack Presentation Classification Error Rate (APCER) and the Bona fide Presentation Classification Error Rate (BPCER) are equal. The APCER indicates the proportion of attack presentations incorrectly classified as bona fide (false positives), while BPCER denotes the proportion of bona fide presentations incorrectly classified as attacks (false negatives). A lower D-EER value reflects the higher accuracy and robustness of the detection system. D-EER is widely used in biometric systems and forgery detection tasks due to its balanced assessment of both types of error rates, providing a comprehensive measure of system performance.

3.3 Feature Extraction

Several feature extraction techniques based on handcrafted features have been employed to distinguish between bona fide and digitally manipulated images [14]. In this study, five frequency handcrafted feature extraction methods were used individually and in combination to improve the detection of manipulated faces: Color (RGB), which is represented by the pixel values, Discrete Cosine Transform (DCT), Steganalysis Rich Model (SRM), Discrete Fourier Transform (DFT), Error Level Analysis (ELA), and Singular Value Decomposition (SVD).

---

[1] https://www.iso.org/standard/79520.html.

All features were extracted from grayscale versions of the images using to emphasize structural and frequency domain characteristics, except for color pixel values of RGB images.

Discrete Cosine Transform (DCT). DCT is a widely used technique in image processing that transforms spatial domain information into the frequency domain. It decomposes an image into a sum of cosine functions oscillating at different frequencies, which helps detect hidden artifacts introduced during manipulations, particularly in compressed images, as it is a core component of popular formats like JPEG that exploit frequency information for efficient compression.

In this study, DCT was applied to the entire image as well as to sub-blocks of varying sizes, specifically 8×8, 12×12, 16×16, 20×20, and 24×24 pixels, with the 20×20 configuration proving to be the most effective.

Steganalysis Rich Model (SRM). SRM is a feature extraction technique commonly used in digital forensics to detect hidden modifications in images. It focuses on capturing high-frequency noise patterns that arise from manipulation processes. In this study, an SRM filter using a kernel described in Eq. 1 was applied to the grayscale images to enhance edge detection and expose subtle alterations.

$$SRM = \begin{bmatrix} 0.0 & 1.0 & 0.0 \\ 1.0 & -4.0 & 1.0 \\ 0.0 & 1.0 & 0.0 \end{bmatrix} \tag{1}$$

This filter emphasizes discrepancies in the high-frequency domain by highlighting regions where pixel intensities exhibit irregular patterns, which are often indicative of tampering.

The Discrete Fourier Transform (DFT). DFT converts an image from the spatial domain to the frequency domain, representing it in terms of sinusoidal components. This transformation helps analyze periodic patterns and identify inconsistencies introduced by generative models or post-processing operations. It is particularly useful for detecting manipulation artifacts that manifest as unnatural frequency distributions.

Error Level Analysis (ELA). ELA is a forensic technique used to detect areas of an image that have undergone different levels of compression. Repeated compression of an image and comparison with the original reveal discrepancies in compression artifacts, which can indicate tampered regions. In this study, it was applied to grayscale images to identify potential manipulation traces based on differences in compression levels across various regions of the image. Areas with significant discrepancies often correspond to edited portions, making ELA a valuable tool for forgery detection.

Singular Value Decomposition (SVD). SVD is a matrix factorization technique that decomposes an image into three matrices, representing its intrinsic structure in terms of singular values and orthogonal components. It is effective in identifying structural changes caused by manipulations, as alterations typically disrupt an image's natural rank and singular value distribution. This study applied SVD to grayscale images to capture global structural inconsistencies with a component of 50. Figure 2, shows an example of the frequency features extracted.

Fig. 2. Feature extraction example for a manipulated image.

3.4   Models

Preprocesing. The preprocessing pipeline begins by cropping faces and adding a 50% padding around each crop. This wider margin exposes background context that face-swap and similar attacks typically leave unaltered, allowing the network to contrast manipulated pixels with their unmodified surroundings. Subsequently, the images are resized to a fixed resolution of 384 × 384 pixels.

The resized images serve as inputs to either EfficientNetV2 B0 [13] or MobileViT-S [8] models, both of which are initialized using ImageNet pre-trained weights. EfficientNetV2-B0 was selected due to its well-balanced trade-off between computational efficiency and performance, making it suitable for deployment in resource-constrained environments. MobileViT-S, chosen for its compact size and rapid inference capability, leverages transformer-like attention mechanisms to capture detailed feature interactions through self-attention maps.

For handcrafted models, images are converted to grayscale. Several data augmentation techniques were employed during training to enhance model robustness, including horizontal flipping, random contrast adjustment, random brightness variation, random hue shifts, random saturation changes, and random JPEG compression, which were applied exclusively to manipulated images. This choice is motivated since GAN generated images often lack of compression artifacts, random JPEG compression was applied to manipulated samples to prevent the model from relying on this pattern and instead focus on manipulation-related traces.

Model weights were optimized using the AdaGrad algorithm with a minibatch size of 32 and an initial learning rate of $1e - 4$. Training was conducted for up to 225 steps, equivalent to approximately 65 epochs. For the MobileViT-S architecture, a patch size of 2 was explicitly adopted. All training was performed using an NVIDIA A100 GPU.

Fusion at Score Level. The fusion of scores involves combining the outputs of different models (RGB and Frequency) based on specific aggregation rules. The fusion strategies considered in this experiment include weighted fusion, where models contribute based on assigned importance; minimum fusion, which selects the lowest score among the models; mean fusion, which computes the average score; and maximum fusion, which takes the highest score from each model.

4   Experiments

Three experiments were proposed to show and compare the results with different frequency filters.

4.1   Experiment 1: Handcrafted Features Benchmark

All filters were trained and evaluated using the datasets mentioned in Table 1 to measure the impact of each one.

Table 2. D-EER % for the different handcrafted features.

		Intra	Cross		
		FF++	Celeb-DF	Dephy	Defacto
Effv2b0	RGB	6.20	35.87	11.72	30.00
	SRM	5.75	51.53	14.11	61.10
	DCT	3.18	46.96	8.19	48.37
	ELA	8.58	45.96	11.61	38.92
	DFT	35.23	46.54	36.30	48.00
	SVD	31.69	44.77	17.63	33.07
MobileViT-S	RGB	1.37	36.76	7.05	31.97
	SRM	13.93	53.98	25.83	53.50
	DCT	5.03	49.18	20.02	50.50
	ELA	13.59	46.11	9.44	33.50
	DFT	40.27	48.67	45.85	48.03
	SVD	35.43	41.56	20.02	32.50

Observing the cross-dataset performance in Table 2, the results are generally suboptimal. These findings suggest that mismatched bona fide distributions due to dataset-specific conditions such as varying image acquisition settings, compression methods, and quality negatively affect model generalization. Experiment 2 explores a potential solution by using a single, consistent source of bona fide images from FaceForensics++(FF++).

4.2   Experiment 2: Handcrafted Features Benchmark Using FF++ Bona Fides

Due to significant discrepancies in the distribution of bona fide images across the public datasets used in this study (see Fig. 3), only bona fide images from the FF++ dataset were employed for final model evaluation. This decision is supported by several considerations.

- **Heterogeneous capture conditions.** The datasets differ markedly in their image sources: some contain "in-the-wild" pictures taken under uncontrolled settings, whereas others include images acquired in controlled or studio environments. This mismatch produces considerable variation in visual characteristics and overall image quality.
- **Compression and format inconsistencies.** Differences in compression schemes, file formats, and orientations introduce additional divergence, making cross-dataset comparisons more difficult.
- **Shortage of bona fide samples.** Several datasets provide only a small number of bona fide images, or none at all, which limits representativeness and reduces statistical reliability during evaluation.

Figure 3 shows the different images from bona fide subsets.

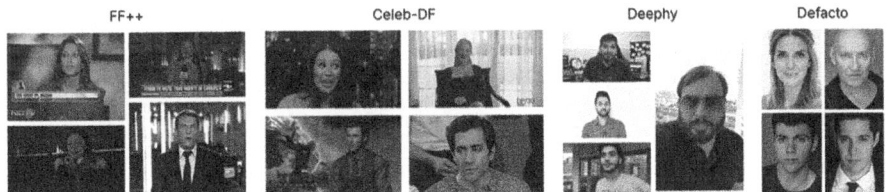

Fig. 3. Bona fide samples distribution across datasets.

4.3   Experiment 3: Fusions at Score Level

Building upon the findings from Experiment 2 in Table 3, RGB and DCT demonstrated superior performance. This experiment focuses on these two feature set. The DCT-based model exhibits strong detection capabilities for identity face swapping but performs poorly in full-face synthesis detection, whereas the RGB-based model shows the opposite trend. The objective is to leverage the strengths of both spatial and frequency domains to enhance detection performance by exploring various fusion strategies.

It is essential to emphasise that model calibration prior to score fusion significantly impacts overall performance. Since FF++ served as the common source of bona fide images for each dataset, calibration was conducted by targeting a BPCER value. This approach ensures a consistent thresholding strategy across datasets.

In Table 4, the default protocol refers to the configuration obtained directly from training without applying any threshold calibration. For the RGB Efficient-Net v2 b0 and DCT EfficientNet v2 b0 models, the "Default" model configuration yielded a BPCER of 19.27% and 7.71%, respectively. In contrast, the RGB MobileViT-S and DCT MobileViT-S models achieved BPCERs of 4.02% and 11.07%, respectively, under the default conditions. "Protocol I" refers to each model being calibrated to BPCER 2.00% before the fusion. "Protocol II" means that each model has been calibrated to BPCER 5.00% before the fusion.

Table 3. D-EER % for the different handcrafted features using FF++ bona fides for every dataset. The highlighted numbers in bold indicate the best performance observed across the dataset.

		Intra	Cross					Avg.
		FF++	Celeb-DF	Df-Timit	Dephy	Defacto	Swan DF	
Effv2b0	RGB	6.20	17.45	**6.39**	9.43	29.36	15.58	14.07
	SRM	5.75	17.12	36.25	22.95	18.43	24.17	20.78
	DCT	3.18	**3.86**	50.49	12.38	**9.93**	29.35	18.20
	ELA	8.58	8.22	13.75	14.57	44.13	33.21	20.41
	DFT	35.23	24.00	61.41	53.14	62.44	69.29	50.92
	SVD	31.69	35.61	10.92	25.53	42.89	36.04	30.78
Mobile ViT-S	RGB	**1.37**	**4.03**	27.18	**6.58**	11.40	**8.72**	9.21
	SRM	13.93	19.13	36.73	34.57	19.11	27.01	25.91
	DCT	**5.03**	7.40	49.35	25.14	**9.53**	34.27	21.79
	ELA	13.59	10.40	17.47	17.81	48.18	46.31	25.63
	DFT	40.27	36.06	47.82	52.66	64.10	66.48	51.57
	SVD	35.43	34.87	18.77	25.72	38.93	41.95	32.78

Table 4. D-EER % for the different fusions by a minimum score between RGB and DCT with the designed protocol. The highlighted numbers in bold indicate the best performance observed across the dataset.

		Intra	Cross				
Model	Protocol	FF++	Celeb-DF	Df.TIMIT	Dephy	Defacto	SwanDF
RGB Effv2b0 + DCT Effv2b0	Default	2.01	5.54	**7.69**	5.52	13.28	15.40
	Protocol I	2.03	4.73	8.74	5.52	11.76	15.97
	Protocol II	1.99	4.49	9.22	**5.33**	11.44	16.44
RGB MobileViT-S + DCT MobileViT-S	Default	1.34	2.98	38.43	8.57	9.80	11.74
	Protocol I	1.34	2.52	34.55	7.05	8.23	9.73
	Protocol II	1.34	2.98	38.43	8.57	9.80	11.74
RGB MobileViT-S + DCT Effv2b0	Protocol I	**1.17**	**2.56**	36.57	5.99	**7.73**	**9.06**

5   Conclusions

Minimum score fusion between spatial and frequency-domain features achieved the best performance. These findings suggest that integrating handcrafted frequency features with deep learning models enhances manipulation detection, demonstrating the effectiveness of this hybrid approach in improving robustness against various manipulation techniques.

Acknowledgements. This work was supported by Facephi, R&D department and the German Federal Ministry of Education and Research and the Hessian Ministry of Higher Education, Research, Science and the Arts within their joint support of the National Research Center for Applied Cybersecurity ATHENE.

References

1. Dong, S., et al.: Implicit identity leakage: the stumbling block to improving deep-fake detection generalization. In: Proceedings of the IEEE/CVF Conference on Computer Vision and Pattern Recognition, pp. 3994–4004 (2023)
2. Korshunov, P., Marcel, S.: DeepFakes: a new threat to face recognition? Assessment and Detection (2018)
3. Korshunov, P., et al.: Vulnerability of automatic identity recognition to audio-visual deepfakes. In: IEEE International Joint Conference on Biometrics (2023)
4. Li, H., et al.: FreqBlender: enhancing deepfake detection by blending frequency knowledge. In: The Thirty-Eighth Annual Conference on Neural Information Processing Systems (2024)
5. Li, Y., et al.: Celeb-DF: a large-scale challenging dataset for deepfake forensics. In: Proceedings of the IEEE/CVF Conference on Computer Vision and Pattern Recognition, pp. 3207–3216 (2020)
6. Luo, Y., et al.: Generalizing face forgery detection with high-frequency features. In: Proceedings of the IEEE/CVF Conference on Computer Vision and Pattern Recognition, pp. 16317–16326 (2021)
7. Mahfoudi, G., et al.: Defacto: image and face manipulation dataset. In: 27Th European Signal Processing Conference (EUSIPCO), pp. 1–5. IEEE (2019)
8. Mehta, S., Rastegari, M.: MobileViT: light-weight, general purpose, and mobile-friendly vision transformer. In: International Conference on Learning Representations (2022)
9. Narayan, K., et al.: Deephy: on deepfake phylogeny. In: IEEE International Joint Conference on Biometrics (IJCB), pp. 1–10. IEEE (2022)
10. Rahaman, N., et al.: On the spectral bias of neural networks. In: International Conference on Machine Learning, pp. 5301–5310. PMLR (2019)
11. Rossler, A., et al.: Faceforensics++: learning to detect manipulated facial images. In: Proceedings of the IEEE/CVF International Conference on Computer Vision, pp. 1–11 (2019)
12. Tan, C., et al.: Frequency-aware deepfake detection: improving generalizability through frequency space domain learning. In: Proceedings of the AAAI Conference on Artificial Intelligence, vol. 38, no. 5 (2024)
13. Tan, M., Le, Q.: Efficientnetv2: smaller models and faster training. In: International Conference on Machine Learning, pp. 10096– 10106. PMLR (2021)

14. Tapia, J.E., Busch, C.: Face feature visualisation of single morphing attack detection. In: 11th International Workshop on Biometrics and Forensics (IWBF), pp. 1–6. IEEE (2023)
15. Tolosana, R., et al.: Deepfakes and beyond: a survey of face manipulation and fake detection. Inf. Fusion **64** (2020)
16. Wang, B., et al.: Frequency domain filtered residual network for deepfake detection. Mathematics **11**(4), 816 (2023)

Fast and Accurate 3D Face Reconstruction from Multiple Uncalibrated Images

Hassan Lhallabi[1,2]([✉]) [iD], Sylvie Chambon[1] [iD], Géraldine Morin[1] [iD],
Simone Gasparini[1] [iD], Xavier Naturel[2] [iD], and Jérôme Guenard[2] [iD]

[1] Institut de recherche en informatique de Toulouse (IRIT), Toulouse, France
{hassan.lhallabi,sylvie.chambon,geraldine.morin,simone.gasparini}@irit.fr
[2] Fittingbox, Labège, France
{hassan.lhallabi,xavier.naturel,jerome.guenard}@fittingbox.com

Abstract. This work aims to achieve high-precision 3D face reconstruction from a sequence of uncalibrated multi-view images acquired directly by the user. The purpose is to compute optical measurements on the face. We therefore focus on reconstructing the face in a neutral expression using a learning-based approach, initialized by a model, which allow for predictions beyond the model's statistics. We tailor an algorithmic pipeline to directly estimate the 3D face model using a multi-view stereovision method. We demonstrate that using a 3DCNN to predict surface proximity enables the network to generalise to unseen and uncalibrated datasets while training on calibrated data. We enhance performance through three proposed improvements: using a dense algorithm to retrieve the initial 3D face, applying image pre-processing, and introducing data augmentation. To demonstrate this behaviour, our method is evaluated on three datasets of uncalibrated images. Beyond generalising to unseen datasets, our approach outperforms state-of-the-art methods in terms of accuracy while maintaining a fast computation time.

Keywords: 3D Face · Multi-view stereovision · Deep learning

1 Introduction

Face reconstruction is an active area of research due to its numerous applications in augmented or virtual reality, gaming, or medicine. The work proposed in this paper is targeted at opticians to capture and measure face geometry to better fit new glasses on the user's face. We are therefore challenged to be efficient while using common devices such as smartphones to capture images or videos of the face. We ask and require the user to maintain the same neutral expression throughout the capture. Our goal is then to estimate an accurate 3D model of the face with sub-millimeter precision to capture fine geometric details (texture retrieval is not necessary). This accuracy requirement strongly determines the proposed method. The camera is not calibrated, either internally or externally, so the camera parameters are estimated through the reconstruction process.

M. Castrillón-Santana et al. (Eds.): CAIP 2025, LNCS 15621, pp. 79–90, 2026.
https://doi.org/10.1007/978-3-032-04968-1_7

For the method to be used by optician customers, the objective is to maintain high accuracy while obtaining the reconstruction within a few minutes at most. Depending on the number of images, sensor characteristics, lighting, and scene, different 3D reconstruction techniques have been introduced. Single-view reconstruction from one in-the-wild image [29] is considered the most difficult context, as this problem is ill-posed. Whereas recent single-view methods are improving accuracy, they remain insufficient for our application. Multi-view approaches thus seem more appropriate. They can be divided into three kinds of techniques:

- *Optimization methods* [1,19] aim to explain model parameters by minimizing an error function in image space at inference.
- *Learning-based methods* [4,13,23,24] intend to learn how to reconstruct a face from images, using large datasets for training.
- *Hybrid methods* [2,6] leverage the non-linear representation ability of neural networks and use optimization at inference time.

Face parametrisation reduces computation time at the cost of representational ability, as it becomes more difficult to capture finer details, while a model-free approach maximizes this ability with higher computing time and less robustness. Optimization and hybrid techniques naturally counterbalance the robustness cost, but execution can take minutes to hours. Conversely, learning-based methods can be less than a second for inference and maximize representation ability, at the cost of robustness. For these reasons, we adopt a learning-based method. To the best of our knowledge, the work presented in [24] is the state-of-the-art 3D face reconstruction using well-calibrated multi-view RGB images. They obtain a very good accuracy, around 0.2 mm. In our preliminary experiments, however, the network showed poor performance on uncalibrated datasets not seen during training. In this paper, we introduce a method for training a multi-view stereovision neural network, using calibrated training data, to achieve reconstruction accuracy beyond the state-of-the-art on different, uncalibrated datasets. The major contributions of the proposed workflow are:

1. We demonstrate that using a 3D Convolutional Neural Network (3DCNN) directly instead of a Multilayer Perceptron (MLP) followed by a 3DCNN post-regularizer enables the network to generalize.
2. The proposed pipeline improves performance and robustness through:
   - Using a dense initialization method, improving initial reconstruction;
   - Data augmentations, adding diversity in image lighting and noise;
   - Image pre-processing, reducing the gap between training and testing datasets by removing background and cropping images around the face.

The paper is structured as follows: Sect. 2 reviews the state of the art on 3D face reconstruction, multi-view stereovision, and explains our choice of multi-view stereovision against other approaches, Sect. 3 presents the algorithmic pipeline we built to reconstruct the face in an uncalibrated setting. The evaluation protocol is presented in Sect. 4, including comparison to other methods on unseen datasets and the gain in performance of our contributions before presenting our analysis, in Sect. 5, and our conclusions in Sect. 6.

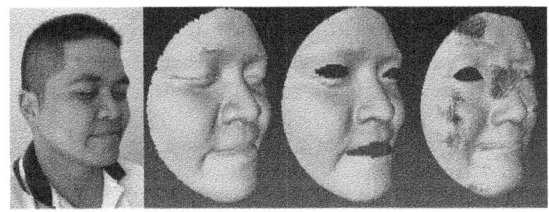

Fig. 1. Example of an incorrect inference from [24]. From left to right: a sample image, 3D ground truth, our prediction, Xiao *et al.* [24] prediction.

2   3D Face Reconstruction

Recent advances in 3D face reconstruction include four method families: 3D Morphable Models, neural implicit representations, depth, and 3D estimation multi-view stereovision. Each class involves trade-offs between reconstruction accuracy, computational cost, and the ability to capture facial details. We review representative approaches to motivate the interest in our method.

2.1   Presentation of Existing Approaches

By leveraging a priori facial knowledge, *3D Morphable Models, 3DMM* [3,12] have long been used for 3D face reconstruction and remain widely used [2,19,23].

3DMMs are constructed from datasets of topologically uniform face meshes, from which data reduction techniques extract a low-dimensional parameter space to simplify the reconstruction process.

The facial model obtained is realistic but, due to the lack of variability in 3DMM, the reconstructed geometry and facial details are limited. Therefore, to achieve fine accuracy, we investigate 3D reconstruction methods that can consider faces outside the 3DMM space.

Neural implicit methods, including neural radiance fields [14], neural surface representations [22], and more recently *Gaussian splatting*, provide a general framework for synthesizing novel views from multi-view images and can be used for 3D face reconstruction.

In general, the focus is on evaluating the ability to synthesize new views and not on the accuracy of the face geometry. Some articles compare computed 3D face surfaces with ground truth scans [6,28]. These methods can model fine facial details and are robust thanks to optimization at inference time. However, they require high computing time, from several minutes to hours, which is impractical for our use case.

The strength of approaches based on *multi-view stereovision depth estimation* is their ability to accumulate information from multiple images. Recently, classical approaches [8] are being outperformed by learning-based methods [21,26]. These methods usually follow a common general structure: they first learn features from images through a 2D Convolutional Neural Network (2DCNN), then they construct a cost volume by projecting 3D depth hypotheses on feature maps, and finally, they regularize the volume with a 3DCNN. Depth predicting multi-view stereovision methods have been adapted to the specific setting of 3D face reconstruction [1,13]. These methods can model fine facial geometry in a short computing time. However, estimating depth on each image independently leads to a lack of coherence between the depths of corresponding pixels in different images. As a result, the point cloud is noisy, making it challenging to obtain an accurate mesh. Depth fusion reduces noise in the point cloud by selecting only 3D points with consistent depth across multiple views. The final point cloud quality is improved, but noise persists and holes appear.

Recent works on *multi-view stereovision 3D estimation* [10,15] skip the intermediate depth representation by directly estimating a 3D representation based on the Truncated Signed Distance Function (TSDF) [16].

In this approach, the 3D space is discretized using voxels. These methods focus on reconstructing indoor scenes and achieve better results than depth-estimating methods while directly predicting a 3D model. Two recent articles perform face reconstruction using end-to-end 3D methods. First, Bolkart *et al.* [4] use a spatial transformer to localize the face in 3D space and reconstruct the face using a 3DCNN through voxel representation. However, they use structured light images as inputs, which are not compatible with our application. Second, Xiao *et al.* [24] reconstruct a coarse face with a bundle adjustment using 68 face landmarks while not optimizing the calibrated intrinsics and extrinsics. Then, 3D hypotheses along vertex normals are given independently to a Multilayer Perceptron (MLP) responsible for predicting a scalar indicating how close the point is to the surface. Finally, a 3DCNN processes the resulting volume to predict final displacements along normals. Multi-view stereovision methods that directly estimate 3D outperform those that estimate depth, resulting in a simpler pipeline because the fusion step is not required. The mesh is also used once per object instead of once per image.

2.2   State of the Art Analysis

Multi-view methods with available code consists of three 3DMM based approaches: *MVFNET* [23], *INORig* [2], *3DI* [19] and the neural implicit approach *Sira++* [6]. We compare our method against them in Sect. 5. Our motivation for designing the proposed approach is threefold:

1. 3DMM-based methods accuracy is limited by their statistical aspect.
2. Neural implicit and Gaussian splatting methods can reconstruct the face with fine accuracy, but are too computationally intensive for our application.
3. Depth-based multiview stereovision methods need depth fusion, which results in a noisy point cloud, making it hard to retrieve the surface.

For these reasons, we focus on multi-view stereovision 3D estimation. The method in [24] reconstructs the face with an accuracy compatible with our requirements (around 0.2 mm) in an end-to-end manner. However, the network is not able to generalise to datasets other than a test split of the training dataset

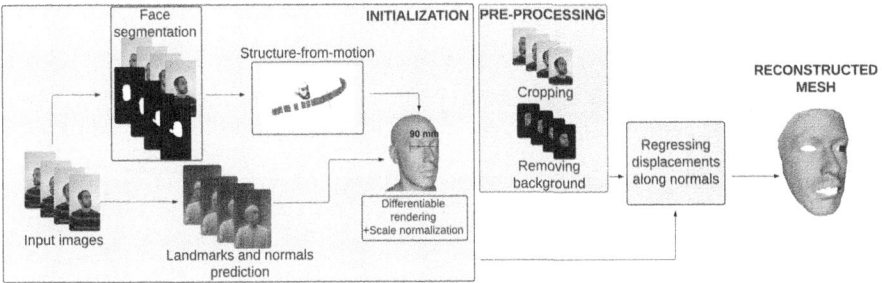

Fig. 2. Overview of the proposed pipeline. All the parts highlighted in red correspond to our contributions and adjustments to the uncalibrated context. (Color figure online)

Facescape [25], see Fig. 1. Building upon [24], we present an algorithmic pipeline able to perform face reconstruction in such an uncontrolled context and generalize to unseen datasets. The next section presents the contributions introduced in this pipeline, resulting in better generalization and performance.

3   Proposed Pipeline for Fine 3D Face Reconstruction

Our proposed pipeline follows the same steps as [24], and Fig. 2 shows our contributions highlighted in red. In this section, we present our improvements that reduce over-fitting on the training data. Further implementation details are given in the supplementary material.

3.1   Initialization

Structure from motion – Since in our application the users can record themselves moving their head instead of the camera, we mask the background and other static parts of the scene, such as the chest and neck, by applying a face segmentation network[1] [27] to detect the face. We then use COLMAP [20] to estimate camera parameters and poses from the masked regions. As 3D hypotheses are made using a uniform step (Sect. 3.4), the estimated calibration is required to reflect the scale of a face (or the steps between each 3D hypothesis along the normals will be too small/too big). Hence, we normalize the eye distance of the initialized 3D face model to 90 mm.

3D Face initialization – In [24], the face is initialized using a bundle adjustment algorithm, using 68 landmarks detected on each view, optimizing for 3DMM parameters and a rotation, translation, and scale applied to the 3D face. This initialization is limited by the sparse aspect of its input data. We propose to use a dense approach (optimizing for the same parameters). This improves the initial face shape and position in 3D space, thus making the network prediction

---

[1] https://github.com/zllrunning/face-parsing.PyTorch.

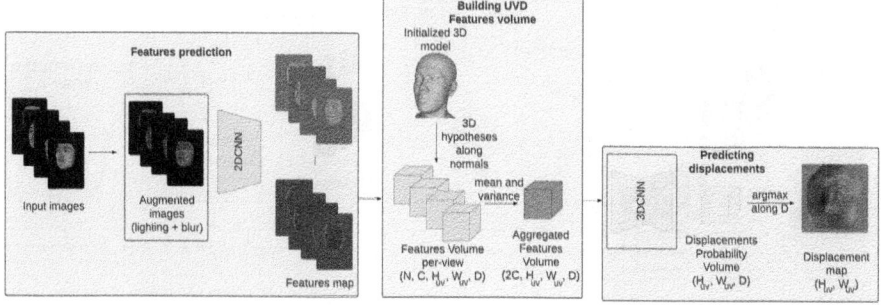

Fig. 3. Overview of the network architecture. All the parts highlighted in red correspond to our contributions. (Color figure online)

easier. To this end, we adapt FOUND [5], an optimization framework for 3D foot reconstruction, to be used with faces. It predicts normals from each view using a neural network and then uses differentiable rendering to recover the 3D shape. We use Sapiens [11] to predict face normals and FLAME [12] as 3DMM. We remove silhouette loss and uncertainty from normals and keypoints losses, as Sapiens does not predict uncertainty. Moreover, we replace the registration step with the one from [24], allowing for a faster and more robust prediction of initial rotation, translation, and scale.

3.2   Images Preprocessing

Removing background information – As 3D hypotheses are made along normals, some may be projected into the background region of an image. Since the background is similar in training data images, this can lead to overfitting. To address this, the initial reconstruction is used to segment the face region in each image. This is done by rendering the mesh with the estimated camera parameters and poses, and then thresholding the depth. By using these masks instead of those computed through segmentation, it's possible to only retain the reconstructed part of the face (excluding hair and ears) and remove the cap that each subject wears in the training dataset.

Cropping images – Training dataset images are well zoomed on the face, which is not the case in the testing data. Naively resizing (downscaling or upscaling) images can lead to a difference in the size of the face between training and testing sets. The locality of features computed by the 2DCNN is dependent on the distribution of face sizes in the training data. Hence, instead of resizing the images, we crop a same-size image centered on the masks used for the Structure-from-Motion step. Cropping increases the size of the face on testing images, making it closer to the training ones.

3.3   Data Augmentations

Noise – As training images are acquired in a well-calibrated context, they do not have the same noise level as testing images, where the camera moves during acquisition. Random Gaussian and motion blur are applied to each input image independently.

Lighting – Training images are acquired in the same lighting context, but we want our method to be able to reconstruct the face in any environment. Random changes in brightness and contrast as well as random gamma correction, are applied to input images. We apply the same change in lighting between multiple images of the same person to simulate a different environment.

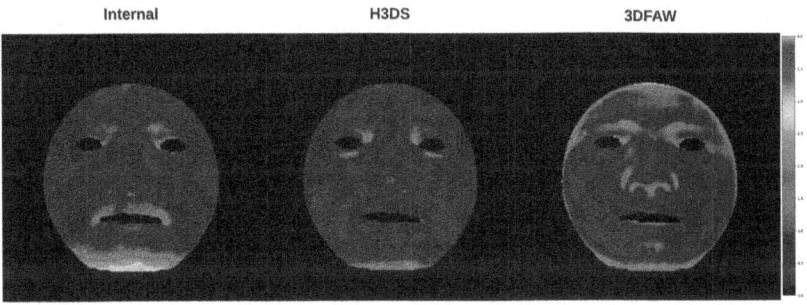

Fig. 4. Error maps reporting mean error for each point in the UV map across every person in a dataset. Errors in mm.

3.4   Networks Architecture

Pre-processed images with corresponding camera parameters and poses are then used to reconstruct the face. The network architecture follows the same steps as [24], see Fig. 3, where our contributions are highlighted in red. Our main improvement is to predict how close each hypothesis is to the surface with a 3DCNN instead of an MLP.

Features backbone – A 2DCNN network takes images, noted $I \in \mathbb{R}^{N \times 3 \times H \times W}$ as input and computes features map $F_{\mathrm{maps}} \in \mathbb{R}^{N \times C \times H/2 \times W/2}$ with N being image number, H, W image height and width and C feature channel number. Camera parameters are scaled accordingly for the next steps.

3D formulation – The 3D reconstruction problem consists of predicting displacements along normals from the initial mesh to the true face surface, which are transferred in the UV space. To this end, 3D hypotheses are made along normals with a uniform step $s \in \mathbb{R}$. Then, these 3D points are projected in the feature map using the estimated camera parameters and poses, resulting in a volume

$V_{ft} \in \mathbb{R}^{N \times C \times H_{uv} \times W_{uv} \times D}$ with H_{uv}, W_{uv} being the height, width of the UV map and D the number of step of size s.

Aggregating information between views – To build a network that is independent of the number of views and their poses, it is necessary to aggregate information between views. Mean and variance along the dimension of the views are computed and concatenated, resulting in a volume $V_{in} \in \mathbb{R}^{2C \times H_{uv} \times W_{uv} \times D}$.

Predicting closeness to the surface – Closeness to the surface is defined by a normalized Gaussian distribution centered on the ground truth displacement:

$$\hat{s}(d) = e^{-\frac{(d-\hat{d})^2}{2\sigma}} \tag{1}$$

Instead of using an MLP that sees each 3D hypothesis independently, we use a U-Net 3DCNN. The network outputs a volume $V_{out} \in \mathbb{R}^{H_{uv} \times W_{uv} \times D}$ on which we apply a sigmoid activation function. The cost function is the mean squared error between V_{out} and the ground truth labels.

Inference – The displacement map $M \in \mathbb{R}^{H_{uv} \times W_{uv}}$ is obtained by computing the arg max on V_{out} last dimension.

Visibility – Bolkart *et al.* [4] show that using visibility when fusing features across multiple views improves performance. We tried adding a visibility-aware feature fusion step, but that did not lead to any improvement in accuracy in our context. This may be caused by a low resolution of the UV map (128×128), which does not provide a sufficiently accurate visibility.

Table 1. Comparison against state of the art results. Results are reported as mean | median in mm.

Method	Internal dataset	H3DS	3DFaw-video
MVFNET [23]	1.42 | 1.08	1.42 | 1.12	1.42 | 1.09
INORig [2]	1.28 | 0.99	1.16 | 0.93	1.46 | 1.07
3DI [19]	1.45 | 1.06	1.47 | 1.11	1.34 | 1.02
Sira++ [6]	unavailable	0.84 | 0.66	unavailable
Our	**0.63 | 0.46**	**0.69 | 0.49**	**0.64 | 0.46**

4   Evaluation Protocol

Datasets – We use the same dataset as [24], *Facescape* [25] for training our network (*cf.* the supplementary material for more details). It includes 3D scans and calibrated multi-view images of 359 individuals, each one performing 20 different expressions. For inference, to evaluate generalization ability, we use three other datasets:

1. *3DFaw-video* dataset [17] composed of 26 people filmed with the back camera of a smartphone (1920 × 1080) by another person, from left to right.
2. *H3DS* [6] dataset, composed of 60 people filmed with a camera from left to right and up and down, images are in lower resolution (512 × 512) than smartphones.
3. We also collected a new dataset of 53 people, each person filmed themselves with a smartphone (1920 × 1080), at a bent arm distance, keeping the camera still and turning their head left to right.

Protocol – To evaluate the 3D reconstruction, we compute the reconstruction error between the ground truth and reconstructed 3D model with a similar protocol as in [18]: we first align the two 3D models using Procrustes [9] with 7 annotated landmarks. Then we use a robust Iterative Closest Point (ICP) algorithm to refine the alignment. We use a threshold of 3 mm to filter out outlier correspondences. We crop the two meshes with a sphere of 80 mm radius centered on the tip of the nose of the ground truth mesh, as all methods do not reconstruct the whole face. We compute a final alignment with the same ICP using the cropped meshes. The errors are computed using the vertex-to-surface distance, from ground truth to prediction. As our method outputs a mesh with holes in the eyes and mouth regions, we close holes using *Meshlab* [7]. We first determine the mean or median error for each face, and then average these values to obtain an overall error across each dataset. We filtered out outliers, one capture from our internal dataset and three from *H3DS*: due to poor registration during structure-from-motion, only half of the face was reconstructed.

Table 2. Ablation study results. Results are reported as mean | median in mm.

Model Variant	Internal dataset	H3DS	3DFaw-video
MLP+3DCNN regularizer [24]	1.96 | 1.37	2.46 | 1.69	1.87 | 1.31
3DCNN (facescape 3DMM)	0.75 | 0.53	0.84 | 0.58	0.69 | 0.50
3DCNN (flame 3DMM)	0.66 | 0.48	0.82 | 0.57	0.70 | 0.51
Without augmentations	0.64 | **0.45**	0.72 | 0.50	0.68 | 0.48
Without pre-processing	0.66 | 0.49	0.74 | 0.51	0.66 | 0.48
Without dense initialization	0.68 | 0.49	0.75 | 0.53	0.69 | 0.51
Complete model	**0.63** | 0.46	**0.69 | 0.49**	**0.64 | 0.46**

5   Results and Analysis

Comparison with the state of the art – As justified in Sect. 2.2, we compare our pipeline with other publicly available methods: *MVFNET* [23], *INORig* [2], *3DI* [19] and *Sira++* [6] on the same three datasets. We also compare our method with Xiao *et al.* [24] on a test split of *Facescape* and show qualitative results in Fig. 5. Quantitative results reported in Table 1 show that our method outperforms other 3DMM-based methods, namely *MVFNET*, *INORig*, and *3DI* by a large margin, showing that reconstructing the face outside of the possible faces parametrized by 3DMMs improves the performance. The last method, Sira++, does not provide code but only results (and alignment code) using different numbers of input images on the *H3DS* dataset. For comparison purposes, we use their results with the most images (32) and our alignment procedure that outperforms theirs (see details in the supplementary material). Our overall method performs better, with a highly reduced computing time; their inference time for 3 images (191 seconds) is higher than ours using all images (they do not provide the time for 32 images).

Ablation study – We evaluate several versions of our method to show the benefit of each proposed step. Table 2 reports these results. The main performance improvement comes from the change in architecture from MLP to 3DCNN. This shows that processing the whole face at once drastically reduces overfitting. Data augmentations have a neutral impact on our internal dataset, but they improve accuracy on *H3DS* and *3DFaw-video*. Augmentations help generalise, particularly by accounting for differences in the acquisition context. For example, some images of *H3DS* are taken in poor lighting conditions, while those from *3DFaw-video* are affected by motion blur. Pre-processing steps improve performance for all datasets. Changing the initialisation algorithm has a greater impact, emphasising the significance of the initial normal positions. We also report error maps for our complete model in Fig. 4.

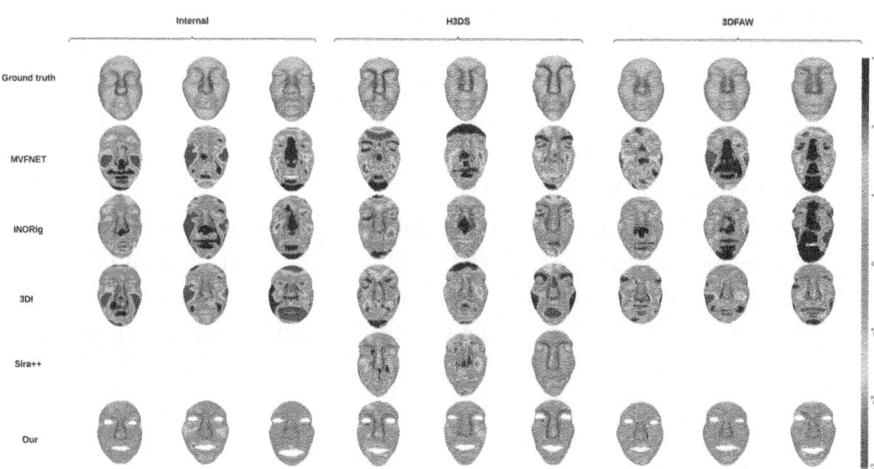

Fig. 5. Qualitative comparison between state of the art methods. Errors are reported in mm

6 Conclusions and Future Work

We propose a novel method for reconstructing a 3D face model from uncalibrated images using end-to-end 3D estimation multi-view stereovision. Building on the work of Xiao et al. [24], we develop a network that generalises to unseen and uncalibrated datasets while learning on calibrated data. Our method outperforms other publicly available multi-view approaches in reconstructing the face in a neutral expression, which is crucial for our application. We showed that 3DMM-based methods are limited by the statistical aspect of their model as they reconstruct the face with lower accuracy. Moreover, our method outperforms a neural implicit one while maintaining a lower computing time. To further improve the generalisation capacities and robustness of the proposed method, we plan to add training data based on the rendering of 3D face scans or real images, estimating 3D ground truth with depth sensors.

Acknowledgments. This research was supported by the CIFRE grant ANRT contract #2022/0356, funding the collaboration between the French company Fittingbox and Toulouse INP–IRIT laboratory. The experiments presented in this paper were carried out using the OCCIDATA platform, funded and administered by IRIT and supported by CNRS.

Disclosure of Interests. The authors have no competing interests to declare concerning the content of this article.

References

1. Agrawal, S., Pahuja, A., Lucey, S.: High accuracy face geometry capture using a smartphone video. In: WACV (2020)
2. Bai, Z., Cui, Z., Liu, X., et al.: Riggable 3D face reconstruction via in-network optimization. In: CVPR (2021)
3. Blanz, V., Vetter, T.: A morphable model for the synthesis of 3D faces. In: SIGGRAH (2002)
4. Bolkart, T., Li, T., Black, M.: Instant multi-view head capture through learnable registration. In: CVPR (2023)
5. Boyne, O., Bae, G., Charles, J. et al.: Found: foot optimization with uncertain normals for surface deformation using synthetic data. In: WACV (2024)
6. Caselles, P., Ramon, E., Garcia, J., et al.: Implicit shape and appearance priors for few-shot full head reconstruction. TPAMI (2025)
7. Cignoni, P., Callieri, M., Corsini, M., et al.: MeshLab: an open-source mesh processing tool. In: Eurographics Italian Chapter Conference (2008)
8. Galliani, S., Lasinger, K., Schindler, K.: Massively parallel multiview stereopsis by surface normal diffusion. In: ICCV (2015)
9. Gower, J., Dijksterhuis, G.: Procrustes problems. Procrustes Problems, Oxford Statistical Science Series (2005)
10. Ju, J., Tseng, C.W., Bailo, O., et al.: DG-recon: depth-guided neural 3D scene reconstruction. In: ICCV (2023)
11. Khirodkar, R., Bagautdinov, T., Martinez, J., et al.: Sapiens: foundation for human vision models. In: ECCV (2024)

12. Li, T., Bolkart, T., Black, M., et al.: Learning a model of facial shape and expression from 4D scans. TOG (2017)
13. Liu, Y., Li, L., An, S., et al.: 3D face reconstruction with mobile phone cameras for rare disease diagnosis. In: AJCAI (2022)
14. Mildenhall, B., Srinivasan, P.P., Tancik, M., et al.: NeRF: representing scenes as neural radiance fields for view synthesis. Commun. ACM (2021)
15. Murez, Z., Van As, T., Bartolozzi, J., et al.: Atlas: end-to-end 3D scene reconstruction from posed images. In: ECCV (2020)
16. Newcombe, R.A., Fitzgibbon, A., Izadi, S., et al.: KinectFusion: real-time dense surface mapping and tracking. In: ISMAR (2011)
17. Pillai, R.K., Jeni, L.A., Yang, H., et al.: The 2nd 3D face alignment in the wild challenge (3DFAW-video): dense reconstruction from video. In: ICCV Workshops (2019)
18. Ramon, E., Triginer, G., Escur, J., et al.: H3d-net: few-shot high-fidelity 3D head reconstruction. In: ICCV (2021)
19. Sariyanidi, E., Zampella, C.J., Schultz, R.T., et al.: Inequality-constrained 3D morphable face model fitting. TPAMI (2024)
20. Schönberger, J.L., Frahm, J.M.: Structure-from-motion revisited. In: CVPR (2016)
21. Wang, F., Galliani, S., Vogel, C., et al.: Patchmatchnet: learned multi-view patchmatch stereo. In: CVPR (2021)
22. Wang, P., Liu, L., Liu, Y., et al.: Neus: learning neural implicit surfaces by volume rendering for multi-view reconstruction. In: NeurIPS (2021)
23. Wu, F., Bao, L., Chen, Y., et al.: MVF-NET: multi-view 3D face morphable model regression. In: CVPR (2019)
24. Xiao, Y., Zhu, H., Yang, H., et al.: Detailed facial geometry recovery from multi-view images by learning an implicit function. In: AAAI (2022)
25. Yang, H., Zhu, H., Wang, Y., et al.: FaceScape: a large-scale high quality 3D face dataset and detailed Riggable 3D face prediction. In: CVPR (2020)
26. Yao, Y., Luo, Z., Li, S., et al.: MVSNet: depth inference for unstructured multi-view stereo. In: ECCV (2018)
27. Yu, C., Wang, J., Peng, C., et al.: Bisenet: bilateral segmentation network for real-time semantic segmentation. In: ECCV (2018)
28. Zheng, M., Zhang, H., Yang, H., et al.: NeuFace: realistic 3D neural face rendering from multi-view images. In: CVPR (2023)
29. Zielonka, W., Bolkart, T., Thies, J.: Towards metrical reconstruction of human faces. In: ECCV (2022)

Seeing Through Wearables: A Comprehensive Face Recognition Dataset from Body Worn Cameras

Sameer Hans[1]([✉]), Jean-Luc Dugelay[1], and Mohd Rizal Mohd Isa[2]

[1] EURECOM, Sophia-Antipolis, France
sameer.hans@eurecom.fr
[2] Universiti Pertahanan Nasional Malaysia (UPNM), Kuala Lumpur, Malaysia

Abstract. Body worn cameras (BWCs) have become more and more popular over the last decade. They are becoming one of the essential tools for law enforcement officers to carry with them for surveillance purposes. Generally, videos captured by BWCs are used a posteriori through visual inspection in case of major problems between police officers and citizens. Limited academic research has been conducted on image and video processing using BWCs. There are extremely few datasets available that are based on BWCs. For this objective, we introduce FALEBface: a novel dataset for face detection and recognition using BWCs. We also provide some baseline experiments on the proposed dataset. This work includes two distinct insights: (1) introduction of a dataset specific to body cameras with the applications of facial recognition, and (2) evaluation of models in different environments. The experiments are carried out in three environments: indoor, outdoor, and dark which includes a variation of expressions for the subjects, and a comparative study is also done to check the performance of the models across environments and spectra. To facilitate further research in this domain, the entire dataset can be obtained upon request from the authors (The dataset can be obtained by visiting https://faleb.eurecom.fr/).

Keywords: Body Worn Camera · Face Recognition · Multimodal Dataset · Near-infrared spectrum image · Visible spectrum image

1 Introduction

Body worn cameras (BWCs) have become increasingly prevalent in various sectors. They have been implemented in different parts of the world, where they serve as a critical tool in the areas of law enforcement for enhancing transparency [8], accountability [20], and evidence collection [21].

Face biometric authentication systems have made significant progress and are now used in a variety of applications such as identifying criminals, surveillance, security systems, and even social networks. Very high performance is achieved for such systems using deep learning-based approaches for feature extraction. In

© The Author(s), under exclusive license to Springer Nature Switzerland AG 2026
M. Castrillón-Santana et al. (Eds.): CAIP 2025, LNCS 15621, pp. 91–101, 2026.
https://doi.org/10.1007/978-3-032-04968-1_8

recent years, the Convolutional Neutral Network (CNN) [17] has become a very popular and successful method for facial recognition. The CNN automatically extracts a variety of features of the image and has good robustness to complex environments. Their achievements have been fueled by the huge amount of data accessible online and the enormous efforts made by the research community to produce vast labeled datasets like CASIAWebFace [24] and VGGFace2 [7].

Law enforcement professionals have recently shown interest in using face recognition with BWCs to protect officers, enable situational awareness, and provide evidence for trial. This function could be useful for the suspect too (if the officer has inappropriate behaviour). In this particular domain, there are currently a limited number of studies. Most previous studies focus on body camera placement or relied on traditional machine learning methods like Principal Component Analysis (PCA) and Linear Discriminant Analysis (LDA) [3], which were widely used before the rise of deep learning techniques.

This work introduces FALEBface, a novel dataset for face detection and recognition using body worn cameras. The dataset contains 485 videos from 97 subjects for each environment for facial recognition: indoor, outdoor, and dark. The videos are classified according to the discussion context, considering expressions of happy, sad, angry, and neutral emotion per subject. We evaluate deep learning models based on VGG-Face [16], Inception ResNet [18], and Vision Transformer (ViT) [9], along with a comparative study on different environments and spectra[1] to identify the individuals across diverse conditions and variations.

The paper is organized as follows. In Sect. 2 we survey previous work related to BWCs. In Sect. 3, we introduce the steps followed in the data collection for the activity. We report our implementation, experiments and results in Sect. 4. Finally, the conclusions and future work follow in Sect. 5.

2   Previous Work

While there is a considerable volume of work in the area of egocentric vision, the use of BWCs for image processing tasks in real-world surveillance settings remains relatively limited and underexplored. BWC footage presents unique challenges like first-person viewpoint, low resolution due to camera limitations, unbalanced data distribution across activities, privacy concerns over identifiable information, and limited annotated training data. Even in this limited work, most of the work in the literature is focused on egocentric vision and actions [12,14] and not on scenarios specific to law like recognition of a suspect at night or matching the face from a different environment.

There exists some work done on facial recognition using body cameras [3,4]. The authors [3] worked with 20 subjects, and evaluated with traditional methods for facial image representation. They obtained accuracy in the range [68%, 75%]. This study had very limited data size to perform further evaluations.

In the most prominent study for facial recognition using body cameras [4], the authors tried to implement a dataset that is based on cropped images in

---

[1] The camera generates images and videos in visible and near-infrared spectra.

indoor and outdoor environments. They also introduced some subjects in the dark environment, but it is very limited (only 2 subjects) to perform any evaluations. Although they introduce facial images of 132 subjects, the dataset is limited by a lack of expression diversity (many of the subjects are present with a neutral expression) and lack of proper environmental conditions. They also include some tests based on ResNet architecture with different loss functions. For future work, they proposed to evaluate some lightweight CNN models.

EgoFace [10] presents a novel lightweight framework for face performance capture and videorealistic reenactment using a single egocentric RGB camera. The method estimates facial expressions from a single oblique view using a deep encoder, and synthesizes a frontal videorealistic face with an adversarially trained network. The system is trained in a supervised manner and handles diverse lighting, movement, and facial expressions. While EgoFace focuses on videorealistic facial reenactment from egocentric views, our dataset targets identity recognition under surveillance-like conditions with head motion, varying environments, and expression diversity, providing a benchmark more aligned with real-world face recognition tasks in mobile contexts.

An in-depth study [12] focuses on action recognition using BWCs. By focusing on actions relevant to law enforcement scenarios, they address the challenges of egocentric motion, dynamic camera perspectives, and the difficulty of recognizing common actions of individuals recorded by law enforcement officers. State-of-the-art models, along with novel approaches are benchmarked, with results highlighting the need for robust fine-tuning strategies and domain adaptation across different scenarios. This work offers a significant step forward for real-world action recognition from wearable devices.

The authors of [11] try to identify users through egocentric motion captured by BWCs. The study benchmarks several deep learning models, along with a two-stream I3D architecture, combining RGB and optical flow inputs, which achieved the best performance, underscoring the complementary strengths of appearance and motion cues in identifying users. The paper also highlights the challenges of domain shift. This work is one of the first to explore user identification from egocentric motion using real-world BWC data, establishing a valuable benchmark for biometric recognition under mobile, first-person conditions.

3   Data Collection

For the data collection, students from a university volunteered. The subjects were recorded using Cammpro[2] I826 Body camera, as shown in Fig. 1. The recording took place over different sessions spread across a week. The camera was fixed on the middle of the chest of the user as recommended in [6]. All the recordings were done with a video resolution of 2304 × 1296 pixels at 30 fps.

For facial recognition, we had 97 volunteers. We recorded 5 videos per subject, each showing them talking for 10–15 seconds. These videos specified the

---

[2] https://www.cammpro.com/.

Fig. 1. Cammpro I826 Body Camera.

expressions of neutral, happy, angry, and sad. Each participant was provided a script before the recording session for consistency. The script included example sentences specifically designed to elicit the target emotions. Participants were instructed to act out these sentences, focusing on both facial expressions and vocal tones to convey the intended emotions. As the camera is positioned on the chest, we record the subject from the head to the torso. This was done in three different environments: indoor, outdoor, and dark. The indoor environment was well-lit with a uniform background and lighting conditions, and the dark environment was in the same place with the lights switched off. The outdoor environment had natural sunlight conditions with varying intensities of light. The distance between the user and subject was kept around 5–6 feet according to the reactionary[3] gap [2]. The recorded videos for indoor and outdoor environments lie in the visible[4] spectrum. The recording in the dark environment was done using the infrared[5] feature of the camera. This data is useful for experiments of matching Near-Infrared (NIR) face images to Visible spectrum (VIS) face images, which is a very challenging task. Figure 2 shows the sample images from the VIS and NIR spectrum. These recordings captured the facial expressions/emotions, speech, hand gestures, and head movements of each subject.

4   Preliminary Assessment of the Dataset

4.1   Preprocessing

The videos are converted into frames and organized according to the expressions. After getting the frames, faces are detected and cropped using the Dlib library

---

[3] The Reactionary Gap is a concept based on the rule that distance equals time. The gap or distance you stay away from a suspect provides time for you to respond.

[4] The visible spectrum (VIS) is the region perceivable by the human eye, which includes wavelengths from 400nm to 700nm.

[5] The camera produces near-infrared (NIR) images. The NIR region spans wavelengths ranging from 780 nm to 2500 nm.

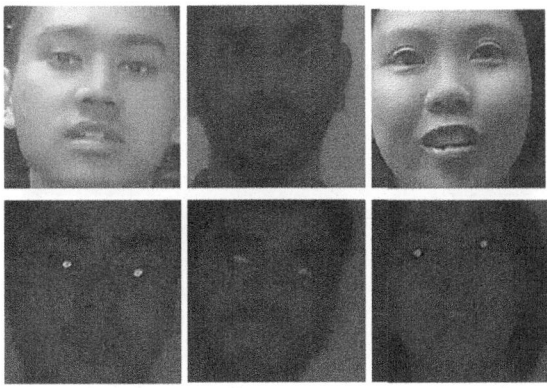

Fig. 2. Examples of VIS images (top row) and NIR images (bottom row).

[13]. The frames are resized into 224×224 to ensure uniformity and compatibility with various model architectures. To improve the contrast and brightness of the frames (and achieve consistent illumination across the frames) [15], histogram equalization is applied to the resized frames. Finally, these frames are selected according to sharpness metrics using Laplace variance. Figure 3 shows some samples obtained after preprocessing. For cross-environment analysis, we perform experiments on two sets of frames: histogram equalized and normal (RGB) frames. The frames for the experiments are selected if their sharpness is higher than a threshold (fixed as 0.01 for equalized and 0.002 for RGB frames). The frames are selected such that we have samples from all the expressions for diversity. We divide the training, validation, and test set in the ratio of 70:15:15.

4.2   Implementation

We used the VGG16 architecture, Inception Resent V1, and BEiT. These models were chosen for their respective strengths: VGG16 as a baseline model, Inception ResNet V1 for its popularity and proven performance, and Bidirectional Encoder representation from Image Transformers (BEiT) for its novelty and recent advancements in the field.

– **VGG16:** We implement a VGG-Face model [16] architecture, a 16-layer CNN that is trained on over 2 million celebrity images. The face weights are loaded into the implemented architecture and using this model, we extract image features from the output of the fc-6 layer and use them in our subsequent classification stage. Each image is represented by a 4096-dimensional feature vector. For the activation function, we use **softmax**, which is useful in dealing with multi-class classification problems. During the training, we use the Adam optimizer and Sparse Categorical Cross-entropy loss. The learning rate is fixed as 0.001.

Fig. 3. Samples of acceptable frames in each environment. The first row represents normal, cropped, and equalized frame in the indoor environment. The second row represents normal, cropped, and equalized frame in the outdoor environment. The last row represents normal, cropped, and equalized frame in the dark environment.

- **Inception Resent V1:** We experiment with Inception ResNet (V1) [18] architecture pretrained on CASIAWebFace and VGGFace2. The process typically starts by loading the pretrained weights and setting up a training loop that includes a data loader, Adam optimizer, and learning rate scheduler. The model is trained over 50 epochs, with each epoch consisting of forward passes through the data, loss computation using Cross-entropy loss, and backpropagation to update the model weights.
- **BEiT:** The BEiT [5] model is a Vision Transformer, pretrained in a self-supervised fashion on ImageNet-21k [1], a dataset comprising of 14 million images and 21,841 classes. Images are presented to the model as a sequence of fixed-size patches (resolution 16×16), which are linearly embedded. The model uses relative position embeddings and performs classification of images by mean-pooling the final hidden states of the patches. The fine-tuning takes place by placing a linear layer on top of the pretrained encoder. The learning rate is fixed as 0.001.

4.3   Experiments and Results

Self-environment Analysis. For the analysis, we consider 20 subjects randomly chosen with a size of 140 frames per subject. The selected frames are normal (not equalized) to see the performance of different models on the videos as captured by the camera. Table 1 shows the performance of different models when fine-tuned on the subjects (environment-wise). BEiT model performs the best among all the models giving accuracy of 98.1% and 99.3% for indoor and outdoor environments respectively. We also receive high accuracies in the range of [96%, 99%] for the indoor and outdoor environments when considering the models of VGG16 and Inception ResNet. We record accuracy value of 97.95% for dark environment with BEiT model.

Table 1. Environment-wise analysis.

Model	Pretrained Dataset	Environ-ment	Validation Accuracy [%]	Rank-1 Accuracy [%]
VGG16	VGGFace2	Indoor	99.25	96.51
		Outdoor	99.25	98.5
		Dark	97.5	95.25
Inception ResNet	VGGFace2	Indoor	99.76	97.84
		Outdoor	99.7	99.26
		Dark	98.7	97.75
Inception ResNet	CASIA- Webface	Indoor	99.5	97.64
		Outdoor	99.7	99.26
		Dark	99	97.7
BEiT	Imagenet-21K	Indoor	99.7	**98.1**
		Outdoor	99.78	**99.3**
		Dark	98.5	**97.95**

Cross-Environment Analysis. It is essential to ensure that our model can generalize well across different environments. For this experiment, the training is done in one particular environment and tested on different validation and test environment to evaluate the model's ability to generalize to unseen environments, which is crucial for real-world deployment. We select 20 subject and perform experiments on 3 sizes, where S1 contains 20 frames per subject, S2 contains 40 frames per subject, and S3 contains 140 frames per subject respectively. For each size, the training, validation, and test set are in the ratio 70:15:15. For this experiment, we consider separate sets from both the frames: equalized and not equalized. Table 2 shows the performance of the VGG-Face model (best model in this experiment) when fine-tuned on the subjects. Rank-1 and Rank-5 accuracy values are recorded. The first test is training on indoor and testing on outdoor environment. Although there is a drop in the accuracy from environment-wise testing, we get a remarkable Rank-1 accuracy in the outdoor test set (87.5%) for the non-equalized frames. When the training set is outdoor, there is around 10-point drop again in the Rank-1 accuracy as compared with the previous test. At Rank-5, we achieve high accuracy value of 95.83% when training environment is indoor. The accuracy values are higher if the training sets are more focused on the indoor environment as it has a uniform background and lighting conditions, controlled environment, and reduced variability.

Table 2. Cross-Environment comparison. We experiment with VGG-Face model, where training is done on one environment, and testing for the other environment.

Training Environment	Test Environment	Equalized	Size	Rank-1 [%]	Rank-5 [%]
Indoor	Outdoor	no	S1	78.33	90
			S2	**87.5**	**95.83**
			S3	77	93.25
		yes	S1	73.33	91.67
			S2	75	**95**
			S3	76	92.75
Outdoor	Indoor	no	S1	73.33	88.33
			S2	72.5	**95.13**
			S3	75.75	91.77
		yes	S1	73.33	91.67
			S2	73.33	85
			S3	75.5	**91.75**

Cross-Spectrum Analysis. For the dark environment, as we shoot using the infrared feature of the camera, we have samples from the NIR spectrum. In our experiment, we selected 20 subjects, and the training set has images from both VIS and NIR spectra for fine-tuning the model. This approach aims to train the model on diverse conditions, potentially improving its robustness to variations. We experiment with the models described earlier. The training set consists of 98 images per subject (49 VIS and 49 NIR spectrum images). For testing, we create validation and test sets from both spectra (VIS and NIR) and see the performance of the models on both these spectra separately. The sizes of the validation and test sets are based on the ratio 70:15:15. Table 3 shows the performance of the models in a cross-spectrum environment. We get comparable results from existing experiments [19] done on datasets with traditional cameras. On the VIS test set, we obtain accuracy of 96.56% and on the NIR test set, we obtain accuracy of 93.36% with Inception ResNet model. Recent advancements in domain adaptation methods [22,23] offer promising opportunities to bridge the performance gap between VIS and NIR spectra.

Table 3. Accuracy of models in Cross-Spectrum environment.

Model	Test Spectrum	Accuracy [%]
VGG16	VIS	93.52
	NIR	91.25
Inception ResNet (VGGFace2)	VIS	94.8
	NIR	92.6
Inception ResNet (CASIA-Webface)	VIS	**96.56**
	NIR	**93.36**
BEiT	VIS	94.26
	NIR	80.5

5 Conclusion

This work introduces FALEBface, a novel dataset for image processing using body worn cameras. By focusing exclusively on BWCs, the dataset provides a unique benchmark for law enforcement applications. For the preliminary experiments, a comparative analysis is done on the dataset. Fine-tuning the model on the dataset produces high recognition accuracy of 99.25%, 99.75%, and 96.67% respectively for indoor, outdoor, and dark environments. The high performance observed in these evaluations is indicative of the progress in existing algorithms rather than a lack of complexity in the dataset. However, when transitioning from self-environment analysis to cross-environment evaluations, a notable performance drop was observed. Rank-1 accuracy showed a decline of around 10 points when testing across different environments, indicating that while environment-specific performance is strong, cross-environment generalization remains challenging. Comparable results are also obtained for cross-spectrum environments (VIS and NIR spectra). For future work, we aim to extend the dataset by incorporating images captured with standard cameras and comparing them with those obtained from BWCs. This approach reflects real life scenarios where law enforcement often has access to images of suspects captured by standard cameras in their databases (gallery), while relying on BWCs during street-level operations. Apart from face recognition, there is limited work in the literature combining action recognition with BWCs. While action recognition has been extensively studied in the community, its application using BWCs remains relatively limited; creating an exciting avenue for new research and advancements in the field. As a part of our ongoing research on BWCs, we aim to experiment with advanced action recognition techniques using BWCs, with actions that are specifically relevant to law enforcement scenarios.

References

1. Imagenet. https://www.image-net.org/. Accessed 22 Feb 2025
2. Safe distance. https://www.officer.com/home/article/10248804/safely-handling-suspicious-person-stops. Accessed 22 Feb 2025
3. Al-Obaydy, W., Sellahewa, H.: On using high-definition body worn cameras for face recognition from a distance. In: Vielhauer, C., Dittmann, J., Drygajlo, A., Juul, N.C., Fairhurst, M.C. (eds.) BioID 2011. LNCS, vol. 6583, pp. 193–204. Springer, Heidelberg (2011). https://doi.org/10.1007/978-3-642-19530-3_18
4. Almadan, A., Krishnan, A., Rattani, A.: BWCFace: open-set face recognition using body-worn camera. arXiv preprint arXiv:2009.11458 (2020)
5. Bao, H., Dong, L., Piao, S., Wei, F.: BEit: BERT pre-training of image transformers. In: International Conference on Learning Representations (2022). https://openreview.net/forum?id=p-BhZSz59o4
6. Bryan, J.: Effects of movement on biometric facial recognition in body-worn cameras. Ph.D. thesis, Purdue University Graduate School (2020). https://doi.org/10.25394/PGS.12227372.v1
7. Cao, Q., Shen, L., Xie, W., Parkhi, O.M., Zisserman, A.: Vggface2: a dataset for recognising faces across pose and age. In: 2018 13th IEEE International Conference on Automatic Face amp; Gesture Recognition (FG 2018), pp. 67–74. IEEE Computer Society, Los Alamitos, CA, USA (2018). https://doi.org/10.1109/FG.2018.00020. https://doi.ieeecomputersociety.org/10.1109/FG.2018.00020
8. Choi, S., Michalski, N.D., Snyder, J.A.: The "civilizing" effect of body-worn cameras on police-civilian interactions: examining the current evidence, potential moderators, and methodological limitations. Criminal Justice Rev. **48**(1), 21–47 (2023). https://doi.org/10.1177/07340168221093549
9. Dosovitskiy, A., et al.: An image is worth 16×16 words: transformers for image recognition at scale. In: International Conference on Learning Representations (2021). https://openreview.net/forum?id=YicbFdNTTy
10. Elgharib, M., et al.: Egoface: Egocentric face performance capture and videorealistic reenactment (2019). https://doi.org/10.48550/arXiv.1905.10822
11. Hans, S., Dugelay, J., Isa, M.R.M.: Identifying individuals through egocentric motion: a study using body worn cameras. In: IEEE (ed.) IWBF 2025, International Workshop on Biometrics and Forensics, Munich, Germany, 24–25 April 2025 (2025)
12. Hans, S., Dugelay, J., Isa, M.R.M., Khairuddin, M.A.: Action recognition in law enforcement: a novel dataset from body worn cameras. In: Proceedings of the 14th International Conference on Pattern Recognition Applications and Methods - ICPRAM, pp. 605–612. INSTICC, SciTePress (2025). https://doi.org/10.5220/0013151900003905
13. King, D.: Dlib-ML: a machine learning toolkit. J. Mach. Learn. Res. **10**, 1755–1758 (2009). https://doi.org/10.1145/1577069.1755843
14. Meng, Z., Sánchez, J., Morel, J.-M., Bertozzi, A.L., Brantingham, P.J.: Ego-motion classification for body-worn videos. In: Tai, X.-C., Bae, E., Lysaker, M. (eds.) IVLOPDE 2016. MV, pp. 221–239. Springer, Cham (2018). https://doi.org/10.1007/978-3-319-91274-5_10
15. Mustafa, W., Kader, M.: A review of histogram equalization techniques in image enhancement application. In: Journal of Physics: Conference Series, vol. 1019, p. 012026 (2018). https://doi.org/10.1088/1742-6596/1019/1/012026

16. Nakada, M., Wang, H., Terzopoulos, D.: ACFR: active face recognition using convolutional neural networks. In: 2017 IEEE Conference on Computer Vision and Pattern Recognition Workshops (CVPRW), pp. 35–40 (2017). https://doi.org/10.1109/CVPRW.2017.11

17. O'Shea, K., Nash, R.: An introduction to convolutional neural networks. arXiv preprint arXiv:1511.08458 (2015)

18. Schroff, F., Kalenichenko, D., Philbin, J.: Facenet: a unified embedding for face recognition and clustering. In: Proceedings of the IEEE Conference on Computer Vision and Pattern Recognition (CVPR) (2015)

19. Siddiqui, N.J.: Novel approach for face recognition using cross-spectral environment. Int. J. Res. Eng. Sci. (IJRES) **09**(10), 70–76 (2021)

20. Cubukcu, S., Sahin, N., Tekin, E., Topalli, V.: The effect of body-worn cameras on the adjudication of citizen complaints of police misconduct. Justice Q. **40**(7), 999–1023 (2023). https://doi.org/10.1080/07418825.2023.2222789

21. Todak, N., Gaub, J.E., White, M.D.: Testing the evidentiary value of police body-worn cameras in misdemeanor court. Crime Delinquency **70**(4), 1249–1273 (2024). https://doi.org/10.1177/00111287221120185

22. Wen, G., Chen, H., Cai, D., He, X.: Improving face recognition with domain adaptation. Neurocomputing **287** (2018). https://doi.org/10.1016/j.neucom.2018.01.079

23. Yang, Y., Hu, W., Lin, H., Hu, H.: Robust cross-domain pseudo-labeling and contrastive learning for unsupervised domain adaptation NIR-VIS face recognition. IEEE Trans. Image Process. **32**, 5231–5244 (2023). https://doi.org/10.1109/TIP.2023.3309110

24. Yi, D., Lei, Z., Liao, S., Li, S.Z.: Learning face representation from scratch. arXiv preprint arXiv:1411.7923 (2014)

SMSCI: Simultaneous Modeling of Social and Contextual Interactions for Multi Pedestrian Trajectory Prediction

Mayssa Zaier[1], Hazem Wannous[1(✉)], and Hassen Drira[2]

[1] IMT Nord Europe, Institut Mines-Télécom, Univ. Lille, Centre for Digital Systems, 59000 Lille, France
{mayssa.zaier,hazem.wannous}@imt-nord-europe.fr
[2] University of Strasbourg, Strasbourg, France
hdrira@unistra.fr

Abstract. Recent advances in pedestrian trajectory prediction aim to capture the complexity of human behavior in autonomous systems. Prior methods often rely solely on positional data, limiting their ability to model social and environmental interactions. We propose *SMSCI*, a novel model that integrates recurrent sequence modeling with generative adversarial networks (GANs) to generate realistic trajectory distributions. By incorporating both historical motion and environmental context, *SMSCI* effectively models social dynamics and scene constraints, achieving higher accuracy and improved collision avoidance. Experimental results confirm the model's ability to produce socially and physically consistent trajectories, with strong implications for applications in self-driving vehicles and social robotics. The code is available at this link.

Keywords: Pedestrian · trajectory prediction · motion analysis

1 Introduction

Predicting pedestrian trajectories is vital for applications such as autonomous driving, robotics, and surveillance systems. Anticipating human trajectories in dynamic environments, such as streets or sports arenas, is complex. Pedestrians interact with their surroundings and each other, requiring the consideration of social behavior, environmental constraints, and multiple path options (Fig. 1). These challenges arise from:

Accurate prediction requires modeling how individuals navigate shared spaces while adapting to obstacles and responding to others' actions. However, this task is challenging due to the variability in social dynamics and physical layout across different scenarios. Early methods employed hand-crafted features, but their generalization was limited. Deep learning-based models, especially RNN-based approaches like Social-LSTM [1] and Social-GAN [2], introduced social pooling to capture interactions, but these models typically assign equal importance to all agents and neglect long-range dependencies. To address this, attention-based and

© The Author(s), under exclusive license to Springer Nature Switzerland AG 2026
M. Castrillón-Santana et al. (Eds.): CAIP 2025, LNCS 15621, pp. 102–113, 2026.
https://doi.org/10.1007/978-3-032-04968-1_9

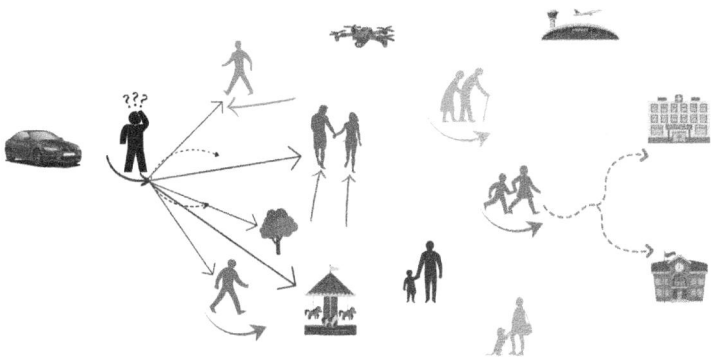

Fig. 1. Challenges in predicting pedestrian paths: human-environment interaction, social interaction, and multi-modal behavior.

graph-based methods [3,4] have been proposed, improving the handling of agent relationships but often simplifying the complex social context or ignoring scene information. Scene-aware approaches [5,6] integrate visual cues to better model physical constraints, typically using CNNs to extract features from static images. However, treating scene and social interactions separately overlooks their joint influence, and static contexts may miss temporal dynamics critical for prediction. Recent work [7,8] has emphasized the need for unified frameworks that address both social and environmental factors over time, incorporating multi-modal predictions to account for uncertainty in motion choices. Finally, Zaier et al. [9,10] employ Transformer networks with a self-attention mechanism to exploit visual context features for modeling complex spatio-temporal interactions. Their approach incorporates both pedestrian trajectory data and contextual information. However, it does not explicitly address social-scene interactions involving multiple trajectories.

In this paper, we introduce *SMSCI*, a GAN-based framework that jointly models social and physical interactions using attention mechanisms. Our model: (1) integrates scene context and social dynamics for improved predictions, (2) uses attentive pooling to prioritize relevant agents, (3) generates diverse trajectories with a recurrent GAN, and (4) outperforms state-of-the-art baselines on benchmark datasets such as UCY, ETH, and SDD.

2   Related Work

Trajectory prediction models are typically categorized based on how they model interactions between humans and the environment (human-space) or among humans themselves (human-human). More recent work attempts to unify these perspectives under social-scene interaction modeling.

Human-Space Interaction. Scene-aware methods rely on contextual cues such as road geometry or obstacles to constrain pedestrian paths. Approaches like

DESIRE [5] and SoPhie [6] use CNNs and attention to integrate scene semantics into trajectory forecasting. Extensions such as Trajectron++ [11] and STGAT [3] refine this by modeling spatial attention and dynamic features. However, many scene-aware methods still decouple scene and social influence, limiting their expressiveness.

Human-Human Interaction. Socially aware models often rely on sequence modeling and pooling. Social-LSTM [1] and Social-GAN [2] use spatial pooling to incorporate nearby agents' states. Attention-based models [12,13] assign dynamic weights to influential neighbors, though they often neglect asymmetric influence and distant agents. Graph-based techniques [4,7] allow for more flexible interaction modeling, yet may still simplify pedestrian relationships through uniform graph structures.

Our work leverages an attention mechanism that computes relative influence between agents while encoding spatial and temporal dependencies using ResNet and Transformer components. This enables learning of complex interaction patterns and variation in influence strengths, improving realism in predicted paths.

Generative Modeling. Human behavior is inherently uncertain, and multimodal forecasting is key to realistic trajectory prediction. GAN-based models [2,6,7] enable diverse path generation by sampling from learned distributions. Variants like PECNet [14] and M2P3 [15] refine end-point selection or enforce physical realism, though often without unified social-scene modeling.

Social-Scene Interaction. Integrating social and physical cues improves accuracy and robustness, as seen in SoPhie [6], which attends to both agent dynamics and scene layout. However, existing models often struggle to capture temporal continuity or prioritize relevant influences over time. Our approach addresses these gaps by jointly modeling scene and social context with time-aware attention and generating diverse, plausible trajectories via GANs.

3   Proposed Method

We propose an encoder-decoder framework with attention mechanisms to predict pedestrian trajectories. Our model integrates each agent's motion history, interactions with others, and environmental context to forecast future paths. As shown in Fig. 2, the architecture consists of four modules: Generator (G), Dynamic Context Encoder (DCE), Attentive Social Pooling Module (ASPM), and Discriminator (D).

The Generator encodes agents' past trajectories (X_i) into historical embeddings (H_i^t). The DCE extracts spatial and temporal features from videos using ResNet50 and a Transformer to focus on feasible future paths. ASPM pools social information into a vector P_i for each individual by emphasizing influential agents. The decoder predicts trajectories from historical data, pooled vectors, and the scene embedding. Finally, the Discriminator distinguishes real from generated trajectories, refining the model's predictions. *SMSCI* employs GANs and recurrent models to generate diverse and accurate trajectory distributions.

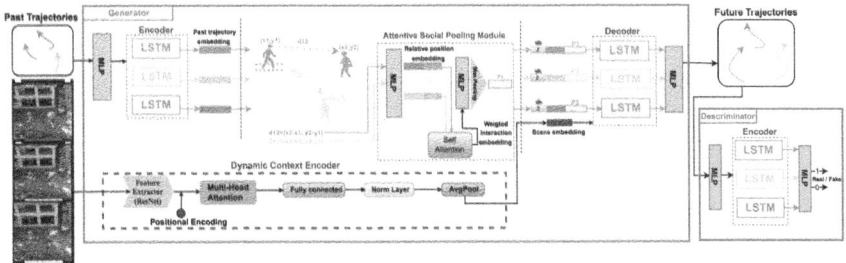

Fig. 2. An overview of *SMSCI* architecture.

3.1 Problem Formulation

We observe n pedestrians over $t = 1$ to T_{obs}, with 2D positions x_i^t and corresponding video clips providing environmental context. The aim is to collectively predict future trajectories $\hat{Y} = \{\hat{Y}_1, \ldots, \hat{Y}_n\}$ from $t = T_{obs} + 1$ to T_{pred}, informed by social norms and visual scene cues.

Let $X = \{X_1, \ldots, X_n\}$ and $V = \{V_1, \ldots, V_n\}$ denote observed positions and corresponding video inputs. The task is to predict $T_{pred}^{(i)} = \{(x_t^{(i)}, y_t^{(i)})\}_{t=T_{obs}+1}^{T_{pred}}$ using both the trajectories T_{obs} and visual context V_{obs}, accounting for physical constraints and social interactions.

3.2 Dynamic Context Encoder (DCE)

The DCE captures environmental features from video inputs, combining a ResNet to extract spatial structure with a Transformer to model temporal dynamics via Multi-head Attention:

$$MultiAtt(x) = (Att(x) \oplus \ldots \oplus Att(x))W^O, \quad \text{where } Att(x) = \sigma\left(\frac{Q_i K_i}{d_k}\right)V_i$$

Here, Q_i, K_i, V_i are projections of x into a shared feature space of dimension d, enabling temporal attention over meaningful regions. A causal mask ensures each prediction relies only on past and present data. Q/K/V are computed from pedestrian and environment embeddings, facilitating relevance-weighted attention for trajectory generation.

3.3 Attentive Social Pooling Module

Modeling social interactions requires sharing information across agents. Prior approaches like grid-based pooling [1] or max-pooling [2] are limited by their inability to account for distant agents or variable influence.

Our ASPM overcomes this by computing relative positions between the target agent and others, concatenating these with hidden states to form an input tensor. A self-attention mechanism ranks the importance of others' states, which are processed by an MLP and pooled to produce the vector P_i, as illustrated in Fig. 2. This is repeated per agent to reflect personalized social context.

3.4   LSTM-Based Generative Adversarial Network

Trajectory forecasting is inherently multimodal, as pedestrians may choose among multiple plausible future paths. Minimizing only the L2 loss tends to produce averaged, unrealistic trajectories. To address this, we adopt a Generative Adversarial Network (GAN) framework [16], which enables learning a distribution over feasible paths.

Generator. Trajectory forecasting is inherently multimodal, as pedestrians may choose among various plausible paths. Minimizing L2 loss alone often leads to unrealistic, averaged trajectories. To address this, we adopt a conditional Generative Adversarial Network (GAN) [16], which enables learning a diverse distribution over future trajectories.

In our generator, each pedestrian's observed 2D position is first embedded using a single-layer MLP and then encoded with an LSTM to model temporal dependencies. The resulting hidden states are passed through the Attentive Social Pooling Module (ASPM) to capture the influence of surrounding pedestrians. This pooled social context is then fused with scene features extracted by the Dynamic Context Encoder (DCE) and a stochastic latent variable sampled from a Gaussian distribution. Together, they form the input to the decoder LSTM, which generates future trajectories in an autoregressive manner. A final MLP maps the decoder's outputs to predicted spatial coordinates.

This setup enables the generator to capture the uncertainty, diversity, and complex interactions that characterize human motion in dynamic scenes.

Discriminator. To encourage socially and physically realistic predictions, we introduce an LSTM-based discriminator. It receives either real or generated trajectories, combined with the observed history, and processes them through an LSTM followed by an MLP to output a probability score. The discriminator learns to differentiate between authentic and synthetic trajectories by identifying violations of plausible social behaviors or environmental constraints.

3.5   Loss Functions

We jointly optimize the generator and discriminator using a combination of adversarial and regression losses. The adversarial loss ensures that generated trajectories resemble real ones in distribution, while the L2 loss penalizes large deviations from ground truth:

$$V = \arg \min_G \max_D \left(\mathcal{L}_{\text{GAN}}(G, D) + \lambda \mathcal{L}_{L2}(G) \right) \tag{1}$$

The adversarial loss $\mathcal{L}_{\text{GAN}}$ drives the generator to produce realistic samples, while the L2 loss $\mathcal{L}_{L2}$ ensures numerical accuracy. The weighting factor λ balances realism and precision, encouraging the model to generate diverse yet accurate trajectories.

3.6   Implementation Details

We trained the Generator and Discriminator for 200 epochs using Adam (lr = 0.001) and a batch size of 64. Encoder/decoder LSTM sizes are 16/32 respectively; input embeddings are 16-dimensional. The Discriminator uses an LSTM with a hidden size of 64. We consider up to $N_{\max} = 32$ agents per scene (with zero-padding if fewer). We use $\lambda = 1$ following prior works such as SoPhie, and implemented our model in PyTorch on an NVIDIA TITAN RTX GPU with 24GB RAM.

4   Experiments

4.1   Datasets

ETH/UCY. These datasets feature real-world top-view pedestrian trajectories annotated in meters, capturing diverse behaviors such as collisions, avoidance, and group movement. Data was collected via a fixed camera at 2.5 Hz with annotations every 0.4 s. ETH [17] and UCY [18] include five scenes: ETH, HOTEL (ETH), and UNIV, ZARA1, ZARA2 (UCY), with 2D positions provided in world coordinates.

Stanford Drone Dataset (SDD). Collected from a drone above a university campus, SDD [19] contains 20 outdoor scenes and over 11,000 pedestrians with 185,000+ interactions. It includes agent-scene interactions and landmarks (e.g., buildings) that guide pedestrian motion. We follow the standard data splits used in [6,14]. Coordinates are reported in pixels; results are converted to meters for consistency.

4.2   Evaluation

Metrics. We adopt standard metrics [1,2,5–7,11]:

Average Displacement Error (ADE): mean L2 distance between predicted and actual positions over all timesteps.

Final Displacement Error (FDE): L2 distance at the final timestep.

$$ADE = \frac{\sum_{i=1}^{n} \sum_{t=T_{\text{obs}}+1}^{T_{\text{pred}}} ||\hat{Y}_t^i - Y_t^i||}{n * T}, \quad FDE = \frac{\sum_{i=1}^{n} ||\hat{Y}_{T_{\text{pred}}}^i - Y_{T_{\text{pred}}}^i||}{n} \quad (2)$$

Protocol. Following prior works (Table 1), we observe 8 timesteps (3.2 s) and predict the next 12 (4.8 s). A leave-one-out scheme trains on four scenes and tests on the fifth. ETH/UCY results are in meters; SDD results are reported in pixels for compatibility with prior benchmarks. Synchronization of video and trajectory timestamps enables better spatio-temporal learning.

Baseline. We sample 20 predictions per input and select the closest to the ground truth. We compare against prior approaches [2,5–7,11–14,20–22].

We also evaluate ablations: **Ours w/o DCE:** no scene encoding, **Ours w/o AM:** no attention in pooling, **Ours w/o AM, DCE:** no attention or scene context, **Ours:** full model.

4.3   Quantitative Results

ETH/UCY combined. Table 1 reports results across ETH/UCY. Linear baselines fail to capture interactions. Social-LSTM [1] improves via pooling; Social-GAN [2] further improves using a generative approach. Our method exceeds Social-GAN, benefiting from scene context and attention.

SMSCI achieves state-of-the-art FDE (0.32), improving upon MemoNet and EqMotion [23], and obtains the lowest ADE (0.20) across all subsets. To assess social acceptability, we also compute near-collisions (within 0.10 m). Figure 2 summarizes this, showing SMSCI minimizes such incidents effectively.

Table 1. Quantitative results on ETH-UCY. Both ADE and FDE are reported in meters in world coordinates. Each method takes 8 frames as input and generates 12 frames as output. The **bold**/<u>underline</u> font denotes the best/second-best result.

Methods	UNIV	ZARA 1	ZARA 2	Hotel	ETH	AVG
Linear	0.82/1.59	0.62/1.21	0.77/1.48	0.39/0.72	1.33/2.94	0.79/1.59
Social-LSTM [1]	0.67/1.40	0.47/1.00	0.56/1.17	0.79/1.76	1.09/2.35	0.72/1.54
Social-ATTN* [12]	0.33/3.92	0.20/0.52	0.30/2.13	0.29/2.64	0.39/3.74	0.30/2.59
DESIRE [5]	0.59/1.27	0.41/0.86	0.33/0.72	0.52/1.03	0.93/1.94	0.53/1.11
Social-GAN [2]	0.60/1.26	0.34/0.69	0.42/0.84	0.72/1.61	0.81/1.52	0.58/1.18
SoPhie [6]	0.54/1.24	0.30/0.63	0.38/0.78	0.76/1.67	0.70/1.43	0.54/1.15
Trajectron [7]	0.54/1.13	0.43/0.83	0.43/0.85	0.35/0.66	0.59/1.14	0.47/0.92
STGAT [3]	0.52/1.10	0.34/0.69	0.29/0.60	0.35/0.66	0.65/1.12	0.43/0.83
Social Ways [13]	0.55/1.31	0.44/0.64	0.51/0.92	0.39/0.66	0.39/0.64	0.46/0.83
NMMP [24]	0.52/1.11	0.32/0.66	0.29/0.61	0.33/0.63	0.61/1.08	0.41/0.82
Social-STGCNN [20]	0.44/0.79	0.34/0.53	0.30/0.48	0.49/0.85	0.64/1.11	0.44/0.75
Causal-STGAT [25]	0.52/1.10	0.32/0.64	0.28/0.58	0.30/0.54	0.60/0.98	0.40/0.77
Causal-STGCNN [25]	0.49/0.81	0.34/0.53	0.32/0.49	0.38/0.45	0.64/1.00	0.43/0.66
PECNet [14]	0.35/0.60	0.22/0.39	0.17/0.30	0.18/0.24	0.54/0.87	0.29/0.48
M2P3 [15]	0.64/1.34	0.45/0.95	0.37/0.79	0.54/1.13	1.04/2.16	0.60/1.27
RSBG [26]	0.68/1.39	0.42/0.89	0.35/0.71	0.35/0.71	0.79/1.47	0.52/1.03
ST-LSTM [4]	0.38/0.84	0.31/0.47	0.33/0.51	0.41/0.73	0.57/1.02	0.40/0.72
SGCN [27]	0.37/0.70	0.29/0.53	0.25/0.45	0.32/0.55	0.63/1.03	0.37/0.65
Trajectron++ [11]	0.30/0.54	0.25/0.41	0.18/0.32	0.18/0.28	0.67/1.18	0.32/0.55
SGN LSTM [28]	0.48/1.08	0.30/0.65	0.26/0.57	0.63/1.01	0.75/1.63	0.48/0.99
SIT [29]	0.27/0.47	0.19/0.33	0.16/0.29	0.14/0.22	0.39/0.62	0.23/0.38
DynGroupNet [8]	0.24/0.44	0.19/0.34	0.15/0.28	0.13/<u>0.20</u>	0.42/0.66	0.23/0.38
EqMotion [23]	<u>0.23</u>/0.43	0.18/0.32	<u>0.13</u>/<u>0.23</u>	<u>0.12</u>/<u>0.18</u>	0.40/<u>0.61</u>	<u>0.21</u>/<u>0.35</u>
Ours (SMSCI)	0.26/**0.34**	**0.16/0.28**	**0.11/0.20**	0.17/0.27	**0.32/0.52**	**0.20/0.32**

SDD. We evaluated our method on the challenging SDD dataset and compared it with 12 baseline methods (Table 3). As expected, linear models underperform, while S-LSTM and S-GAN benefit from social modeling. CAR-Net improves further via physical attention, and DESIRE introduces a generative perspective. Incorporating both social and scene context, as in SoPhie, significantly enhances performance.

Table 2. Average collision percentage per frame in ETH/UCY scenes. Collision is detected if distance is less than 0.10 m.

Methods	UNIV	ZARA 1	ZARA 2	Hotel	ETH	Average
Linear*	1.24	3.78	3.63	1.57	3.14	2.67
Social-GAN [2]	0.56	1.75	2.02	1.75	2.51	1.72
SoPhie [6]	0.62	1.03	1.46	1.94	1.76	1.36
ST-LSTM [4]	0.58	0.87	**0.98**	1.23	1.25	0.98
Ours (SMSCI)	**0.49**	**0.72**	1.05	**0.63**	**0.81**	**0.74**

Our method outperforms all baselines, achieving an ADE of 8.83 and FDE of 13.41, surpassing notable methods like Social-GAN, SoPhie, Social-STGCNN, PECNet, and Trajectron++. These results validate the strength of our attention-based design in modeling complex interactions and constraints across diverse scenarios.

Table 3. ADE and FDE in pixels of various models in the SDD.

Methods	Linear	Social Force	Social-LSTM	Social-GAN	DESIRE	CAR-Net	SoPhie	Social-STGCNN	NMMP	PECNet	Trajectron++	SIT	Ours
ADE	37.11	36.48	31.19	27.246	19.25	25.72	16.27	20.60	14.67	9.96	19.30	9.13	**8.83**
FDE	63.51	58.14	56.97	41.44	34.05	51.8	29.38	33.10	26.72	15.88	32.70	15.42	**13.41**

Ablation Study. We conducted ablation experiments to assess the contributions of the attention mechanism in ASPM and the Dynamic Context Encoder (DCE) (Tables 4 and 5). The DCE captures spatial and temporal environmental features, while ASPM models pedestrian interactions. Their combination yields notable gains over baselines.

Effect of the Self-attention Mechanism in ASPM. Without attention (ours w/o AM), the model lacks focus, treating all agent interactions equally. This limits the representation of varying social influences. With self-attention, the model adaptively emphasizes relevant interactions, improving performance in dense or complex scenarios. In ETH/UCY, ADE/FDE improve from 0.30/0.52 to 0.20/0.32. In SDD, errors reduce from 11.76/16.53 to 8.83/13.41. These gains confirm attention's role in refining interaction modeling via adaptive weighting.

Effect of the Dynamic Context Encoder (DCE). Omitting the DCE (ours w/o DCE) excludes critical physical context, hindering understanding of how environments influence pedestrian behavior. Including the DCE, which uses multi-head attention over video frames, improves trajectory realism and spatial compliance. Performance improves from 0.35/0.60 to 0.20/0.32 (ETH/UCY) and from 12.89/17.61 to 8.83/13.41 (SDD). The DCE enables the model to interpret scene structure (e.g., walls, turns), reducing errors and enhancing generalization in varied environments.

Effect of Removing Both AM and DCE (Ours w/o AM, DCE). Removing both components significantly degrades performance. Without AM, the model cannot prioritize influential agents. Without DCE, it fails to account for environmental constraints. This leads to higher error: 0.58/1.18 in ETH/UCY and 27.25/41.44 in SDD. These results confirm that AM and DCE are complementary: the former captures nuanced social behavior, and the latter encodes physical constraints. Their synergy is critical to achieving strong performance.

Table 4. Enhancements in model performance by combination of DCE and ASPM on ETH/UCY

Methods	UNIV	ZARA 1	ZARA 2	Hotel	ETH	AVG
Ours w/o AM	0.38/0.65	0.32/0.52	0.14/0.25	0.25/0.45	0.43/0.73	0.30/0.52
Ours w/o DCE	0.45/0.76	0.33/0.62	0.16/0.29	0.30/0.52	0.51/0.79	0.35/0.60
Ours w/o AM, DCE	0.60/1.26	0.34/0.69	0.42/0.84	0.72/1.61	0.81/1.52	0.58/1.18
Ours	0.26/0.34	0.16/0.28	0.11/0.20	0.17/0.27	0.32/0.52	**0.20/0.32**

Table 5. Enhancements in model performance by combination of DCE and ASPM on SDD

Methods	Ours w/o AM	Ours w/o DCE	Ours w/o AM, DCE	Ours
ADE	11.76	12.89	27.25	**8.83**
FDE	16.53	17.61	41.44	**13.41**

4.4   Qualitative Results

We visualize predicted trajectories in Fig. 3 to demonstrate *SMSCI*'s qualitative performance on ETH/UCY and SDD. Predictions closely follow ground truth, illustrating accurate modeling of motion trends and interactions.

For instance, in the Hyang scene, a pedestrian veers off a straight path due to an obstacle. *SMSCI* correctly predicts this deviation, capturing both the obstacle and the behavioral adaptation. This reflects the model's strength in dynamically modeling agent-environment interactions, especially in complex outdoor settings like SDD with dense layouts and varying paths.

Fig. 3. Qualitative examples of trajectory prediction. Yellow, green, and red dots correspond to observations, ground truth, and predictions respectively. (Color figure online)

5   Conclusion

This work addresses the challenge of predicting pedestrian trajectories by modeling complex human-human and human-environment interactions. Our GAN-based encoder-decoder framework, *SMSCI*, outperforms recent baselines across multiple datasets. Key components include an attentive social pooling module, a dynamic context encoder for scene understanding, and an attention mechanism to emphasize impactful interactions. Ablation studies validate the contribution of each module, highlighting the robustness of our design. *SMSCI* consistently delivers accurate predictions by effectively leveraging both social and environmental cues.

Acknowledgement. This work is co-funded by the AI@IMT program of the Agence Nationale de la Recherche (ANR) and the region Hauts-de-France in France.

References

1. Alahi, A., Goel, K., Ramanathan, V., Robicquet, A., Fei-Fei, L., Savarese, S.: Social LSTM: human trajectory prediction in crowded spaces. In: Proceedings of the IEEE Conference on Computer Vision and Pattern Recognition (CVPR), pp. 961–971 (2016)

2. Gupta, A., Johnson, J., Fei-Fei, L., Savarese, S., Alahi, A.: Social GAN: socially acceptable trajectories with generative adversarial networks. In: IEEE/CVF Conference on Computer Vision and Pattern Recognition, pp. 2255–2264 (2018)

3. Huang, Y., Bi, H., Li, Z., Mao, T., Wang, Z.: STGAT: modeling spatial-temporal interactions for human trajectory prediction. In: 2019 IEEE/CVF International Conference on Computer Vision (ICCV), Seoul, Korea (South), pp. 6271–6280. IEEE (2019)

4. Xiong, D.: Spatial-temporal block and LSTM network for pedestrian trajectories prediction. arXiv preprint arXiv:2009.10468, vol. 2 (2020)

5. Lee, N., Choi, W., Vernaza, P., Choy, C.B., Torr, P.H.S., Chandraker, M.: DESIRE: distant future prediction in dynamic scenes with interacting agents. In: IEEE Conference on Computer Vision and Pattern Recognition (CVPR), pp. 2165–2174 (2017)

6. Sadeghian, A., Kosaraju, V., Sadeghian, A., Hirose, N., Rezatofighi, S.H., Savarese, S.: SoPhie: an attentive GAN for predicting paths compliant to social and physical constraints. In: Proceedings of the IEEE Computer Society Conference on Computer Vision and Pattern Recognition, CVPR 2019, Institute of Electrical and Electronics Engineers, United States of America, pp. 1349–1358. IEEE (2019)

7. Ivanovic, B., Pavone, M.: The trajectron: probabilistic multi-agent trajectory modeling with dynamic spatiotemporal graphs. In: IEEE/CVF International Conference on Computer Vision (ICCV), pp. 2375–2384 (2019)

8. Xu, C., Wei, Y., Tang, B., Yin, S., Zhang, Y., Chen, S.: Dynamic-group-aware networks for multi-agent trajectory prediction with relational reasoning. arXiv preprint arXiv:2206.13114 (2022)

9. Zaier, M., Wannous, H., Drira, H., Boonaert, J.: Cross-modal attention for accurate pedestrian trajectory prediction. In: Accepted at the 34rd British Machine Vision Conference, BMVC 2023 (2023)

10. Zaier, M., Wannous, H., Drira, H., Boonaert, J.: A dual perspective of human motion analysis-3D pose estimation and 2D trajectory prediction. In: Proceedings of the IEEE/CVF ICCV, pp. 2189–2199 (2023)

11. Salzmann, T., Ivanovic, B., Chakravarty, P., Pavone, M.: Trajectron++: dynamically-feasible trajectory forecasting with heterogeneous data. arXiv:2001.03093 [cs] (2021)

12. Vemula, A., Muelling, K., Oh, J.: Social attention: modeling attention in human crowds. In: 2018 IEEE International Conference on Robotics and Automation (ICRA), pp. 4601–4607 (2018)

13. Amirian, J., Hayet, J.-B., Pettré, J.: Social ways: learning multi-modal distributions of pedestrian trajectories with GANs. In: Proceedings of the IEEE/CVF Conference on Computer Vision and Pattern Recognition Workshops (2019)

14. Mangalam, K., et al.: It is not the journey but the destination: endpoint conditioned trajectory prediction. In: Vedaldi, A., Bischof, H., Brox, T., Frahm, J.-M. (eds.) ECCV 2020. LNCS, vol. 12347, pp. 759–776. Springer, Cham (2020). https://doi.org/10.1007/978-3-030-58536-5_45

15. Poibrenski, A., Klusch, M., Vozniak, I., Müller, C.: M2P3: multimodal multi-pedestrian path prediction by self-driving cars with egocentric vision. In: ACM Symposium on Applied Computing, Brno, Czech Republic, pp. 190–197. ACM (2020)

16. Goodfellow, I., et al.: Generative adversarial nets. In: Advances in Neural Information Processing Systems, vol. 27. Curran Associates, Inc., Montreal, QC, Canada (2014)

17. Pellegrini, S., Ess, A., Schindler, K., van Gool, L.: You'll never walk alone: modeling social behavior for multi-target tracking. In: IEEE 12th International Conference on Computer Vision, Kyoto, pp. 261–268. IEEE (2009)

18. Lerner, A., Chrysanthou, Y., Lischinski, D.: Crowds by example. Comput. Graph. Forum **26**(3), 655–664 (2007)

19. Robicquet, A., Sadeghian, A., Alahi, A., Savarese, S.: Learning social etiquette: human trajectory understanding in crowded scenes. In: Leibe, B., Matas, J., Sebe, N., Welling, M. (eds.) ECCV 2016. LNCS, vol. 9912, pp. 549–565. Springer, Cham (2016). https://doi.org/10.1007/978-3-319-46484-8_33

20. Mohamed, A., Qian, K., Elhoseiny, M., Claudel, C.: Social-STGCNN: a social spatio-temporal graph convolutional neural network for human trajectory prediction. In: Proceedings of the IEEE/CVF Conference on Computer Vision and Pattern Recognition, pp. 14 424–14 432 (2020)

21. Shafiee, N., Padir, T., Elhamifar, E.: Introvert: human trajectory prediction via conditional 3D attention. In: 2021 Conference on Computer Vision and Pattern Recognition (CVPR), Nashville, TN, USA, pp. 16810–16820. IEEE (2021)

22. Giuliari, F., Hasan, I., Cristani, M., Galasso, F.: Transformer networks for trajectory forecasting. In: 2020 25th International Conference on Pattern Recognition (ICPR), pp. 10 335–10 342. IEEE (2021)

23. Xu, C., et al.: Eqmotion: equivariant multi-agent motion prediction with invariant interaction reasoning. In: Proceedings of the IEEE/CVF Conference on Computer Vision and Pattern Recognition, pp. 1410–1420 (2023)

24. Hu, Y., Chen, S., Zhang, Y., Gu, X.: Collaborative motion prediction via neural motion message passing. In: Proceedings of the IEEE/CVF Conference on Computer Vision and Pattern Recognition, pp. 6319–6328 (2020)

25. Chen, G., Li, J., Lu, J., Zhou, J.: Human trajectory prediction via counterfactual analysis. In: 2021 IEEE International Conference on Computer Vision (ICCV), Montreal, QC, Canada, pp. 9804–9813. IEEE (2021)

26. Sun, J., Jiang, Q., Lu, C.: Recursive social behavior graph for trajectory prediction. In: Proceedings of the IEEE/CVF Conference on Computer Vision and Pattern Recognition, pp. 660–669 (2020)

27. Shi, L., et al.: SGCN: sparse graph convolution network for pedestrian trajectory prediction. In: Proceedings of the IEEE/CVF Conference on Computer Vision and Pattern Recognition, pp. 8994–9003 (2021)

28. Zhang, L., She, Q., Guo, P.: Stochastic trajectory prediction with social graph network. CoRR, vol. abs/1907.10233 (2019)

29. Shi, L., et al.: Social interpretable tree for pedestrian trajectory prediction. In: Proceedings of the AAAI Conference on Artificial Intelligence, vol. 36, no. 2, pp. 2235–2243 (2022)

SPL-BEV: Soccer Player Localization and Birds-Eye-View Estimation

Ivar Persson[1]([✉])[iD], Håkan Ardö[2][iD], and Mikael Nilsson[1][iD]

[1] Division of Computer Vision and Machine Learning, Centre for Mathematical Sciences, Lund University, Lund, Sweden
{ivar.persson,mikael.nilsson}@math.lth.se
[2] Spiideo, Malmö, Sweden
hakan.ardo@spiideo.com

Abstract. In this work we present SPL-BEV, a method to localize soccer players on a pitch from a monocular RGB camera. SPL-BEV features a network with few parameters that does not need to make any explicit object detection before localization is made. With SPL-BEV we show increased performance on the Spiideo SoccerNet SynLoc dataset compared to the best provided baseline result. The SPL-BEV system samples features from the U-Net feature space using bi-linear interpolation, guided by camera calibration, to generate features at grid points across multiple planes in a 3D world coordinate system. This forms a voxel feature space, which is then processed into grid cell detections on the ground plane, with final location refinement through x/y correction. The code for SPL-BEV is also published open source on GitHub.

Keywords: Birds-Eye-View estimation · 2D positioning · U-Net · Sports Analytics · Soccer

1 Introduction

In the area of automated sport broadcasts and sport analysis an intermediate problem is to detect and position all players on the field. This can be used in live broadcasts to, for example, estimate a suitable camera direction based on player positions, but also in post-game analysis of how players have moved during the game and in specific game scenarios. An intuitive way to display this is through a Birds-Eye-View (BEV) of all players that give a clear understanding of the current game state (Fig. 1).

A clear limitation in what type of methods that can be used, especially for live broadcasts, is the quick inference time needed as well as the need to be able to run these systems on devices with limited memory at the pitch location, not on remote servers. In this work we present a possible solution with a BEV estimation network with few parameters aimed at cost-efficient deployment. The work is engineered first and foremost to detecting players on a soccer pitch but the system is also aimed to be transferable to other sports as well as to other orientation and planning tasks. The main contributions of this paper are:

M. Castrillón-Santana et al. (Eds.): CAIP 2025, LNCS 15621, pp. 114–123, 2026.
https://doi.org/10.1007/978-3-032-04968-1_10

Fig. 1. The SPL-BEV system estimates a Bird-Eye-View of the soccer pitch and the world coordinate of each player with no prior object detection.

- We present SPL-BEV, a system with few parameters, evaluated on the synthetic dataset Spiideo SoccerNet SynLoc [2], achieving state-of-the-art performance.
- We present an ablation study showing the performance differences from different experimental setups.
- The code to train and run the networks will be provided at: https://github.com/IvarPersson/SPL-BEV

2   Related Work

Creating a BEV of a scene is an important step in different perception tasks and has been an active field for many decades [13]. The idea is to use the information in one (or more) ordinary RGB images to construct an overview, a BEV of the scene in question. The survey study [12] divides popular methods into four major categories of BEV methods; Homograph-based [5], Depth-based [7,16], MLP-based [18] and Transformer-based [10]. Some of these methods combine their method with polar coordinate representations [18,20], thus representing points in a natural way with projective camera models.

In the depth-based category of studies one could combine the regular images with other modalities as a guidance, such as LIDAR [11] where a complete 3D structure makes the conversion to BEV simple. When not supplementing with e.g. LIDAR, a common method is to create a voxel space across the scene and then fill it features from the image or feature extractors [7,16]. Other related problems to monocular BEV estimation can also share methodologies with other problems, such as monocular depth estimation [9] as well as pose och shape estimation [8,15], where complete systems that estimate depth and pose can be used to easily infer a BEV.

A lot of the earlier works in the BEV estimation field have been focused on automated driving by evaluating their methods on datasets such as KITTI [14] and NuScenes [4] or have had the focus on other robotics connected views [13]. The number of works concerning only humans are fewer. One work by Dai et al. come with cameras already in a close to BEV position focused on social distancing compliance during the COVID-19 pandemic [5], however the distances are shorter than in this work. In sport settings, there has been more focus on pose estimation of human joints in different key moments [22] or with a video stream [1]. These existing methods may detect players first with an object detection network [21] possibly from television broadcasts with limited field of view [6]. In order to avoid using parameter expensive object detectors, we are going to base our method on Simple-BEV [7]. Common object detectors include different versions of YOLO [17], used in e.g. [2,21]. This will avoid explicit object detection and thus keeping the number of parameters low. Given the conclusions in Simple-BEV, that the performance decrease was very limited compared to other methods, that method seems promising. This work will also not feature a polar coordinate system, like [18,20], but a rectangular voxel in global coordinates as we want precision to be equal across the field, even though distances may hinder this.

3   Method

The method to estimate the BEV is in its original form based on Simple-BEV [7]. First we calculate a feature-map by using different of-the-shelf feature extractor structures before we bi-linearly sample values from this map onto a rendered voxel grid over the soccer pitch. An illustration of this can be seen in Fig. 2. The extractor transforms the original input image of size $H_{in} \times W_{in} \times 3$ to a feature map, see step i in Fig. 2, $H \times W \times C$ of, possibly, different size H and W than H_{in} and W_{in}, where C is the number of features in the feature extractor. This is described in detail in Sect. 3.1.

Step ii in Fig. 2 is done by projecting the voxel positions, in global coordinates, into the camera plane by using the known camera calibration, i.e. camera matrix and distortion model. These positions are then bi-linearly sampled. The initialized voxel grid with bi-linearly sampled features covers the entire soccer pitch and several layers into the air. As the BEV reconstruction concerns humans a reasonable conclusion is to consider a total height of the voxel grid to be three meters to also capture jumping players. The voxel grid is sampled of size $X \times Y \times Z$ to cover the entire pitch, with a sampling density of s, in all three directions. As C features are sampled in each voxel the total grid become of size $X \times Y \times Z \times C$. A simple restructuring of data creates a voxel grid size of $X \times Y \times ZC$.

A feature reduction, seen as step iii in Fig. 2, is done by a 2D 3×3 convolution that reduces the possibly high number of features into a more manageable size of $X \times Y \times C$.

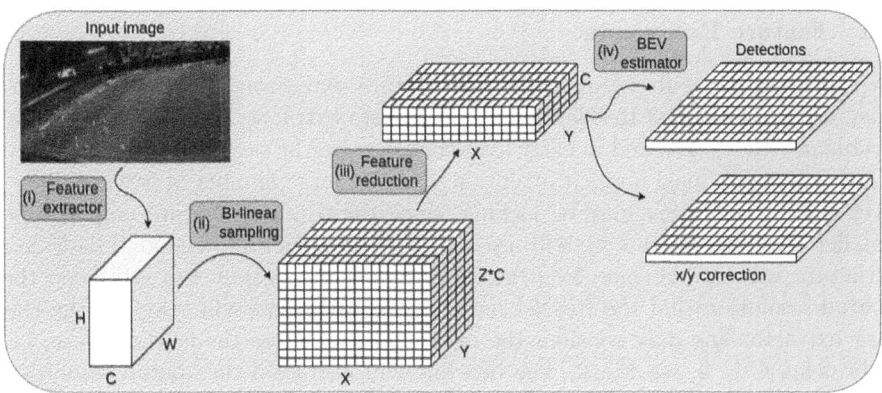

Fig. 2. An image is fed through the feature extractor. The subsequent feature map is used to do bi-linear sampling along rays from the camera center to each voxel position. This voxel space is then used to create the actual BEV-estimation, including the x/y correction.

Next the BEV estimation, step *iv*, consists of two ResNet blocks where the final output layer consists of a grid with the same X and Y size as the sampled grid but only on the ground plane. The first output, denoted *Detections* after step *iv* in Fig. 2 is a binary estimation of whether a player is detected within the grid cell in question or not. The ground truth is built up by considering all grid cell points within r meters of the position to be a positive hit of a soccer player. The loss for the first output is a Binary Cross Entropy (BCE), with an added weight due to the uneven class ratio on the dataset. The second and third outputs represent the local x and y offsets respectively of the voxel position relative to the exact ground-truth location, serving as a correction in x/y, denoted *x/y correction* in the Figure. This is analogous to the bounding box adjustments in the image plane used in the original YOLO detector loss [17], but here the corrections are applied in world coordinates on a 3D plane. This correction will be evaluated for all voxel points within the same r meters from the ground truth with a simple L1-loss.

The final detections are made by a Soft Non-Maximum Suppression (Soft-NMS) [3], with hyper parameter tweaks to fit the studied problem better. First, the detections lack a bounding box that is usually used, however here a circle is added to all detections in order to calculate IoU. Second, the penalty for overlapping is not excessively large, allowing players in close proximity to all be detected. Non-overlapping circles are of course not effecting each other. The detected players' positions are then corrected with the x/y correction. The idea is thus that the grid detects approximate positions of players before the x/y correction finds the exact position.

3.1   Feature Extractors

In this work we will examine different feature extractors. In Simple-BEV [7] they use a pre-trained ResNet50 as their feature extractor, however a potential problem with it, observed in early investigation, are the several down-samplings. They will create local implicit attention along its large number of features, however spatial resolution may be lacking, since the resulting feature map will be smaller. To mitigate this we will use a U-Net [19] structure instead. The U-Net structure will enable some local attention while the output will still keep the spatial resolution that the ResNet lacks. As a baseline we will also skip the feature extractor and only consider the input RGB image as the feature space, i.e. only using $C = 3$, see Fig. 2. The baseline will also keep the spatial resolution for the bi-linear sampling but may be lacking in descriptive features and local attention. The choice of structure for the feature extractor will be investigated later in the ablation studies in Sect. 4.2. None of the feature extractors will be pre-trained, however you should be able to use a pre-trained extractor. A possible risk is that it will require a more parameter intensive one than our specially trained extractors.

Two different sizes of U-Net will be used, the structure of them will be the same but they will differ in the number of features. Two convolutions are used at each feature level with standard skip-connections across the U-shape. The feature extractor denoted just U-Net have $8, 16, 32$ features and a final 16 features output while the smaller extractor, denoted U-Net small have $4, 8, 16$ features and a final 8 output features.

3.2   BEV-Estimator

The BEV-estimation used is in its standard form consists of two ResNet-blocks which outputs the estimation and x/y correction. An alternative used in the ablation is to skip the BEV-estimator based on ResNet blocks and just add a 1D convolution in the features. In effect, this is a classical logistic regression in the three outputs. This alternative will drastically reduce the number of parameters in the network if it is used.

4   Experiments and Results

As described in the method, the loss consists of a standard weighted BCE-loss for the detection and L1-loss for the x/y corrections. The BCE-loss thus become:

$$\mathcal{L}_{\text{BCE}} = \frac{1}{|V|} \sum_{x,y} w[V(x,y)log(\hat{V}(x,y))$$
$$+ (1 - V(x,y))log(1 - \hat{V}(x,y))],$$

(1)

where w is the weight calculated over the entire training set as the fraction of pixels without a player over the number of pixels with a player. V denotes the grid of binary classification and $\hat{V}$ the estimated grid. The L1-loss for the x/y corrections can be written as:

$$\mathcal{L}_{\text{pos}} = \sum_{x,y} \mathbb{1}^p_{x,y} |\hat{dx}(x,y) - dx(x,y)|)$$
$$+ \sum_{x,y} \mathbb{1}^p_{x,y} |\hat{dy}(x,y) - dy(x,y)|), \tag{2}$$

where $\mathbb{1}^p_{x,y}$ is the indicator function if a player is present in the (x,y) position or not. These losses are weighed as

$$\mathcal{L} = \mathcal{L}_{\text{BCE}} + \lambda\mathcal{L}_{\text{pos}}. \tag{3}$$

In the cases of training with both the detection and the x/y correction, $\lambda = 10$, a value determined by minor experimentation but may be improved by hyper-parameter optimization. When training without the x/y correction $\lambda = 0$.

The Soft-NMS [3] differ from a standard NMS since it still allows overlap between objects, or in the case of this work, allows players to be close to each other. A standard NMS would automatically hinder player detections close to each other by canceling them but the Soft-NMS does not. Here players were given a circle of 1 m radius to be used when considering overlap instead of the normal bounding boxes.

The parameter r that controlled the ground truth creation of which voxels were considered to contain a player was chosen to be 0.5 m. The sample density s of the voxel grid may, just like r, impact the results, however to narrow the scope of this study a spatial sampling granularity of 0.5 m was chosen to correspond with ground truth creation.

4.1   Dataset and Metrics

The dataset used in this work is the Spiideo SoccerNet SynLoc [2], a synthetic dataset directed at detecting and position players. It consists of real-world images from soccer stadiums where realistic soccer players are rendered in different parts of the field. The images are high resolution (3840*2160) showing approximately one half of the field at a time. The paper also introduces a metric called mAP-LocSim [2]. It is a mean Average Precision metric with the IoU similarity measure replaced replaced with the LocSim similarity measure,

$$\text{LocSim} = e^{ln(0.05)\frac{d^2}{\tau^2}}, \tag{4}$$

where d is the distance between the true and estimated positions. A player is considered detected if the LocSim is greater than 0.5. The amount of images with all soccer players detected and no extra detections will be shown in Table 1 as the metric Frame accuracy. The cameras have known calibration and are positioned at different heights depending on the format of the stadium, thus this is something our model has to learn as well. The standard train and validation splits from Spiideo SoccerNet SynLoc is used [2].

Table 1. Quantitative results on the dataset Spiideo SoccerNet SynLoc. The model YOLOX-m pose is the best performing model of the baselines in [2]. The columns *BEV est* and *x/y corr.* denote different training setups whether the used BEV estimator was the ResNet blocks (ResNet) or logistic regression (log.reg.) and if the x/y correction was trained and used in the evaluation or not. The best scoring values in each column are marked bold.

Feature extractor	BEV est.	x/y corr.	mAP LocSim	Precision	Recall	F1	Frame Acc
YOLOX-m pose [2]			0.793	0.928	**0.890**	0.907	0.316
No extractor	ResNet	x	0.440	0.732	0.640	0.683	0.050
U-Net small	log.reg.		0.405	0.801	0.790	0.798	0.162
U-Net small	log.reg.	x	0.731	0.913	0.820	0.864	0.180
U-Net small	ResNet	x	0.832	0.966	0.870	0.915	0.295
U-Net	log.reg.		0.526	0.431	0.410	0.420	0.012
U-Net	log.reg.	x	0.646	0.915	0.850	0.881	0.287
U-Net	ResNet	x	**0.863**	**0.974**	0.880	**0.925**	**0.324**

4.2   Results and Ablation Study

The results are shown in Table 1 where the largest U-Net with both the ResNet blocks as well as the x/y correction performed best in four out of five metrics. It is also clear that both of these additions, the ResNet blocks and the x/y correction, improved the results across all metrics. Comparing e.g. the three tests with the full U-Net we see that adding x/y correction almost doubles Precision and Recall with further improvements when adding the ResNet blocks. When comparing the best result in the table with the second best it seems like the number of features in the U-Net affected the results less than the addition of the ResNet blocks and the x/y correction.

Of the major building blocks in our system we focus the ablations to the two blocks with the largest amount of parameters, the feature extractor and the BEV estimator. The feature extraction is tested with no feature extraction at all and two sizes of U-Net, denoted U-Net and U-Net small, all tests are shown in Table 1.

In Fig. 3 we can see two example results from the validation set. The two U-Net variants perform in a similar manner but the smaller variant makes a spurious detection in *Ex 1*. Both systems fail to detect the two players close to each other in the center of the visible field. The Baseline with no explicit feature extraction makes several spurious detections and also fails to detect several players, this is extra clear in *Ex 2*. The visual results in Fig. 3 is in line with the results in Table 1.

U-Net   U-Net small   No feature extractor

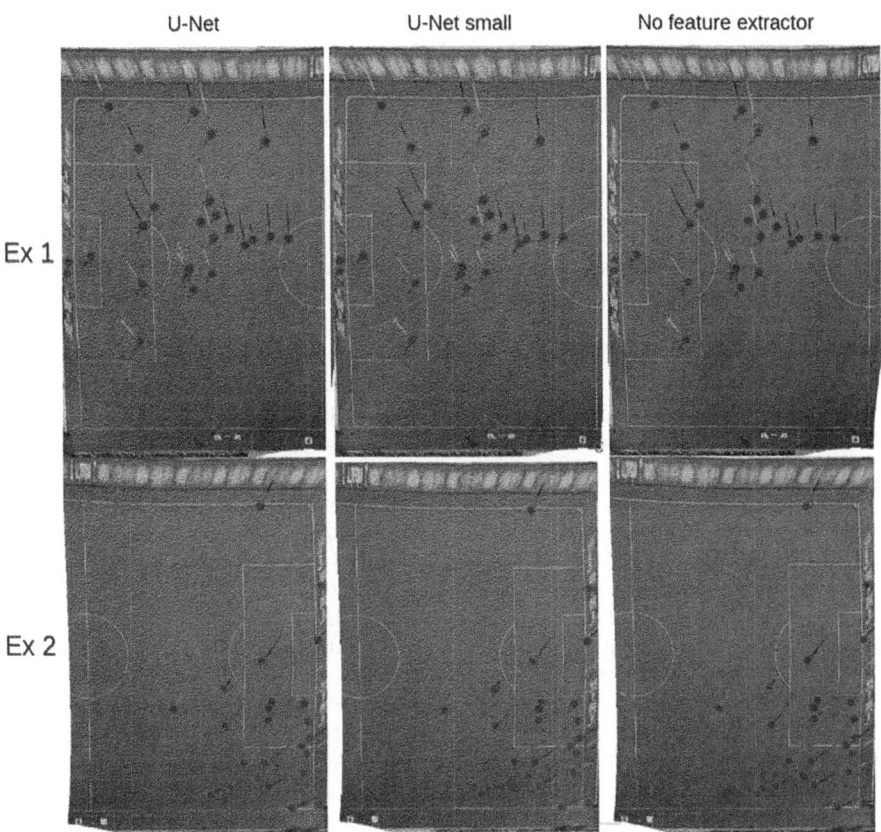

Fig. 3. Qualitative comparison of the results for three of the tested methods. This shows two random examples (Ex 1, Ex 2) from the validation set, one for each side of the field. All of these three methods feature ResNet-blocks as BEV estimator and x/y correction but different feature extractors. Blue discs are the ground truth position and red circles are the estimated positions. (Color figure online)

5   Conclusion

In this paper we present SPL-BEV that achieves state-of-the-art results on almost all metrics on the Spiideo SoccerNet SynLoc dataset. An ablation study shows the importance of the different blocks. As future work, we consider, among other things, several hyperparameters that may have improved results further and also synthetic-to-real generalization.

Acknowledgements. The machine learning computations were enabled by resources provided by the National Academic Infrastructure for Supercomputing in Sweden (NAISS), partially funded by the Swedish Research Council through grant agreement no. 2022-06725.

References

1. Arbués-Sangüesa, A., Martín, A., Fernández, J., Rodríguez, C., Haro, G., Ballester, C.: Always look on the bright side of the field: merging pose and contextual data to estimate orientation of soccer players. In: 2020 IEEE International Conference on Image Processing (ICIP), pp. 1506–1510 (2020)
2. Ardö, H., et al.: Spiideo soccernet synloc - single frame world coordinate athlete detection and localization with synthetic data. In: In Proceedings of the 20th International Joint Conference on Computer Vision, Imaging and Computer Graphics Theory and Applications - Volume 2: VISAPP, pp. 278–285. INSTICC, SciTePress (2025)
3. Bodla, N., Singh, B., Chellappa, R., Davis, L.S.: Soft-NMS – improving object detection with one line of code. In: Proceedings of the IEEE/CVF International Conference on Computer Vision (ICCV), pp. 5561–5569 (2017)
4. Caesar, H., et al.: nuscenes: a multimodal dataset for autonomous driving. In: CVPR (2020)
5. Dai, Z., Jiang, Y., Li, Y., Liu, B., Chan, A.B., Vasconcelos, N.: BEV-net: assessing social distancing compliance by joint people localization and geometric reasoning. In: Proceedings of the IEEE/CVF International Conference on Computer Vision (ICCV), pp. 5401–5411 (2021)
6. Gupta, M., Singh, N., Gajawada, R., Dambekodi, S.: Hawk eye: automatic birds eye view registration of sports videos (2023). https://nihal111.github.io/hawk_eye/. Accessed 28 Nov 2024
7. Harley, A.W., Fang, Z., Li, J., Ambrus, R., Fragkiadaki, K.: Simple-BEV: what really matters for multi-sensor BEV perception? In: arXiv:2206.07959 (2022)
8. Lee, H.J., Kim, H., Choi, S.M., Jeong, S.G., Koh, Y.J.: Baam: monocular 3D pose and shape reconstruction with bi-contextual attention module and attention-guided modeling. In: 2023 IEEE/CVF Conference on Computer Vision and Pattern Recognition (CVPR), pp. 9011–9020 (2023)
9. Li, Z., Bhat, S.F., Wonka, P.: Patchfusion: an end-to-end tile-based framework for high-resolution monocular metric depth estimation. In: Proceedings of the IEEE/CVF Conference on Computer Vision and Pattern Recognition (CVPR), pp. 10016–10025 (2024)
10. Li, Z., et al.: BEVFormer: learning bird's-eye-view representation from multi-camera images via spatiotemporal transformers. In: Avidan, S., Brostow, G., Cissé, M., Farinella, G.M., Hassner, T. (eds.) Computer Vision - ECCV 2022, pp. 1–18. Springer, Cham (2022)
11. Ma, X., Wang, Z., Li, H., Zhang, P., Ouyang, W., Fan, X.: Accurate monocular 3D object detection via color-embedded 3D reconstruction for autonomous driving. In: Proceedings of the IEEE/CVF International Conference on Computer Vision (ICCV) (2019)
12. Ma, Y., et al.: Vision-centric BEV perception: a survey. IEEE Trans. Pattern Anal. Mach. Intell. **46**(12), 10978–10997 (2024)
13. Mallot, H.A., Bülthoff, H.H., Little, J.J., Bohrer, S.: Inverse perspective mapping simplifies optical flow computation and obstacle detection. Biol. Cybern. **64**, 177–185 (1991)
14. Menze, M., Geiger, A.: Object scene flow for autonomous vehicles. In: Conference on Computer Vision and Pattern Recognition (CVPR) (2015)

15. Persson, I., Ahrnbom, M., Nilsson, M.: Monocular estimation of translation, pose and 3D shape on detected objects using a convolutional autoencoder. In: 17th International Joint Conference on Computer Vision, Imaging and Computer Graphics Theory and Applications (VISAPP), vol. 5, pp. 390–395 (2022). https://doi.org/10.5220/0010826600003124

16. Philion, J., Fidler, S.: Lift, splat, shoot: encoding images from arbitrary camera rigs by implicitly unprojecting to 3D. In: Vedaldi, A., Bischof, H., Brox, T., Frahm, J.-M. (eds.) ECCV 2020. LNCS, vol. 12359, pp. 194–210. Springer, Cham (2020). https://doi.org/10.1007/978-3-030-58568-6_12

17. Redmon, J., Divvala, S., Girshick, R., Farhadi, A.: You only look once: unified, real-time object detection. In: Proceedings of the IEEE Conference on Computer Vision and Pattern Recognition, pp. 779–788 (2016)

18. Roddick, T., Cipolla, R.: Predicting semantic map representations from images using pyramid occupancy networks. In: IEEE/CVF Conference on Computer Vision and Pattern Recognition (CVPR) (2020)

19. Ronneberger, O., Fischer, P., Brox, T.: U-net: convolutional networks for biomedical image segmentation. In: Navab, N., Hornegger, J., Wells, W.M., Frangi, A.F. (eds.) MICCAI 2015. LNCS, vol. 9351, pp. 234–241. Springer, Cham (2015). https://doi.org/10.1007/978-3-319-24574-4_28

20. Saha, A., Mendez, O., Russell, C., Bowden, R.: Translating images into maps. In: 2022 International Conference on Robotics and Automation (ICRA), pp. 9200–9206 (2022). https://doi.org/10.1109/ICRA46639.2022.9811901

21. SoccerEye: Soccereye: an open-source system for understanding monocular football videos (2023). https://github.com/WWandP/SoccerEye. Accessed 28 Nov 2024

22. Yeung, C., Ide, K., Fujii, K.: Autosoccerpose: automated 3D posture analysis of soccer shot movements. In: Proceedings of the IEEE/CVF Conference on Computer Vision and Pattern Recognition, pp. 3214–3224 (2024)

CANpose: A Cross-Attention Framework for Human Pose Recognition

M. S. Subodh Raj[1]([✉])(ID), Sudhish N. George[2](ID), and Kiran Raja[1](ID)

[1] Department of Computer Science, Norwegian University of Science and Technology,
Gjøvik, Norway
subodh.m.raj@ntnu.no, kiran.raja@ntnu.no
[2] Department of Electronics and Communication Engineering, National Institute
of Technology Calicut, Kozhikode, India
sudhish@nitc.ac.in

Abstract. We propose CANpose, a lightweight model for human pose recognition that integrates convolutional networks with a bi-modal cross-attention mechanism to effectively capture both local and global features. In contrast to conventional CNN-based and hybrid attention approaches, CANpose leverages the complementary nature of pose depth maps and binary silhouettes through cross-attention, enhancing its ability to distinguish between visually similar poses. By employing depth-wise separable convolutions for efficient local feature extraction, CANpose significantly reduces computational complexity–achieving up to 17× fewer parameters and 11× lower FLOPs–while maintaining high recognition accuracy across multiple benchmark datasets. These results position CANpose as a scalable and efficient solution for real-world pose recognition applications.

Keywords: Cross-attention · Depth-wise convolution · Human pose recognition

1 Introduction

Driven by a large number of applications including visual surveillance, gait analysis, human-computer interface, and ego-centric action detection from videos [1,15], human pose recognition (HPR) has become increasingly important [2]. Human poses convey cues through the spatial orientation of human body parts and provide details about a person's actions [8]. HPR has been performed with different data modalities including RGB images/videos in recent applications such as egocentric vision and motion capture (mocap) data in dedicated applications. The rich 3D structural information, geometrical shape information, and inherent privacy preservation capability of depth maps make them an attractive

Supplementary Information The online version contains supplementary material available at https://doi.org/10.1007/978-3-032-04968-1_11.

candidate for pose recognition tasks [12]. Many earlier works have used human pose depth maps for estimating poses and have not explored their potential for pose recognition tasks [13, 25, 27]. We therefore aim to leverage depth maps for HPR to achieve high accuracy while maintaining low computational complexity.

The intricate interactions among human body parts play a critical role in HPR. These interactions encode both local features, such as joint angles and posture nuances, and global features, reflecting the overall spatial configuration of the body. Effective HPR requires capturing both levels of information. While convolutional neural networks (CNNs) are well-suited for local feature extraction, attention mechanisms excel at modeling global dependencies. Recent works have therefore explored hybrid models combining CNNs and attention to leverage both aspects [29, 32]. Building on this idea, we propose a cross-attention-based approach that jointly utilizes pose depth maps and their corresponding binary silhouettes derived from the depth maps. Depth maps encode rich 3D structural information and spatial positioning of body parts, whereas binary silhouettes emphasize body shape and contours. By fusing these modalities through a bi-modal cross-attention mechanism, the model learns to attend to semantically important regions, improving discrimination between visually similar poses.

Although attention mechanisms enhance recognition by focusing on relevant spatial regions, they incur high computational costs. CNNs, while more efficient, may underperform due to limited context modeling and sensitivity to pose variation, especially in smaller datasets. To address these limitations, the proposed model combines depth-wise separable convolution [23] for efficient local feature extraction with bi-modal cross-attention for capturing global pose relationships. Cross-attention explicitly aligns tokens from depth maps and silhouettes, enabling the model to generalize across diverse pose variations. Depth-wise convolution significantly reduces computational overhead, leading to a lightweight architecture with improved accuracy.

In summary, the proposed model offers two core advantages: (i) joint learning of global and local pose-specific features from dual modalities, and (ii) reduced computational cost through depth-wise convolution. These design choices are empirically validated across multiple datasets, demonstrating improvements in both recognition performance and efficiency.

We present a set of relevant works in Sect. 2 and the proposed architecture in Sect. 3 followed by results in Sect. 4.

2   Related Works

While classical machine learning methods have been explored for HPR [6, 24, 34], we focus on deep learning approaches due to their superior performance and alignment with the proposed method.

Related Works with CNN. Transfer learning with CNNs has been widely used for human activity recognition (HAR) [2, 10], leveraging strong local spatial feature extraction. However, these models often lack the ability to capture global pose context, which is critical for HPR. To address this, attention mechanisms have been integrated with CNNs. Yan *et al.* [30] introduced a multi-branch soft attention module on VGG16 to fuse scene-level and region-level features. Similarly, Chapariniya *et al.* [3] proposed a student-teacher framework using self-regulated attention to improve generalization. Zheng *et al.* [33] combined visual and semantic attention for better context modeling in still-image action recognition.

Related Works with Attention. Attention mechanisms are effective for modeling long-range dependencies and enhancing context awareness [14]. Their combination with CNNs has shown improved performance in visual tasks [29,32]. For instance, Hosseyni *et al.* [11] used ResNet50 followed by vision transformers (ViT) to capture both local and inter-region features. Dai *et al.* [5] introduced CoAtNet, integrating convolutions into transformers to retain spatial locality. Li *et al.* [16] proposed UniFormer, blending local and global attention across layers to enhance learning efficiency and reduce data requirements.

Related Works in Pose Estimation. Depth-based pose estimation has also seen notable progress. Shotton *et al.* [25] introduced per-pixel 3D joint prediction from depth images. Later works refined this with improved annotations [7], silhouette-based ridge features [13], and prototype-based pose decomposition [19]. Moon *et al.* [21] proposed voxel-to-voxel prediction to address 2D-to-3D regression limitations, while Wu *et al.* [27] modeled limb uncertainties via depth-encoded representations.

Limitations of Current Works for HPR. Despite their success, existing CNN-attention hybrids often fall short in modeling the spatial hierarchies and inter-part dependencies essential for HPR. CNNs excel at capturing local patterns but struggle with global body configurations, making it difficult to recognize subtle pose variations [30]. Although attention mechanisms improve global reasoning [20], their application to human pose parts remains limited. Furthermore, large labeled datasets are scarce, hindering the generalization of data-hungry models like CNNs and transformers [5]. Combining these architectures also introduces computational overhead, underscoring the need for lightweight designs that balance accuracy and efficiency [31].

3   Proposed Model

To address the limitations of existing methods, we propose a model that integrates pose depth maps and binary silhouettes using a bi-modal cross-attention mechanism for HPR. This approach leverages the complementary nature of depth and silhouette information to enhance recognition accuracy, while depth-wise convolution improves local feature extraction with reduced computational complexity.

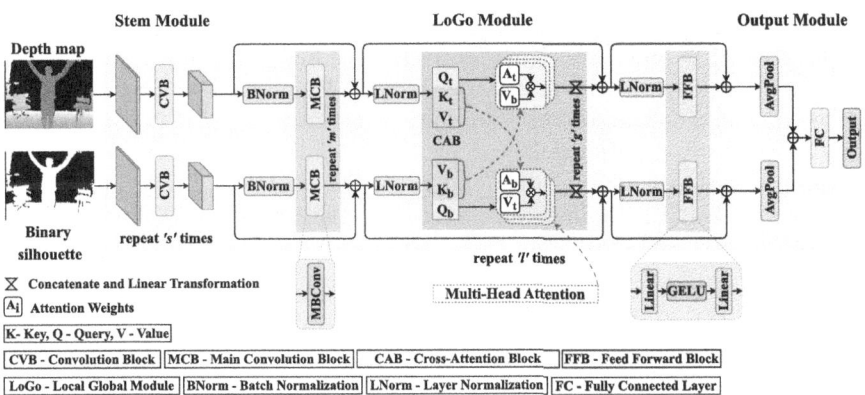

Fig. 1. Overview of the proposed CANpose architecture. The model follows a two-stream design, processing depth maps and corresponding binary silhouettes in parallel.

The proposed cross-attention convolution network for human pose recognition (CANpose), illustrated in Fig. 1, comprises three main components: a stem module, a local-global (LoGo) module, and an output module. Two parallel streams process depth maps and binary silhouettes independently. Their outputs are fused via a cross-attention mechanism before reverting to the two-stream structure. CANpose leverages synchronized depth maps and binary silhouettes through a bi-modal cross-attention framework that performs relative self-attention between each token of the depth map and the corresponding binary silhouette, facilitating global feature learning across modalities. Additionally, depth-wise separable convolution [23] captures local spatial patterns efficiently, contributing to the model's lightweight and effective design. At each stage, spatial dimensions are progressively reduced while channel depth increases, optimizing both performance and computational efficiency.

Stem Module. Human pose frames are typically large (e.g., 640×480), making global feature extraction via relative attention computationally intensive. Direct downscaling risks losing essential spatial details. To address this, the stem module systematically reduces frame dimensions while preserving important features. We employ convolution blocks (CVB), which apply 3×3 convolutions followed by batch normalization and Gaussian error linear unit (GELU) activation to extract features from both depth maps and binary silhouettes. This structure facilitates faster convergence and efficient representation learning.

LoGo Module. The core of the proposed architecture is the LoGo module, composed of three blocks: the main convolution block (MCB), the cross-attention block (CAB), and the feed-forward block (FFB). MCB employs depth-wise convolution to extract local features from each modality, while CAB and FFB collectively implement the attention mechanism to capture global context. The

LoGo module can be repeated ℓ times.[1] Within each repetition, MCB and CAB may be applied m and g times, respectively. The impact of these parameters is analyzed in Sect. 4.2. In the first instance of MCB and CAB during each LoGo repetition, frame downsampling is applied by a factor of 2. MCB performs this using a *MaxPool* operation followed by *Conv*, while CAB uses only *MaxPool*. The FFB introduces non-linearity and helps the model capture complex relationships. Like MCB, it follows an inverted bottleneck design. Top and bottom streams are processed separately through the FFB.

Output Module. After ℓ repetitions of the LoGo module, average pooling is applied to both streams. The pooled features are added and rearranged before being passed to a fully connected (FC) layer for classification:

$$\text{FC}_{input} = AvgPool(\text{FFB}_t) + AvgPool(\text{FFB}_b) \qquad (1)$$

To preserve critical information during feature extraction, pre-activation residual connections are added after each block. Batch normalization is used for MCB, while layer normalization is used for CAB and FFB.

See Supplementary Material for block definitions of CVB, MCB, CAB, and FFB.

4   Experiments

Dataset Specifications. The proposed model is evaluated on three public datasets: MSR Action3D (MSRA) [17], UTD-Multimodal Human Action (UTD) [4], and UTKinect-Action3D (UTK) [28]. These datasets include both clean and cluttered samples, supporting robust evaluation. MSRA contains 567 sequences (320×240) covering 20 actions by 10 subjects, from which 3760 human poses across 7 classes (*forward punch, hand raise, horizontal wave, pick up, side punch, two-hand raise, and vertical stretch*) are sampled. UTD comprises 861 sequences (320×240) representing 27 actions by 8 subjects, yielding 4382 poses across 6 classes (*arm cross, baseball swing, boxing, pick up, sit, and swipe*). UTK consists of 20 multi-action sequences with 10 actions performed by 10 subjects, parsed to extract 5308 poses across 6 classes (*push-pull, clap hands, pick up, sit down, walk, and wave hands*). Subject-disjoint splits for training, validation, and testing are provided in Table 1.

Network Settings. Input pose depth maps and binary silhouettes were resized to 128×128. The model[2] was trained using stochastic gradient descent (learning rate: 0.001) with cross-entropy loss and converged in 35 epochs. Images were progressively downscaled by a factor of two across successive blocks to enhance

---

[1] We empirically set $\ell = 2$ to balance recognition performance and computational cost.

[2] Implemented in PyTorch 1.10 and trained on NVIDIA DGX Station (Tesla V100, 32 GB GPU, CUDA 11.4).

Table 1. Dataset split overview for training, validation, and testing.

Dataset	Training	Validation	Testing
MSRA [17]	2387	762	611
UTD [4]	2595	874	913
UTK [28]	3615	891	802

computational efficiency and enable hierarchical, multi-scale feature learning. During training, data augmentation[3] was applied. Additionally, high inter-frame correlation within sampled sequences provided self-augmentation. For fairness, all baseline models were re-trained under identical settings using the official code bases.

Evaluation Settings. We evaluate multiple configurations of the proposed architecture to highlight the contribution of each component. These configurations are denoted as CANpose-mg, where m and g refer to the number of MCBs and CABs, respectively. Each version has two variants: small (S) and base (B), denoted as CANpose-mg (S) and CANpose-mg (B), differing in the number of output channels per block (Table 3). The architecture uses eight attention heads, each with a dimension of 32.

4.1 Results

We compare CANpose against state-of-the-art CNNs, attention-based transformers, and hybrid architectures. CNN-based baselines include RegNetY-8G [22], ResNet-50 [9], and EfficientNet-B7 [26]. Transformer models include ViT [14], Swin-T [18], Focal-T small (S), and Focal-T base (B) [31]. As hybrid baselines, we consider CoAtNet-0, CoAtNet-4 [5], and Uniformer (S, B) [16] to demonstrate the advantages of our CNN-attention integration. Performance is benchmarked using accuracy, precision, recall, and F1-score.

Accuracy and Computational Cost. Table 2 highlights that CANpose consistently delivers higher accuracy with significantly lower computational cost than existing models. The lightweight variant, CANpose-11 (S), achieves strong performance on the MSRA and UTD datasets using only 0.20M parameters and 0.03B FLOPs—far more efficient than CoAtNet-0, which requires 17M parameters and 1.09B FLOPs. The CANpose-22 (B) variant achieves classification accuracies of 95.1%, 90.0%, and 90.8% on the MSRA, UTD, and UTK datasets, respectively, with just 11.75M parameters and 1.5B FLOPs. In contrast, CoAtNet-4 attains 91.3%, 90.5%, and 84.0%, respectively, while incurring nearly 17× more parameters and 10× higher FLOPs.

---

[3] Rotation: 5° clockwise (3 steps) and counterclockwise (2 steps), increasing the dataset by 5×.

Table 2. Comparison of classification accuracy between the proposed model and state-of-the-art methods. Models marked with (∗) use an input image size of 224 × 224.

Arch.	Method	#Param (M)	FLOPs (B)	Accuracy		
				MSRA	UTD	UTK
CNN	RegNetY-8G [22]	39.09	2.60	0.655	0.768	0.657
	ResNet-50 [9]	23.52	1.35	0.445	0.475	0.495
	EfficientNet-B7 [26]	63.80	0.33	0.908	0.802	0.813
Trans	ViT* [14]	85.65	33.73	0.800	0.494	0.672
	Swin-T* [18]	87.70	15.17	0.882	0.587	0.579
	Focal-T (S) [31]	53.08	3.16	0.885	0.447	0.545
	Focal-T (B) [31]	93.26	5.49	0.876	0.823	0.613
CNN + T	CoAtNet-0 [5]	17.00	1.09	0.881	0.830	0.847
	CoAtNet-4 [5]	202.27	15.85	0.913	0.905	0.840
	UniFormer (S) [16]	21.55	1.12	0.856	0.794	0.849
	UniFormer (B) [16]	49.78	2.54	0.851	0.889	0.840
Proposed	CANpose-11 (S)	0.20↓	0.03↓	0.910	0.919↑	0.771
	CANpose-11 (B)	4.15↓	0.75↓	0.945	0.909	0.818
	CANpose-22 (S)	0.56↓	0.08↓	0.945	0.894	0.848
	CANpose-22 (B)	11.75↓	1.50↓	0.951	0.900	0.908↑
	CANpose-33 (S)	0.91↓	0.12↓	0.938	0.905	0.814
	CANpose-33 (B)	19.34↓	2.26↓	0.969↑	0.913	0.839

Notably, CANpose-33 (B) offers the best accuracy on MSRA (96.9%) and UTD (91.3%) with only 19.34M parameters and 2.26B FLOPs. Additionally, CANpose achieves higher accuracy than ViT and Swin-T models, despite operating on smaller inputs (128 × 128 vs. 224 × 224), demonstrating its efficient learning capability. On the challenging UTK dataset with cluttered scenes, CANpose-22 (B) achieves the highest accuracy (90.8%), confirming its robustness in complex environments.

Precision, Recall, and F1-Score Analysis. The average precision, recall, and F1-scores of the CANpose variants across datasets are compared with baseline models in Fig. 2. The figure highlights the superior average performance of the CANpose-22 (B) model compared to all other methods. It consistently outperforms in precision, recall, and F1-score—achieving the highest precision for accurate positive predictions and excelling in recall by effectively capturing true positive cases. These gains are achieved with only 11.75M parameters and 1.5B FLOPs, highlighting the efficiency of the model. Interestingly, hybrid models that combine CNNs and attention mechanisms outperform standalone CNNs and transformers, demonstrating the benefit of combining local and global feature learning. However, CANpose surpasses all hybrid baselines, validating the effectiveness of the proposed bi-modal cross-attention mechanism.

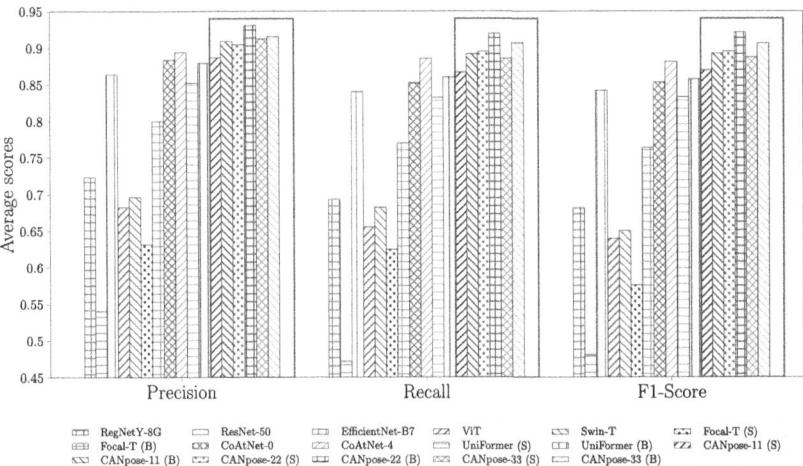

Fig. 2. Average precision, recall, and F1-score comparison across methods. The red box highlights the performance of different CANpose variants, demonstrating consistent improvements over baseline models. (Color figure online)

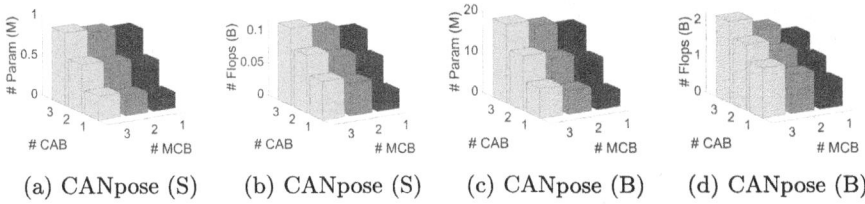

(a) CANpose (S)   (b) CANpose (S)   (c) CANpose (B)   (d) CANpose (B)

Fig. 3. Variation in the number of parameters and FLOPs with respect to the number of MCB and CAB blocks in the proposed CANpose (S) and CANpose (B) models. The figure illustrates how architectural depth impacts computational complexity across model variants.

4.2 Ablation Study

A series of ablation experiments were conducted to evaluate the contribution of individual components in the proposed architecture.

Ablation on Architectural Aspects. We conducted a controlled ablation study by independently varying the number of MCBs and CABs while keeping the other fixed. Performance was evaluated on both small and base variants across all three datasets. As shown in Fig. 3 and Fig. 4, increasing the number of blocks improves accuracy up to a point, after which gains diminish. The best performance is achieved with two MCBs and two CABs, highlighting the need for a balanced integration of convolutional and attention modules in CANpose.

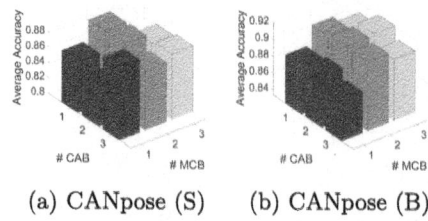

(a) CANpose (S)   (b) CANpose (B)

Fig. 4. Average accuracy variation with respect to the number of MCB and CAB blocks in the proposed CANpose models. The figure highlights the impact of architectural depth on recognition performance for both small and base variants.

Ablation on Hidden Dimension. The main difference between the small and base variants of CANpose is the number of hidden dimensions, i.e., output channels per block (Table 3). As seen in Fig. 4, the base variant consistently performs better, suggesting stronger feature representation with more channels. However, this comes with increased computational cost, as shown in Fig. 3.

Table 3. Number of output channels in different blocks of the small and base variants of the proposed CANpose model. Here, 'C' refers to the CVB, 'M' to the MCB, and 'S' to the CAB blocks.

Variant	Type	#Channels
Small	[C, M, S, M, S]	[8, 16, 32, 64, 128]
Base	[C, M, S, M, S]	[64, 96, 192, 384, 768]

Ablation on Head Size. We assess the CANpose-22 (B) variant with different head sizes (Table 4). While a head size of 16 slightly reduces performance, sizes 32 and 64 show comparable gains. However, size 64 increases computational overhead. Thus, a head size of 32 offers the best balance of accuracy and efficiency and is used in all CANpose configurations.

Table 4. Average scores for different head sizes in the multi-head attention mechanism of the CANpose-22 (B) model.

Head Size	#Param (M)	FLOPs (B)	Average Scores			
			Accuracy	Precision	Recall	F1-Score
16	10.94	1.4	0.887	0.912	0.887	0.890
32	11.75	1.5	0.920	0.930	0.920	0.921
64	13.34	1.71	0.920	0.929	0.923	0.923

5   Conclusions

This work presents CANpose, a hybrid model that combines the strengths of convolutional and attention-based mechanisms for HPR. The model adopts a bi-modal strategy, utilizing both pose depth maps and corresponding binary silhouettes to enhance recognition accuracy. Local features are extracted using depth-wise separable convolution, while global dependencies are captured through a cross-attention network. Evaluations on three widely used human action datasets demonstrate that CANpose outperforms state-of-the-art CNNs, transformers, and hybrid models, delivering superior accuracy with significantly reduced computational cost.

While the proposed model demonstrates strong performance across diverse human poses, it exhibits limitations when handling poses with similar visual characteristics. Misclassifications often occur when distinguishing between poses that differ only in subtle aspects of body orientation or limb positioning. To address this, future work can explore the integration of features that emphasize local structures and spatial dependencies inherent in human poses, potentially through multi-scale CNN architectures or advanced spatial encoding techniques. Additionally, refining attention mechanisms to better focus on pose-specific details may further enhance recognition accuracy. However, such enhancements may come at the cost of increased computational complexity, particularly when introducing deeper networks or more sophisticated attention layers. Exploring model explainability also presents a valuable direction, offering insights into the decision-making process and improving the transparency of pose recognition systems.

Acknowledgments. Funded by the European Union. Views and opinions expressed are however those of the author(s) only and do not necessarily reflect those of the European Union or the European Research Executive Agency. Neither the European Union nor the granting authority can be held responsible for them.

References

1. Chaaraoui, A.A., Climent-Pérez, P., Flórez-Revuelta, F.: Silhouette-based human action recognition using sequences of key poses. Pattern Recogn. Lett. **34**(15), 1799–1807 (2013)
2. Chakraborty, S., Mondal, R., Singh, P.K., Sarkar, R., Bhattacharjee, D.: Transfer learning with fine tuning for human action recognition from still images. Multimedia Tools Appl. **80**(13), 20547–20578 (2021). https://doi.org/10.1007/s11042-021-10753-y
3. Chapariniya, M., Barazande, S.V., Ashrafi, S.S., Shokouhi, S.B.: Attention transfer in self-regulated networks for recognizing human actions from still images. In: 2022 12th International Conference on Computer and Knowledge Engineering (ICCKE), pp. 036–041. IEEE (2022)

4. Chen, C., Jafari, R., Kehtarnavaz, N.: UTD-MHAD: a multimodal dataset for human action recognition utilizing a depth camera and a wearable inertial sensor. In: 2015 IEEE International Conference on Image Processing (ICIP), pp. 168–172. IEEE (2015)

5. Dai, Z., Liu, H., Le, Q.V., Tan, M.: CoAtNet: marrying convolution and attention for all data sizes. In: Advances in Neural Information Processing Systems, vol. 34, pp. 3965–3977. Curran Associates, Inc. (2021)

6. Delaitre, V., Laptev, I., Sivic, J.: Recognizing human actions in still images: a study of bag-of-features and part-based representations. In: BMVC 2010-21st British Machine Vision Conference (2010)

7. D'Eusanio, A., Pini, S., Borghi, G., Vezzani, R., Cucchiara, R.: Manual annotations on depth maps for human pose estimation. In: Ricci, E., Rota Bulò, S., Snoek, C., Lanz, O., Messelodi, S., Sebe, N. (eds.) ICIAP 2019. LNCS, vol. 11751, pp. 233–244. Springer, Cham (2019). https://doi.org/10.1007/978-3-030-30642-7_21

8. Guo, G., Lai, A.: A survey on still image based human action recognition. Pattern Recogn. 47(10), 3343–3361 (2014)

9. He, K., Zhang, X., Ren, S., Sun, J.: Deep residual learning for image recognition. In: Proceedings of the IEEE Conference on Computer Vision and Pattern Recognition, pp. 770–778 (2016)

10. Herath, S., Fernando, B., Harandi, M.: Using temporal information for recognizing actions from still images. Pattern Recogn. 96, 106989 (2019)

11. Hosseyni, S.R., Taheri, H., Seyedin, S., Rahmani, A.A.: Human action recognition in still images using ConViT. arXiv preprint arXiv:2307.08994 (2023)

12. Kamel, A., Sheng, B., Yang, P., Li, P., Shen, R., Feng, D.D.: Deep convolutional neural networks for human action recognition using depth maps and postures. IEEE Trans. Syst. Man Cybern. Syst. 49(9), 1806–1819 (2018)

13. Kim, Y., Kim, D.: A cnn-based 3d human pose estimation based on projection of depth and ridge data. Pattern Recogn. 106, 107462 (2020)

14. Kolesnikov, A., et al.: An image is worth 16x16 words: transformers for image recognition at scale. In: The International Conference on Learning Representations (2021)

15. Kong, Y., Fu, Y.: Human action recognition and prediction: a survey. Int. J. Comput. Vision 130(5), 1366–1401 (2022)

16. Li, K., et al.: UniFormer: unifying convolution and self-attention for visual recognition. IEEE Trans. Pattern Anal. Mach. Intell. (2023)

17. Li, W., Zhang, Z., Liu, Z.: Action recognition based on a bag of 3D points. In: 2010 IEEE Computer Society Conference on Computer Vision and Pattern Recognition-Workshops, pp. 9–14. IEEE (2010)

18. Liu, Z., et al.: Swin transformer: hierarchical vision transformer using shifted windows. In: Proceedings of the IEEE/CVF International Conference on Computer Vision, pp. 10012–10022 (2021)

19. Marin-Jimenez, M.J., Romero-Ramirez, F.J., Munoz-Salinas, R., Medina-Carnicer, R.: 3d human pose estimation from depth maps using a deep combination of poses. J. Vis. Commun. Image Represent. 55, 627–639 (2018)

20. Mazzia, V., Angarano, S., Salvetti, F., Angelini, F., Chiaberge, M.: Action transformer: a self-attention model for short-time pose-based human action recognition. Pattern Recogn. 124, 108487 (2022)

21. Moon, G., Chang, J.Y., Lee, K.M.: V2V-PoseNet: voxel-to-voxel prediction network for accurate 3D hand and human pose estimation from a single depth map. In: Proceedings of the IEEE Conference on Computer Vision and Pattern Recognition, pp. 5079–5088 (2018)

22. Radosavovic, I., Kosaraju, R.P., Girshick, R., He, K., Dollár, P.: Designing network design spaces. In: Proceedings of the IEEE/CVF Conference on Computer Vision and Pattern Recognition, pp. 10428–10436 (2020)
23. Sandler, M., Howard, A., Zhu, M., Zhmoginov, A., Chen, L.C.: MobileNetV2: inverted residuals and linear bottlenecks. In: Proceedings of the IEEE Conference on Computer Vision and Pattern Recognition, pp. 4510–4520 (2018)
24. Sener, F., Bas, C., Ikizler-Cinbis, N.: On recognizing actions in still images via multiple features. In: Fusiello, A., Murino, V., Cucchiara, R. (eds.) ECCV 2012. LNCS, vol. 7585, pp. 263–272. Springer, Heidelberg (2012). https://doi.org/10.1007/978-3-642-33885-4_27
25. Shotton, J., et al.: Efficient human pose estimation from single depth images. IEEE Trans. Pattern Anal. Mach. Intell. **35**(12), 2821–2840 (2012)
26. Tan, M., Le, Q.: EfficientNet: rethinking model scaling for convolutional neural networks. In: International Conference on Machine Learning, pp. 6105–6114. PMLR (2019)
27. Wu, J., Hu, D., Xiang, F., Yuan, X., Su, J.: 3d human pose estimation by depth map. Vis. Comput. **36**, 1401–1410 (2020)
28. Xia, L., Chen, C., Aggarwal, J.: View invariant human action recognition using histograms of 3D joints. In: 2012 IEEE Computer Society Conference on Computer Vision and Pattern Recognition Workshops (CVPRW), pp. 20–27. IEEE (2012)
29. Xiao, T., Singh, M., Mintun, E., Darrell, T., Dollár, P., Girshick, R.: Early convolutions help transformers see better. Adv. Neural. Inf. Process. Syst. **34**, 30392–30400 (2021)
30. Yan, S., Smith, J.S., Lu, W., Zhang, B.: Multibranch attention networks for action recognition in still images. IEEE Trans. Cogn. Dev. Syst. **10**(4), 1116–1125 (2017)
31. Yang, J., et al.: Focal attention for long-range interactions in vision transformers. In: Advances in Neural Information Processing Systems, vol. 34, pp. 30008–30022. Curran Associates, Inc. (2021)
32. Yuan, K., Guo, S., Liu, Z., Zhou, A., Yu, F., Wu, W.: Incorporating convolution designs into visual transformers. In: Proceedings of the IEEE/CVF International Conference on Computer Vision, pp. 579–588 (2021)
33. Zheng, Y., Zheng, X., Lu, X., Wu, S.: Spatial attention based visual semantic learning for action recognition in still images. Neurocomputing **413**, 383–396 (2020)
34. Zhu, H., Hu, J.-F., Zheng, W.-S.: Learning hierarchical context for action recognition in still images. In: Hong, R., Cheng, W.-H., Yamasaki, T., Wang, M., Ngo, C.-W. (eds.) PCM 2018. LNCS, vol. 11166, pp. 67–77. Springer, Cham (2018). https://doi.org/10.1007/978-3-030-00764-5_7

RTFVE: Realtime Face Video Enhancement

Varun Ramesh Jois[✉][iD], Antonella DiLillo, and James Storer

Brandeis University, Waltham, MA 02453, USA
{vjois,dilant,storer}@brandeis.edu

Abstract. There's been a surge in adoption of video conferencing applications for both personal and business use cases. However, the bandwidth limitations faced by many users worldwide may restrict the optimal use of such applications. Although deep learning offers a solution for enhancing low bit rate videos, most models today are either hard to incorporate with modern compression standards or require specialized hardware to run such as significant GPUs making these models impractical. To address these issues, we introduce the Realtime Face Video Enhancement (RTFVE) model which can be easily incorporated with any video decoder and can run in realtime on ordinary CPUs. Experiments show that our model improves perceptual quality over the compressed video baseline at multiple low bitrate settings. The source code will be made available at https://github.com/varun-jois/RTFVE.

Keywords: Face Enhancement · Realtime · Videoconferencing

1 Introduction

Over the last decade videocalling has become one of the most popular forms of communication. With today's technology, everyone has access to a phone with a front-facing camera and an internet connection, often making video calling a daily occurrence for many users. Ever since the lockdowns during the COVID-19 pandemic, the adoption of this technology has grown exponentially with many new companies providing services for it.

While there has been a global trend towards faster internet speeds, there are still millions of people around the world that experience slow internet connections. And while the video compression standards are flexible with handling low bandwidth settings, they come with the cost of low quality video quite often making videocalls untenable. This being the case even though many of these users have the basic hardware to handle a standard videocall.

Deep learning based methods hold promise when it comes to improving the perceptual quality of faces. There have been advances in a myriad of fields such as face restoration, face deblurring, face super-resolution, etc. However, when it comes to face enhancement for videocalls there have been two problems stymieing the adoption of deep learning: 1) Models that cannot be easily integrated into the

M. Castrillón-Santana et al. (Eds.): CAIP 2025, LNCS 15621, pp. 136–148, 2026.
https://doi.org/10.1007/978-3-032-04968-1_12

Table 1. Relative speed comparison of various methods ranging from classical computer vision algorithms to current state-of-the-art face restoration models on a low cost CPU. For realtime performance an FPS of at least 24 is required.

Model	Type	Frames per Second (↑)
Non-Local Means Denoising [3]	Classical	4.56
Bilateral Filtering [17]	Classical	13.89
Split-Bregman [6]	Classical	17.71
Chambolle [4]	Classical	7.60
Codeformer [23]	Neural Network	0.05
GPEN [21]	Neural Network	2.47
GFP-GAN [19]	Neural Network	2.02
RTFVE (Ours)	Neural Network	24.85

ubiquitous compression standards and 2) Models that require significant GPU and NPU resources that typical users don't have. In this paper, we address both of these issues.

Our contributions are as follows:

1. We present our model **R**eal**T**ime **F**ace **V**ideo **E**nhance-ment (RTFVE) a model for face enhancement that can easily be integrated with the decoder of a video compression standard.
2. For the bandwidth cost of a handful of high-quality reference frames, our model is able to improve the visual quality of faces over the compression standards running in realtime on typical CPUs (e.g. laptops).

2 Related Work

2.1 Face Restoration

Face restoration is the task of producing a high quality image of a face from its low quality counterpart. GPEN [21] works by embedding a trained GAN-prior-network as the decoder of a U-shaped DNN and then fine-tuning the DNN with synthesized low quality face images. CodeFormer [23] looks at the blind face restoration task as a codebook prediction task. They first learn a discrete codebook and decoder to store high quality parts of faces. With the codebook and decoder fixed, they train a transformer based model to predict the code sequence of the low quality input image thereby modeling the properties of the low quality faces. PGDiff [20] introduces the concept of partial guidance to the diffusion process by modeling properties such as face structure and color statistics and applying this guidance during the reverse diffusion process. FRR-Net [14] is a face restoration and lighting network that incorporates a distortion classifier

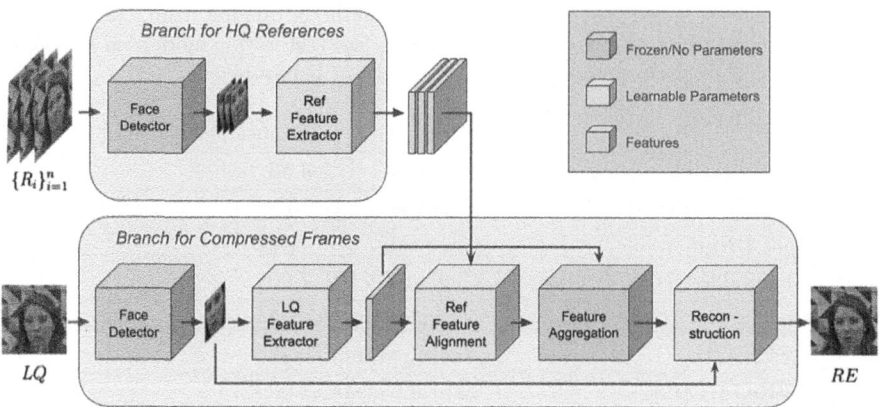

Fig. 1. The Realtime Face Video Enhancement (RTFVE) model. Our model takes the compressed, low-quality frames LQ from a video call, and n high-quality reference frames $\{R_i\}_{i=1}^n$ to produce the enhanced reconstructed frame RE.

to use the class as a prior and dice loss for segmentation masks to only focus on the face region. They propose a new degradation scheme that also include illumination distortions.

2.2 Video Restoration

MFQE [7] leverages a BiLSTM based detector to locate Peak Quality Frames (PQFs) in compressed video and then uses a Multi-Frame Convolutional Neural Network to enhance the quality of compressed video using neighboring PQFs. They use multi scale features and dense connections to improve enhancement performance. In EDVR [18] the authors perform feature alignment using a pyramid, cascading, deformable convolution block (PCD). They then combine the aligned frame features using the temporal and spatial attention module (TSA) to perform attention both temporally and spatially before finally reconstructing the output. BasicVSR [5] uses bidirectional propagation, flow-based alignment, concatenation based aggregation and pixel shuffle to perform restoration. [1] uses keyframes as references and computes multiscale features which are used to produce the final output.

3 RTFVE Framework

Our model is inspired by previous work that leverages reference images such as [9,10] for Super-Resolution and [1] for Face Video Enhancement. The objective of our model is to leverage a handful of high quality reference images to improve the aesthetic of a low quality video stream. An equally important objective of ours is to make our model practical in terms of latency and hardware requirements. A model that is not at least 24 frames per second may diminish the user experience and a model that requires substantial GPU power to run may not be practical for the limited resources of many users.

Our model, named Realtime Face Video Enhancement (RTFVE) is depicted in Fig. 1. It is comprised of five parts; a face detector, feature extraction blocks, an alignment module, a feature aggregation module and a reconstruction module. The model takes a low quality image LQ and n high quality reference images $\{R_i\}_{i=1}^n$ as inputs to produce the enhanced reconstructed frame RE.

3.1   Face Detector

The first part of our model is the face detector f_d which locates and crops the face in the low quality and reference images and passes them along to the rest of the model.

$$LQ^{\text{face}} = f_d(LQ)$$
$$R_i^{\text{face}} = f_d(R_i) \quad \text{for} \quad i = 1...n \tag{1}$$

By working on just the faces, we can considerably reduce the resolution of our inputs thereby reducing latency. In our model, we select a 256×256 size crop. For our work, we used an off the shelf face detector [2] that we fixed. We also add a gaussian blur to the low quality image using a kernel of shape 3×3 and $\sigma = 0.8$.

3.2   Feature Extractor and the Shuffle Unit

After obtaining the face crops, we first perform pixel unshuffling [16] to reduce the resolution before passing them to the feature extractors f_e. This is the reverse of the shuffling operation mentioned in [16] that is a popular method for reducing the resolution size without losing any data by increasing the number of channels. By working in lower resolution space, we greatly increase speed of the model. Here we reduce the resolution by $4\times$. We use different extractors for the LQ and reference images.

$$F_{LQ} = f_e^{LQ}(LQ^{\text{face}})$$
$$F_{R_i} = f_e^R(R_i^{\text{face}}) \quad \text{for} \quad i = 1...n \tag{2}$$

Feature extraction is a necessary step for neural network architectures and deeper features offer better semantic representation. But deep models are computationally expensive making it unfeasible for low latency. To overcome this we designed the extractors, as well as the other modules using shuffle units [12]. These units are analogous to the units in ResNet [8] with three convolutional layers per unit. However, they greatly differ in the number of parameters and speed since the former performs depthwise group convolutions and 1×1 convolutions instead of dense convolutions. The shuffle unit also performs a channel split operation working on half the number of channels in any block thereby further increasing speed. For our features extractors, we used 10 shuffle units giving us deeper features and speed.

3.3   Feature Alignment

To get the most information out of the reference features, they need to be aligned with the low quality image features. We perform the alignment using spatial transformer alignment [10] f_a:

$$F_{R_iA} = f_a(F_{LQ}, F_{R_i}) \quad \text{for} \quad i = 1...n \tag{3}$$

This module works by aligning the features of the high quality reference images $\{F_{R_i}\}_{i=1}^n$ with the features of the low quality frame F_{LQ} in feature space. The original alignment module had the major drawback of two fully connected layers. This drastically increased the number of parameters and computation time. To remedy this, we used downsampling shuffle units within the localization network so the number of parameters given to the fully connected layers is low.

Table 2. Performance scores with respect to video compression codec and compression rate. Higher the CRF, more the compression.

Codec	CRF	Model	PSNR (↑)	SSIM (↑)	LPIPS (↓)
H.264	36	H.264 (baseline)	33.0196	0.9089	0.0826
		RTFVE	33.5172	0.92	0.1073
		RTFVE-gan	33.3745	0.9118	0.0715
	40	H.264 (baseline)	30.833	0.8752	0.1273
		RTFVE	31.6259	0.8961	0.1365
		RTFVE-gan	31.2276	0.8799	0.1076
	44	H.264 (baseline)	28.566	0.8307	0.1872
		RTFVE	29.3205	0.8586	0.1851
		RTFVE-gan	29.0675	0.8433	0.1447
H.265	36	H.265 (baseline)	33.0416	0.9131	0.0779
		RTFVE	33.7089	0.9271	0.0926
		RTFVE-gan	33.1195	0.9086	0.0899
	40	H.265 (baseline)	30.7886	0.8788	0.1226
		RTFVE	31.5147	0.8993	0.133
		RTFVE-gan	31.0642	0.8813	0.1106
	44	H.265 (baseline)	28.5467	0.8353	0.1899
		RTFVE	29.2606	0.8631	0.1876
		RTFVE-gan	28.9606	0.8468	0.1434

3.4   Feature Aggregation

With the aligned reference features $\{F_{R_iA}\}_{i=1}^n$ in hand, we can now aggregate our references to obtain the most useful signals. We adopt the weighted aggregation module f_{agg} in [10] as is: since it's parameter free and efficient.

$$F_{agg} = f_{agg}(F_{LQ}, \{F_{R_i A}\}_{i=1}^n) \tag{4}$$

This module works by giving a greater weight to the reference features that are closest (in terms of L2 distance) to the features of the low-quality frame. This weighting is done for all regions in feature space and so one feature map can be given a greater weight at one region whereas another feature map can be given a greater weight in another region. It should be noted that this module does not contain parameters as it is based on the L2 distance and the softmax function and can be executed in a few steps efficiently.

3.5   Reconstruction Module

We pass the aggregated features F_{agg} to the reconstruction module f_r which produces the residuals that get added to the input image. To match the resolutions, a pixel shuffle [16] is produced at the end of the reconstruction block.

$$RE^{\text{face}} = LQ^{\text{face}} + f_r(F_{agg}) \tag{5}$$

Similar to other papers [1,10] we find producing residuals with the model more effective than directly producing the output. The reconstruction module is composed of 10 shuffle units. Finally, we replace the low-quality face crop with the enhanced face crop RE^{face} to produce the final output RE.

3.6   Enhanced Speed For Subsequent Frames

One integral aspect of the model is the ability to reuse reference features for subsequent frames. Once features have been extracted from the high quality reference images, they can be cached and used throughout the video without having to go through the feature extractor. With these reference features, we can bypass the detector and extraction steps and directly move to the alignment step saving computation over the course of a call.

4   Experiments

4.1   Datasets

Similar to [1] and [10], we use the publicly available DeepFakeDetection dataset [15] to perform our experiments. The dataset is comprised of 363 1080p videos that were made using 28 paid actors. Out of these videos, we used the videos belonging to the categories *"outside talking still laughing"*, *"podium speech happy"* and *"talking against wall"* as these were the ones most similar to a video call setting. The first 22 identities corresponding to 65 videos were used as the training set and the remaining 6 identities corresponding to 17 videos were used as the testing set.

Low Quality
Frame

Enhanced
Frame

Ground
Truth

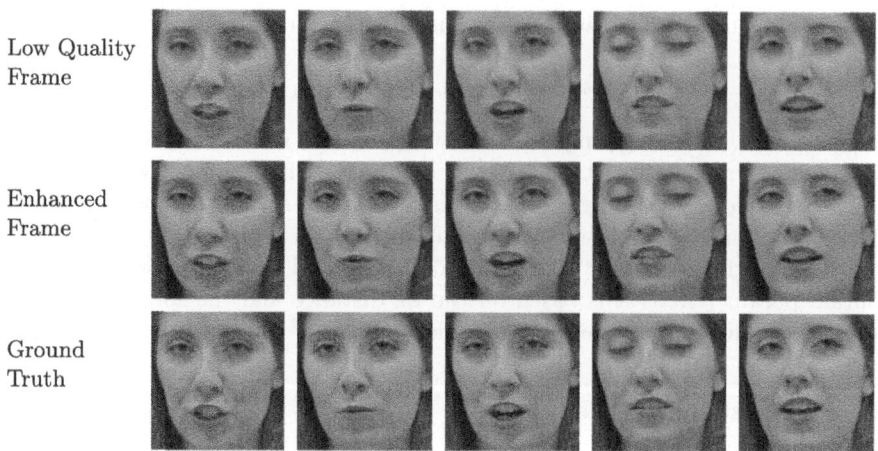

Fig. 2. Stills from videos compressed at CRF 40.

For all videos, we first cropped a 512 × 512 pixel region around the human subject, extracted 5 frames from the first 12 s of the video at a sampling rate of 72 frames (input videos have a frame rate of 24 frames per second) to form our reference images, skipped 3 s and then compressed the rest of the video. We chose this policy as it most resembles a real videocall scenario where the high quality frames can be sent early in the stream, with the rest of the stream being compressed. While we extracted 5 reference frames, we only used 3 in our model. Some frames didn't contain the entire face so we extracted 2 more to have more options to choose from.

Our low quality videos were compressed at three different compression rates using two of the most popular video compression standards - H.264 and H.265. We used ffmpeg with the libx264 encoder for H.264 and libx265 for H.265. The compression rates we chose were a Constant Rate Factor (CRF) of 36, 40 and 44 to mimic a range of low bandwidth settings. For reference, the CRF parameter ranges from 0–51 where 0 is lossless, 18 is considered visually lossless and 51 is the worst quality possible. Generally speaking, a CRF of 17–28 is considered an acceptable range according to the ffmpeg video encoding guide. The reason we chose to experiment with H.264 is because it is still a widely adopted codec for videocalls with the greatest support across a range of devices including older Android and iOS phones. H.264 and its variants are also the codecs used by popular videoconferencing applications such as Zoom and Webex. By including H.265 video in our tests, we demonstrate that our model operates reliably irrespective of the codec used.

To speed up training, instead of training on the entire video, we trained on 20 frames from each video sampled at a rate of 10 frames. We also trained our model on a tighter cropped region around the face with a resolution of 256 × 256. These were obtained using the aforementioned face detector Sect. 3.1.

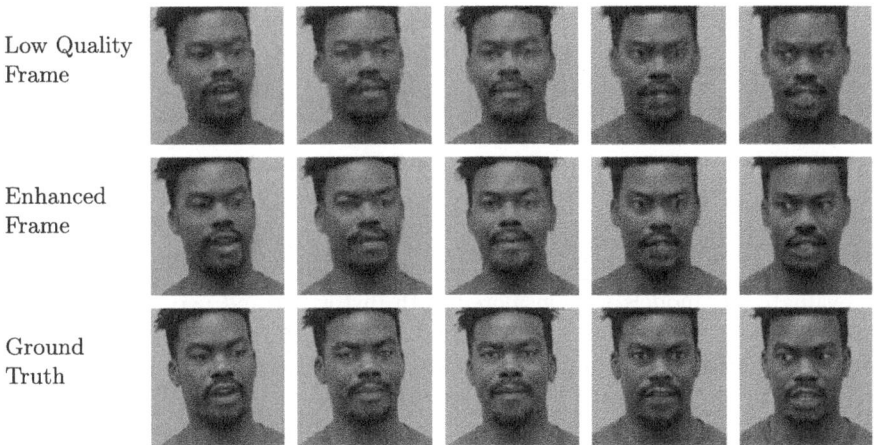

Fig. 3. Stills from videos compressed at CRF 44.

4.2   Training

We trained our model for 1000 epochs using the Adam optimizer with a learning rate of 10^{-4} and the default $\beta_1 = 0.9$, $\beta_2 = 0.999$. To improve the performance of our model, we performed a number of random augmentations including, random horizontal flip, random vertical flip, and a random affine transformation with a rotation in the range $[-5°, 5°]$, translation along the horizontal and vertical axes in the range $[-2\%, 2\%]$ and scaling in the range of $[0.95, 1.05]$. No shearing was performed. Our model can take any number 1 or more reference images; for our experiments, we use 3 reference images.

We compared our model's performance to the baseline input compressed video. While there are numerous works on face restoration and face enhancement, to the best of our knowledge, none of these works can run in realtime (at least 24 FPS) on a typical CPU, and thus we didn't evaluate comparisons against them. In this work we are primarily concerned with face enhancement in constrained resource settings.

The recreation loss l_{rec} we optimized for was the Mean Absolute Error which is given by:

$$l_{rec} = \frac{1}{HWC} \|HQ - RE\|_1 \tag{6}$$

where HQ is the high quality ground truth frame and RE is the reconstructed frame. HWC is the height, width and number of channels in HQ and RE. We also performed adversarial training on our model based on the LS-GAN [13] where discriminator loss l_{dis} and generator loss l_{gen} are given by:

$$l_{dis} = \frac{1}{2}\mathbb{E}_{HQ \sim P(HQ)}\big[(D(HQ) - 1)^2\big] + \\ \frac{1}{2}\mathbb{E}_{LQ \sim P(LQ)}\big[D(G(LQ))^2\big] \tag{7}$$

$$l_{gen} = \frac{1}{2}\mathbb{E}_{LQ \sim P(LQ)}\left[(D(G(LQ)) - 1)^2\right] \tag{8}$$

The total loss l_{total} for our GAN based model is given by:

$$l_{total} = l_{rec} + 0.01 * l_{gen} \tag{9}$$

4.3   Quantitative Results

To evaluate our models we considered both distortion and perceptual metrics. For distortion, we measured Peak Signal to Noise Ratio (PSNR), Structural Similarity (SSIM) and for a perceptual metric, we measured Learned Perceptual Image Patch Similarity (LPIPS) [22]. We used the version of LPIPS that uses the AlexNet [11] backbone. The scores are reported in Table 2. All scores reported on a 256×256 face region. Our RTFVE model, when trained solely with reconstruction loss, achieves PSNR improvements of approximately 0.5, 0.8, and 0.76 dB for CRF values 36, 40, and 44 on the H.264 datasets respectively. Similarly, for videos compressed with the H.265 standard, the model yields PSNR gains of about 0.66, 0.73, and 0.72 dB for CRF 36, 40, and 44 respectively. These results confirm that our model delivers consistent performance improvements regardless of the codec used. Similarly, there is also an increase in SSIM. This is impressive given the severity of degradation found in the CRF 40 and 44 videos. Our GAN model, named RTFVE-gan, is not only able to increase PSNR and SSIM but is also able to reduce LPIPS scores which is considered better for this metric. We found that when the model has fewer parameters to train, it becomes harder to lower LPIPS scores so given the low number of parameters in our model, lowering LPIPS is also impressive.

Table 1 shows the speed of our model on CPU compared to other state-of-the-art face restoration models. Our frame rates include the time taken by the face detector model to find the face. With minimal inference optimizations and no quantization our model runs at a median of 24.85 frames per second on a circa 2018 laptop with an Intel®Core™i7-8550U CPU @ 1.80GHz $\times$ 8. It is reasonable to assume that a typical video call user may have this level of resource or better, and thus in practice our model runs in realtime and is suited for most everyday settings where a GPU or NPU is unavailable. On the other hand, typical face restoration models are not designed for low latency and low compute environments and hence fall well short of the 24 frames per second required for realtime applications. Our model also consumes low memory and is only 190K parameters - less than a megabyte of disk space. When reference features are recalculated for each frame, speed falls to 19.04 fps, which shows how our caching design improves inference speed.

Since our model, to the best of our knowledge, is the only model capable of running in realtime on CPU we couldn't compare our model with other face restoration models. Even classic, handcrafted image processing algorithms such as Non-Local Means Denoising [3], Bilateral Filtering [17] and total variation denoising methods such as Split Bregman [6] and Chambolle [4] do not run in realtime Table 1. We did however compare against low-pass filters that have lower complexity and can be run in realtime. For this experiment we considered H.264 videos compressed at a CRF of 40. As we can see in Table 3, many of these filters produce worse results compared to the given compressed video (baseline). Only the Gaussian Filter was able to marginally improve performance and is well short of our model's output.

Table 3. Various realtime methods on H.264 CRF 40 data.

Model	PSNR (↑)	SSIM (↑)	LPIPS (↓)
H.264 (baseline)	30.833	0.8752	0.1273
Mean Filter	28.2071	0.8787	0.2259
Median Filter	30.5624	0.8766	0.1519
Wavelet Denoising	30.848	0.8762	0.1288
Gaussian Filter	30.964	0.8821	0.1623
RTFVE (Ours)	31.6259	0.8961	0.1365

4.4   Qualitative Results

Figures 2 and 3 compare quality between the compressed and enhanced frames. These stills were taken from videos compressed using the H.264 standard. The reader is encouraged to view these figures realistically enlarged on a screen. Figure 2 shows the output stills of our model for a video compressed at CRF 40. From the first row, we can see that the input low quality frame has many artifacts including unnatural skin texture and blocking artifacts around the mouth. In the second row we can see that the model was able to alleviate these issues by producing smoother looking skin and has considerably reduced the blocking around the mouth, nose and chin. The output produced is also considerably sharper with the individual parts of the face being more well defined. In Fig. 3 we can see the model's performance on a CRF 44 video. Here the input is heavily deformed with large blocking artifacts and speckly skin texture. The model again does a good job smoothing the skin and improving the definition of the face under these difficult conditions. The improved quality is arguably more significant than the improvement in PSNR suggests. A video demo is available at https://sigport. org/documents/rtfve.

5   Conclusion

This paper introduces a novel lightweight model for videocall face enhancement. Unlike previous models for face enhancement, our model can run in realtime on typical user devices (e.g. laptop CPU) making it widely applicable. Our model is able to objectively and subjectively improve the video call experience under constrained circumstances where compression rates are high and visual quality is low. It achieves this by leveraging the information contained in high quality reference frames which are very similar to the low quality frames in the input stream. Our model can be seamlessly integrated with current compression standards where the model would be taking as input the output of the decoder.

For future work, we would like to find ways of further improving quality under constrained circumstances. Obtaining better perceptual quality with small models is challenging due to a lack of modeling ability but we seek to find ways to better exploit high quality reference images. We are also interested in building realtime models for larger resolutions.

Disclosure of Interests. The authors have no competing interests to declare that are relevant to the content of this article.

References

1. Agnolucci, L., Galteri, L., Bertini, M., Bimbo, A.D.: Perceptual quality improvement in videoconferencing using keyframes-based GAN. IEEE Trans. Multimedia **26**, 339–352 (2024). https://doi.org/10.1109/TMM.2023.3264882
2. Bazarevsky, V., Kartynnik, Y., Vakunov, A., Raveendran, K., Grundmann, M.: Blazeface: sub-millisecond neural face detection on mobile GPUs. arXiv (2019). https://arxiv.org/abs/1907.05047
3. Buades, A., Coll, B., Morel, J.M.: A non-local algorithm for image denoising. In: 2005 IEEE Computer Society Conference on Computer Vision and Pattern Recognition (CVPR 2005), vol. 2, pp. 60–65 (2005).https://doi.org/10.1109/CVPR.2005. 38
4. Chambolle, A.: An algorithm for total variation minimization and applications. J. Math. Imaging Vis. **20**, 89–97 (2004). https://doi.org/10.1023/B.0000011325. 36760.1e
5. Chan, K.C., Wang, X., Yu, K., Dong, C., Loy, C.C.: Basicvsr: the search for essential components in video super-resolution and beyond. In: Proceedings of the IEEE/CVF Conference on Computer Vision and Pattern Recognition (CVPR), pp. 4947–4956 (2021)
6. Getreuer, P.: Rudin-Osher-Fatemi total variation denoising using split Bregman. Image Process. Line **2**, 74–95 (2012). https://doi.org/10.5201/ipol.2012.g-tvd
7. Guan, Z., Xing, Q., Xu, M., Yang, R., Liu, T., Wang, Z.: Mfqe 2.0: a new approach for multi-frame quality enhancement on compressed video. IEEE Trans. Pattern Anal. Mach. Intell. **43**(3), 949–963 (2021). https://doi.org/10.1109/TPAMI.2019. 2944806
8. He, K., Zhang, X., Ren, S., Sun, J.: Deep residual learning for image recognition. In: 2016 IEEE Conference on Computer Vision and Pattern Recognition (CVPR), pp. 770–778 (2016). https://doi.org/10.1109/CVPR.2016.90
9. Jiang, Y., Chan, K.C., Wang, X., Loy, C.C., Liu, Z.: Robust reference-based super-resolution via C2-matching. In: 2021 IEEE/CVF Conference on Computer Vision and Pattern Recognition (CVPR), pp. 2103–2112 (2021). https://doi.org/10.1109/ CVPR46437.2021.00214

10. Jois, V.R., DiLillo, A., Storer, J.: Reference-based face super-resolution using the spatial transformer. In: Computer Vision – ACCV 2024: 17th Asian Conference on Computer Vision, Hanoi, Vietnam, 8–12 December 2024, Proceedings, Part IV, pp. 409–425. Springer, Heidelberg (2024). https://doi.org/10.1007/978-981-96-0911-6_24

11. Krizhevsky, A., Sutskever, I., Hinton, G.E.: Imagenet classification with deep convolutional neural networks. Commun. ACM **60**(6), 84–90 (2017). https://doi.org/10.1145/3065386

12. Ma, N., Zhang, X., Zheng, H.T., Sun, J.: Shufflenet v2: practical guidelines for efficient CNN architecture design. In: Ferrari, V., Hebert, M., Sminchisescu, C., Weiss, Y. (eds.) Computer Vision - ECCV 2018, pp. 122–138. Springer, Cham (2018)

13. Mao, X., Li, Q., Xie, H., Lau, R.Y., Wang, Z., Smolley, S.P.: Least squares generative adversarial networks. In: 2017 IEEE International Conference on Computer Vision (ICCV), pp. 2813–2821 (2017). https://doi.org/10.1109/ICCV.2017.304

14. Pouyanfar, S., et al.: FRR-Net: a real-time blind face restoration and relighting network. In: Proceedings of the IEEE/CVF Conference on Computer Vision and Pattern Recognition (CVPR) Workshops, pp. 1240–1250 (2023)

15. Rössler, A., Cozzolino, D., Verdoliva, L., Riess, C., Thies, J., Niessner, M.: Faceforensics++: learning to detect manipulated facial images. In: 2019 IEEE/CVF International Conference on Computer Vision (ICCV), pp. 1–11 (2019). https://doi.org/10.1109/ICCV.2019.00009

16. Shi, W., et al.: Real-time single image and video super-resolution using an efficient sub-pixel convolutional neural network. In: 2016 IEEE Conference on Computer Vision and Pattern Recognition (CVPR), pp. 1874–1883 (2016). https://doi.org/10.1109/CVPR.2016.207

17. Tomasi, C., Manduchi, R.: Bilateral filtering for gray and color images. In: Sixth International Conference on Computer Vision (IEEE Cat. No. 98CH36271), pp. 839–846 (1998). https://doi.org/10.1109/ICCV.1998.710815

18. Wang, X., Chan, K.C., Yu, K., Dong, C., Change Loy, C.: EDVR: video restoration with enhanced deformable convolutional networks. In: Proceedings of the IEEE/CVF Conference on Computer Vision and Pattern Recognition (CVPR) Workshops (2019)

19. Wang, X., Li, Y., Zhang, H., Shan, Y.: Towards real-world blind face restoration with generative facial prior. In: 2021 IEEE/CVF Conference on Computer Vision and Pattern Recognition (CVPR), pp. 9164–9174 (2021). https://doi.org/10.1109/CVPR46437.2021.00905

20. Yang, P., Zhou, S., Tao, Q., Loy, C.C.: PGDiff: guiding diffusion models for versatile face restoration via partial guidance. In: Oh, A., Naumann, T., Globerson, A., Saenko, K., Hardt, M., Levine, S. (eds.) Advances in Neural Information Processing Systems, vol. 36, pp. 32194–32214. Curran Associates, Inc. (2023). https://proceedings.neurips.cc/paper_files/paper/2023/file/661c37f3b098bdee53fd7d9c4ef6964a-Paper-Conference.pdf

21. Yang, T., Ren, P., Xie, X., Zhang, L.: Gan prior embedded network for blind face restoration in the wild. In: Proceedings of the IEEE/CVF Conference on Computer Vision and Pattern Recognition (CVPR), pp. 672–681 (2021)

22. Zhang, R., Isola, P., Efros, A.A., Shechtman, E., Wang, O.: The unreasonable effectiveness of deep features as a perceptual metric. In: Proceedings of the IEEE Conference on Computer Vision and Pattern Recognition (CVPR) (2018)
23. Zhou, S., Chan, K., Li, C., Loy, C.C.: Towards robust blind face restoration with codebook lookup transformer. In: Koyejo, S., Mohamed, S., Agarwal, A., Belgrave, D., Cho, K., Oh, A. (eds.) Advances in Neural Information Processing Systems, vol. 35, pp. 30599–30611. Curran Associates, Inc. (2022). https://proceedings.neurips.cc/paper_files/paper/2022/file/c573258c38d0a3919d8c1364053c45df-Paper-Conference.pdf

Image Segmentation

Walk the Lines 2: Contour Tracking for Detailed Segmentation of Infrared Ships and Other Objects

André Peter Kelm[✉] ⓘ, Max Braeschke, Emre Gülsoylu ⓘ, and Simone Frintrop ⓘ

University of Hamburg, Mittelweg 177, 20148 Hamburg, Germany
{andre.kelm,emre.guelsoylu,simone.frintrop}@uni-hamburg.de

Abstract. This paper presents Walk the Lines 2 (WtL2), a unique contour tracking algorithm specifically adapted for detailed segmentation of infrared (IR) ships and various objects in RGB (parts of this work will also appear in the forthcoming doctoral dissertation of André Kelm at Helmut Schmidt University [11]). This extends the original Walk the Lines (WtL) [12], which focused solely on detailed ship segmentation in color. These innovative WtLs can replace the standard non-maximum suppression (NMS) by using contour tracking to refine the object contour until a 1-pixel-wide closed shape can be binarized, forming a segmentable area in foreground-background scenarios. WtL2 broadens the application range of WtL beyond its original scope, adapting to IR and expanding to diverse objects within the RGB context. To achieve IR segmentation, we adapt its input, the object contour detector, to IR ships. In addition, the algorithm is enhanced to process a wide range of RGB objects, outperforming the latest generation of contour-based methods when achieving a closed object contour, offering high peak Intersection over Union (IoU) with impressive details. This positions WtL2 as a compelling method for specialized applications that require detailed segmentation or high-quality samples, potentially accelerating progress in several niche areas of image segmentation.

Keywords: Object Contour Tracking · Detailed Infrared Ship Segmentation · High-Quality Shape Extraction

1 Introduction

Image segmentation is a fundamental task in computer vision, with applications in medical imaging, robotics, autonomous vehicles, agriculture, and the maritime sector [20,25]; for example, it plays an important role in maritime navigation [13] and traffic safety [7]. Fundamental models such as SAM are already powerful in

---

M. Braeschke-Contributed to this work as part of his bachelor thesis and developed the WtL2 binarization method.

M. Castrillón-Santana et al. (Eds.): CAIP 2025, LNCS 15621, pp. 151–162, 2026.
https://doi.org/10.1007/978-3-032-04968-1_13

the RGB domain [14] and are well suited for ship segmentation since ships are a common category in many benchmarks.

In practice, however, there are many niches where fundamental models do not perform optimally out of the box and require customization. For example, the infrared (IR) domain is widely used in maritime applications, due to its ability to enhance visibility in the maritime environment [3], but there are few methods or datasets available for IR ship segmentation [28]. Challenges increase when applications require a high level of detail [9], which is not uncommon in the maritime and other specialized fields.

To address this, we present 'Walk the Lines 2' (WtL2), a method that complements traditional IR ship segmentation techniques, extracts impressive details, and appears to be even more versatile by extending its application to various objects for color images. Building on the original Walk the Lines (WtL) [12], which was developed for RGB ship segmentation only, to capture the fine details of ship antennas and superstructures, WtL2 extends these capabilities. The adaptation for the IR domain consists of simply replacing its input, the object contour detector designed for color images, with one trained for IR ships.

Contour-based segmentation like the WtLs promises detailed results as it is usually based on a contour detection method, providing detailed predictions (Fig. 1b). Non-maximum suppression (NMS) is often used for thinning or refinement, but often introduces new gaps in the contour (Fig. 1c). To avoid this, WtLs use contour tracking to refine object contours along with details (Fig. 1d), generating closed binary contours (Fig. 1e). Other contour-based methods exist: A graphical network can compute the shape of the object's contour [18], Deep-Snake [22] integrates a learnable active contour into a continuous deep network for better refinement, and E2EC [27] further innovates by a learnable contour initialization architecture, multi-direction alignment, and dynamic matching loss, achieving state-of-the-art (sota) for this type of methods. However, our algorithm segments many details better (Fig. 1f vs. 1g). This paper contains two contributions:

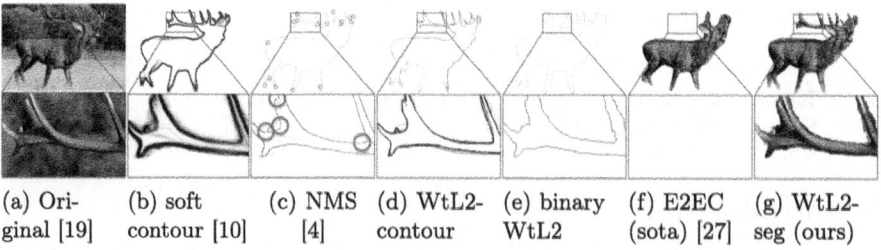

(a) Ori-   (b) soft   (c) NMS   (d) WtL2-   (e) binary   (f) E2EC   (g) WtL2-
ginal [19]   contour [10]   [4]   contour   WtL2   (sota) [27]   seg (ours)

Fig. 1. Visualization of (a) the original, which, when processed with an object contour detector, produces a (b) soft contour with two post-processing options: (c) NMS, which inserts gaps in the contours (marked by red circles), or (d) WtL2-contour, which refines the contours and keeps them connected. Further processing results in the (e) 'perfect' binary WtL2. We visually compare contour-based segmentation, in particular (f) E2EC with our (g) WtL2-seg.

– We demonstrate the applicability of WtL2 for detailed segmentation of IR ship images.
– Our WtL2 is capable of segmenting a variety of objects, including, but not limited to, cars, dogs, and deer. We have almost doubled the IoU from WtL to WtL2, and outperform the sota in contour-based segmentation for images with successfully closed object contours.

While we do not claim that our method consistently outperforms existing benchmarks in terms of overall IoU, it shines in subsets where the algorithm works as intended and a closed shape is recognized. WtL2 achieves high IoU peaks and high-quality segmentations even for complex shapes. This demonstrates the potential of WtL2 for detailed segmentation in a wide range of applications, such as rare object categories or niche areas like IR ship segmentation, as well as other specialized fields not yet fully explored. This ability to address niches makes WtL2 a valuable tool for pushing the boundaries of current computer vision applications.

2   Related Work

In this section, we first review recent advances in IR ship segmentation. We then discuss the current state in contour-based segmentation, closing with a little description to show how our method differs.

2.1   Infrared Ship Segmentation

IR ship segmentation presents unique challenges compared to RGB segmentation due to the scarcity of datasets and the limited number of studies and methods, where model weights are often not publicly available [28].

Progress in IR ship segmentation has been hindered by the unavailability of public datasets. MassMIND [21] provides a significant contribution with more than 2,900 segmented IR images of coastal areas and seven classes (such as sky, water, bridge, obstacle, living obstacle, etc.), but none explicitly for ships. Unfortunately, this and the presence of the coast do not fit with our foreground/background approach, where the ship and its details are more in focus.

Some methods use adversarial domain adaptation [28] to overcome the lack of IR ship segmentation data. Other works also try to overcome data scarcity by proposing weakly supervised or semi-supervised methods [1]. The use of foundational models such as SAM for similar purposes is still an emerging area of research [24], but also requires annotations [6]. In short, despite its critical importance to maritime applications, the maritime IR domain has been underexplored. This highlights the importance of our work in extracting highly detailed ship segmentations from IR imagery.

2.2   Contour-Based Segmentation of Various Objects

Contour-based segmentation was a competitive alternative to traditional mask-based segmentation [2], especially before the era of deep learning (DL). Despite their incredible performance gains, DL methods initially suffered from a rather blob-like mask, and even today some details are difficult to segment [26]. Recent contour-based methods aim to extract even more detail [23], for example by combining advanced object detection with active contours [5] or embedding them in an end-to-end network, such as Deep Snake [22]. E2EC builds on this and shows sota results for contour-based methods using a semantic edge detector as the basis for the active contours [27]. Our approach differs in that we use an object contour detector whose prediction is not improved by an active contour, but innovatively by a tracking CNN that circles the contour around the object.

3   Walk the Lines 2

The Walk the Lines (WtL) algorithm has a very narrow application: detailed segmentation of RGB ships, including antennas and superstructures. WtL2[1] demonstrates that the algorithm is more versatile. It introduces two key contributions: adapting the algorithm for IR ship segmentation and extending its application to various objects in the RGB domain. First, we explain the main function of both WtLs: contour tracking. Then we show the process flow, and discuss the implemented contributions in Sects. 3.1 and 3.2.

The uniqueness of these algorithms is defined by its contour tracking capability. A detailed description of this feature and the flat CNN utilized can be found in the original source [12], as the principle of both WtLs remain consistent. The processes require an object contour detector (we use RefineContourNet (RCN) [10]) and are outlined here as well in Fig. 2: **(0) init: select center point:** randomly select from ground truth (training) or high confidence value from NMS processed soft object contour map (inference) **(1) crop image patch:** centered around the selected point. **(2) rotate patch:** based on directional changes retrieved from the previous center point. **(3) run tracking CNN:** input is always a $7 \times 7 \times 4$ patch, concatenating image and soft contour. **(4) output in degrees:** example output is $-12.054°$. **(5) select new center point:** determine by rounding based on pixel step (stochastically chosen, with a preference for 1 over 2 or 3); for a pixel step of 1, round to $-135°$ (far right), $-90°$ (right), $-45°$ (slight right), $0°$ (straight), $45°$ (slight left), $90°$ (left), $135°$ (far left), or in rare cases, $\pm180°$ (turn around).

The process has been scaled with hundreds of trackers, or thousands depending on the contour, as some have reached dead ends. The pixel-by-pixel contour

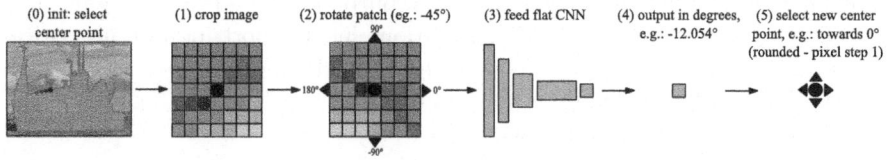

Fig. 2. Diagram illustrating the contour tracking process of WtLs.

---

[1] https://github.com/AndreKelm/WalktheLines2.

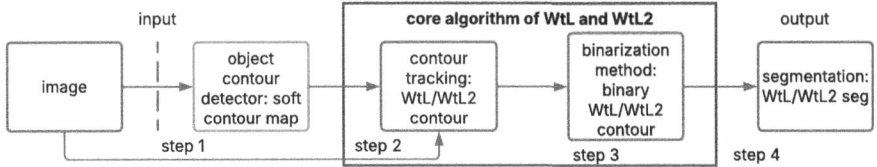

Fig. 3. Overview of WtL and WtL2 processes. Red outlines core algorithm. Modifications for WtL2 are outlined in color: adaptation for IR ship via object contour detector retraining in blue, and extension for various objects in green. (Color figure online)

tracking of each image, with all trackers running simultaneously, typically takes a few tens of seconds to complete.

To contextualize contour tracking and highlight our main contributions Fig. 3 outlines the high-level processes that apply to both WtLs. The changes we made are color-coded for clarity.

It uses the results from an object contour detector [10] to produce a soft contour (ex. in Fig. 1b). For IR ships, it was sufficient to retrain the detector, which allows detailed segmentation in this domain. The already explained unique contour tracking refines its result concatenated with the image (ex. in Fig. 1d). Further processing results in a closed shape (ex. in Fig. 1e) with a 1-pixel-wide contour that encloses an area that can also be used as a mask (ex. in Fig. 1e). To extend to various RGB objects, we retrained the tracking CNN and modified the binarization (which was very specific for ships) allows for its application across various categories, including cars, dogs, deer, giraffes, and many other.

3.1 Infrared Ship Segmentation

Figure 4 illustrates the access to the ship's IR object contour detection. By using two strategies: a highly unbalanced RGB ship dataset, so that other classes are still considered, but the algorithm's response to ship segmentation is overly emphasized, and an adjustment of the IR image intensity channel to better match the typical color channel values of ship images, so that the RGB segmentation becomes more sensitive to IR. A Conditional Random Field (CRF) method [15] occasionally creates meaningful labels from it. Self-training generates an appropriate number of labels. It was stopped at 227 images. Resulting masks are then converted to train an object contour detector for IR data. Specifically, by adapting the RCN [10] model to produce weights for RCN-IR^2.

3.2 Segmentation of Various Objects

To extend WtL to more objects than only ship, we selected 90 COCO *test-dev* [17] images with a typical object focus and manually refined the rough annotations in detail. We call this small subset of COCO: DOC (Detailed Object

---

^{2} https://github.com/AndreKelm/RefineContourNet/tree/master/refinenet-contour-master/model_trained.

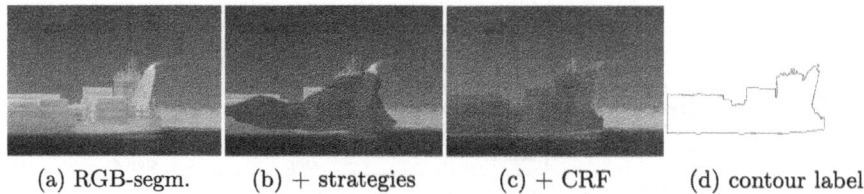

(a) RGB-segm.   (b) + strategies   (c) + CRF   (d) contour label

Fig. 4. Visualization of (a) RGB segmentation method (no segment visible, like original image) (b) + strategies, (c) + CRF, and (d) conversion to contour label; images from [11].

Contour) dataset and publish these together with WtL2's code and our 50 validation images and their ground truth. For details on training the flat tracking CNN and its architecture, see the WtL paper [12].

The WtL binarization method was developed originally for ships, where the longest line from the soft contour map is used to create a starting condition for thresholding, which searches for the lowest value at which the object contour is closed. Thresholding is done by artificially splitting the object contour by a separation line and identifying two neighboring points that, when reconnected, indicate that the object contour is closed. Objects other than ships usually do not have a typical waterline visible from the side. WtL2 uses the same thresholding method as WtL, but starts with a more general assumption for different types of objects. Algorithm 1 takes the WtL contour, a grayscale image, as input and produces a binary WtL2 output. Initialization includes a list of likely object contour pixels. In code, a loop checks the remaining highest pixel intensity to see if its location is suitable for the separation line. The condition is met if the

---

Algorithm 1. *object contour binarization*

---

input: WtL contour
output: binary WtL2
    initialization: create list $L(p, x)$ of pairs (pixel, value) from WtL contour
1: **repeat**
2:    $p_x \leftarrow \max(x)$ in $L(p, x)$
3:    edgegradient(p_x) $\leftarrow$ sobel-filter(gaussian-filter(WtL contour(p_x)))
4:    *separationline* $\leftarrow$ edgegradient(p_x)
5:    optimal *pixel*$_1$, *pixel*$_2$ $\leftarrow$ *separationline*
6:    2 *endpoints* $\leftarrow$ *separationline*
7:    **if** *endpoints* connect object and background **then**
8:        save p_x
9:    **else**
10:       delete p_x from L
11:    **end if**
12: **until** p_x is not deleted
13: binary WtL2 $\leftarrow$ **thresholding**(*pixel*$_1$, *pixel*$_2$, *separationline*)
14: **return** binary WtL2

---

separation line points orthogonally from the background to the object contour or vice versa, and if the end points of the separation line touch the two largest enclosed areas, which are expected to be the background and the largest area in the object. If not, the pixel is discarded and the next one is evaluated. Once a suitable pixel and corresponding separation line are identified, WtL thresholding searches for an optimal value to close the object contour and returns a binary image (see [12] for details).

4   Evaluation

We first compare the two WtL versions using our 50 DOC-val dataset (Sect. 3.2) to show the progress made. For IR ship segmentation, we could not find any public data/methods for a comparable evaluation, so we are releasing complementary data to MASSmind, the Elbe Ship IR and RGB Image Dataset (ESIR-RID[3]), which consists of unregistered RGB and IR images captured in the Lower Elbe River region, along with our self-annotated 10 Detailed IR Ship Contour (DIRSC) Ground Truth (GT); available in the WtL2 GitHub repository to enable comparisons. To investigate the performance of the innovative contour tracking approach in WtL2 we build our own baseline using RefineNet (RN) [16], which we call RN-IR. We used 227 labels generated by self-training (Sect. 3.1) for its supervised fine-tuning. RN-IR uses a ResNet101 backbone with training, similar to the object contour detector, allowing a direct comparison between conventional DL and the innovative WtL2. When we evaluate objects in color, we follow a similar baseline. There are not many methods similar to our approach, so we focused on the sota contour-based segmentation method E2EC.

4.1   WtL vs. WtL2

We compare the two WtL versions using our 50 DOC-val dataset (see Sect. 3.2) to show the progress made, which is clearly visible in the Table 1 for all metrics.

WtL2 creates much more closed shapes of objects, resulting in a higher IoU when these shapes are transformed into masks and used for segmentation.

Table 1. Comparison of NMS, WtL, and WtL2 algorithms on 50 DOC validation images, evaluating closed shapes by absolute number, percentage, and IoU.

method	closed shapes	percentage of closed shape	IoU
NMS	2	4%	-
WTL	28	56%	44.57
WTL2	**40**	**80%**	**75.74**

---

[3] https://cloud.uni-hamburg.de/s/YKN8Lqe58tS2TdR.

4.2 Infrared Ship Segmentation

Some methods focus on their specific and non-public test data and do not always provide model weights [8,28]. So we manually label our own test data, 10 Detailed InfraRed Ship Contour (DIRSC) images, along with our very robust baseline[4], both of which we provide for comparison.

Table 2. IR ship segmentation results on 10 DIRSC validation images, sorted by highest IoU. NMS and binary WtL2 are evaluated for closed object contours. If a closed contour is detected, the row is marked in gray.

method	NMS	binary WtL2	RN-IR			WtL2-seg-IR		
image	closed shape	closed shape	P	R	IoU	P	R	IoU
1	N/A	yes	**98,00**	95,74	93,90	97,24	**98,56**	**95,88**
2	N/A	yes	96,09	**96,39**	**92,75**	**96,44**	95,84	92,57
3	N/A	yes	94,15	**96,13**	90,71	**96,11**	96,00	**92,41**
4	N/A	yes	92,90	96,24	89,65	**93,27**	**97,21**	**90,83**
5	N/A	yes	**90,57**	97,95	**88,89**	87,93	**98,73**	86,95
6	N/A	N/A	**82,38**	**99,04**	**81,73**	0	0	0
7	N/A	N/A	98,19	**78,97**	**77,84**	100	0,05	0,05
8	N/A	N/A	**71,68**	**90,49**	**66,67**	8,89	0,02	0,02
9	N/A	yes	**66,69**	93,48	**63,69**	63,58	**94,72**	61,40
10	N/A	N/A	**23,73**	**80,96**	**22,47**	0,05	0	0
∅	0 %	**60 %**	**81,44**	**92,54**	**76,83**	69,35	58,11	52,01
∅$_{1,2,3,4,5,9}$	0 %	**100 %**	**89,73**	95,99	86,60	89,10	**96,84**	**86,67**

Table 2 shows all 10 val images. In terms of total IoU, WtL2-seg-IR struggles against our robust baseline, which is already at a high level with an average of about 77 IoU. The extremes of the algorithm become visible. Sometimes there are complete failures in IoU, but then there are exceptionally high peaks. Noteworthy is the high recall, a property of WtLs. Focusing only on the images where the algorithm works as intended (marked in gray), it slightly outperforms the comparable baseline. It is noteworthy that only the object contour detector was adapted to IR, not the tracking CNN itself, which indicates a certain robustness of the core algorithm. Considering that our baseline already achieves an average value of 77 IoU, this performance by WtL2-seg-IR (with closed object contours) is remarkable.

The Fig. 5 shows no images where no object contour was closed and therefore no area for segmentation was found. Since these cases are clearly evident from the

---

[4] https://github.com/AndreKelm/Infrared-Ship-and-RGB-Ship-Scene-Segmentation.

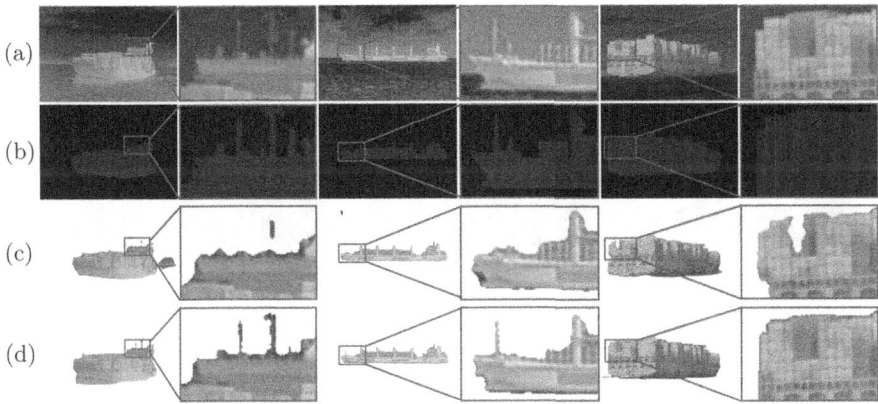

Fig. 5. Visualization of different IR ship segmentation results: (a) Original, (b) Ground Truth, (c) RefineNet-IR (baseline), (d) WtL2-seg-IR (ours); images from [11].

Table 2 with very poor IoU. Figure 5 shows three different IR ships in comparison to our baseline. For the ship on the left, the WtL2-seg-IR does not have a false positive area behind the ship, as the baseline does. WtL2 incorrectly segments the bow wave, but performed significantly better in detecting the antennas. The middle image shows a similar segmentation for both in the first glance, but a better detailed bow (including anchor, stem, and mast) is achieved with WtL2-seg-IR. On the right, the stern of the large, angular container ship is much better segmented. Even the bridge wing is accurately extracted.

4.3 Detailed Segmentation of Various Objects

While recent contour-based segmentation methods are evaluated on COCO, our objective differs. We aim for highly detailed segmentations, which COCO does not provide. Therefore, we updated 50 COCO *test-dev* labels to align with our needs. The first row in Table 3 reveals the disadvantage of the method's robustness, since a further examination shows that for some images there is almost a total failure in terms of IoU. However, the second row shows the advantage when

Table 3. Object segmentation results for 50 refined COCO *test-dev* for our baseline RN and two contour-based methods E2EC and WtL2-seg. The top row shows the overall result for all images. The second row (marked in gray) shows the 40 images for which WtL2 was able to form a closed object contour.

method	RN			E2EC			WtL2-seg		
metric	P	R	IoU	P	R	IoU	P	R	IoU
∅50: all images	**93.14**	79.02	75.04	89.31	**89.71**	**84.50**	89.55	81.02	75.74
∅40: cl. shapes	92.09	80.36	75.95	89.40	89.84	84.06	**93.58**	**94.83**	**89.01**

Fig. 6. Visualization of different segmentation results: (a) Original, (b) Ground Truth, (c) RefineNet (baseline) [16], contour-based segmentation: (d) E2EC (sota) and (e) WtL2-seg (ours); images from COCO [17].

it works and a closed object contour is formed: a high level of detail is given. In this case, the IoU even outperforms the sota in contour-based approaches. For illustration, Fig. 6 shows images from this subset. WtL2 extracts significantly more detail than the baseline and E2EC overall. This includes features such as the motorcyclist's helmet shield and tire treads, the elephant's belly and right tail, and the giraffe's hump, ears, mouth, neck, tail and legs.

5   Conclusion

We have enhanced a unique contour tracking method for contour refinement. To demonstrate the potential of this innovative approach, we adapted it for the niche of IR ship segmentation and generally extended it to various objects beyond ships in the RGB domain. In typical foreground-background scenarios, it outperforms the sota contour-based methods when forming a closed object contour, and it segments common categories such as dogs and cars, as well as rare and unusual categories such as deer, giraffes, and elephants, in a robust and detailed manner. This makes it a compelling method for specialized applications requiring detailed segmentation or the production of high-quality samples, potentially accelerating development in niche areas of computer vision, even with complex object contours or masks.

References

1. Ali Ibrahim, I., Namoun, A., Ullah, S., Alasmary, H., Waqas, M., Ahmad, I.: Infrared ship segmentation based on weakly-supervised and semi-supervised learning. IEEE Access **12**, 117908–117920 (2024)
2. Arbeláez, P., Maire, M., Fowlkes, C., Malik, J.: Contour detection and hierarchical image segmentation. TPAMI **33**(5), 898–916 (2011)
3. Bhattacharya, P., Riechen, J., Zölzer, U.: Infrared image enhancement in maritime environment with CNNs. In: VISAPP, pp. 37–46. SciTePress (2018)
4. Canny, J.: A computational approach to edge detection. TPAMI **6**, 679–698 (1986)
5. Chen, C., Hu, S., Ma, F., Sun, J., Lu, T., Wu, B.: Ship contour: a novel ship instance segmentation method using deep snake and attention mechanism. J. Mar. Sci. Eng. **13**(3) (2025)
6. Chen, J., Bai, X.: Learning to "segment anything" in thermal infrared images through knowledge distillation with a large scale dataset satir (2023). https://arxiv.org/abs/2304.07969
7. Chen, X., Chen, W., Wu, B., Wu, H., Xian, J.: Ship visual trajectory exploitation via an ensemble instance segmentation framework. Ocean Eng. **313**, 119368 (2024)
8. Chen, X., Qiu, C., Zhang, Z.: A multiscale method for infrared ship detection based on morphological reconstruction and two-branch compensation strategy. Sensors **23**(16) (2023)
9. Ke, L., et al.: Segment anything in high quality. In: NeurIPS (2023)
10. Kelm, A.P., Rao, V.S., Zölzer, U.: Object contour and edge detection with RefineContourNet. In: Vento, M., Percannella, G. (eds.) CAIP 2019. LNCS, vol. 11678, pp. 246–258. Springer, Cham (2019). https://doi.org/10.1007/978-3-030-29888-3_20
11. Kelm, A.P.: Extraktion geschlossener Schiffs- und Objektkonturen mit 1-Pixel-Breite zur präzisen Segmentierung in Farb- und Infrarotbildern durch Deep Learning. Ph.D. thesis, Helmut Schmidt University, Holstenhofweg 85, 22043 Hamburg, Germany (2025), doctoral dissertation, expected 2025
12. Kelm, A.P., Zölzer, U.: Walk the lines: object contour tracing CNN for contour completion of ships. In: ICPR, pp. 3993–4000 (2021)
13. Kim, Y., Park, J., Kang, S., Kim, H.: Introducing VaDA: novel image segmentation model for maritime object segmentation using new dataset (2024). https://arxiv.org/abs/2407.09005
14. Kirillov, A., et al.: Segment anything. In: ICCV, pp. 3992–4003 (2023)
15. Krähenbühl, P., Koltun, V.: Efficient inference in fully connected CRFs with gaussian edge potentials. In: Neurips, vol. 24 (2011)
16. Lin, G., Liu, F., Milan, A., Shen, C., Reid, I.: RefineNet: multi-path refinement networks for dense prediction. TPAMI (2019)
17. Lin, T.-Y., et al.: Microsoft COCO: common objects in context. In: Fleet, D., Pajdla, T., Schiele, B., Tuytelaars, T. (eds.) ECCV 2014. LNCS, vol. 8693, pp. 740–755. Springer, Cham (2014). https://doi.org/10.1007/978-3-319-10602-1_48
18. Ling, H., Gao, J., Kar, A., Chen, W., Fidler, S.: Fast interactive object annotation with curve-GCN. In: CVPR, pp. 5252–5261 (2019)
19. Mantell, S.: Free deer image. Online: Pixabay. https://pixabay.com/de/photos/rentier-hirsch-brown-tier-geweih-1323000. Accessed 20 Mar 2023
20. Minaee, S., Boykov, Y., Porikli, F., Plaza, A., Kehtarnavaz, N., Terzopoulos, D.: Image segmentation using deep learning: a survey. TPAMI **44**(7), 3523–3542 (2022)

21. Nirgudkar, S., DeFilippo, M., Sacarny, M., Benjamin, M., Robinette, P.: Massmind: Massachusetts maritime infrared dataset. Int. J. Robot. Res. **42**(1–2), 21–32 (2023)
22. Peng, S., Jiang, W., Pi, H., Li, X., Bao, H., Zhou, X.: Deep snake for real-time instance segmentation. In: CVPR (2020)
23. Xie, E., et al.: Polarmask: single shot instance segmentation with polar representation. In: CVPR, pp. 12190–12199 (2020)
24. Zhang, M., Wang, Y., Guo, J., Li, Y., Gao, X., Zhang, J.: Irsam: advancing segment anything model for infrared small target detection. In: ECCV, pp. 233–249 (2024)
25. Zhang, M., Zhang, Q., Song, R., Rosin, P.L., Zhang, W.: Ship landmark: an informative ship image annotation and its applications. Trans. Intell. Transp. Syst. **25**(11), 17778–17793 (2024)
26. Zhang, Q.L., Yang, Y.B.: A boundary-preserving conditional convolution network for instance segmentation. Pattern Recogn. Lett. **163**, 1–9 (2022)
27. Zhang, T., Wei, S., Ji, S.: E2EC: an end-to-end contour-based method for high-quality high-speed instance segmentation. In: CVPR, pp. 4443–4452 (2022)
28. Zhang, T., et al.: Infrared ship target segmentation based on adversarial domain adaptation. Knowl.-Based Syst. **265**, 110344 (2023)

Dirichlet Process Mixture Model and Markov Random Field for PolSAR Image Segmentation

Wassim Bdiri[1,4(✉)], Nizar Bouhlel[2], Stéphane Méric[3], Eric Pottier[1], and Fathi Kallel[4]

[1] Université de Rennes, IETR UMR CNRS 6164, Campus de Beaulieu, bat 11D,
35000 Rennes Cedex, France
bdiriwassim@gmail.com
[2] Institut Agro Rennes-Angers, Université d'Angers, INRAE, IRHS,
Angers, France
[3] INSA Rennes, IETR UMR CNRS 6164, 20 Av. des Buttes de Coesmes,
35700 Rennes, France
[4] ATMS, Enetcom, Univ Sfax, Sfax Technopark, BP 1163, 3018 Sfax, Tunisia

Abstract. The Dirichlet Process Mixture Model (DPMM) is a Bayesian nonparametric approach commonly used in unsupervised learning, notable for its ability to infer the number of components in the data. In this work, DPMM is integrated with a Markov Random Field (MRF) to tackle the segmentation of Polarimetric Synthetic Aperture Radar (PolSAR) images. The MRF incorporates spatial context, enhancing segmentation accuracy. Class labels are updated using the Expectation Maximization algorithm. The proposed EM-DPMM-MRF model is evaluated on both simulated and real PolSAR images with known ground truth. The experimental results demonstrate strong and consistent performance.

Keywords: Polarimetric synthetic aperture radar (PolSAR) · product model · unsupervised segmentation Dirichlet process mixture model (DPMM) · Expectation-Maximization (EM) · Markov Random Fields (MRF)

1 Introduction

Unsupervised segmentation is a crucial step in understanding Polarimetric Synthetic Aperture Radar (PolSAR) images. These methods partition PolSAR data into different classes by examining their inherent statistical features. For instance, the Wishart distribution has been applied for segmenting low-resolution PolSAR images [14], and the H/α-Wishart method offers another approach [8]. To address the characteristics of high-resolution polarimetric data, non-Gaussian statistical models, including heterogeneous models derived from the

© The Author(s), under exclusive license to Springer Nature Switzerland AG 2026
M. Castrillón-Santana et al. (Eds.): CAIP 2025, LNCS 15621, pp. 163–173, 2026.
https://doi.org/10.1007/978-3-032-04968-1_14

product model [20], have been introduced. Consequently, non-Gaussian clustering algorithms are frequently used for polarimetric segmentation, including the
K-Wishart algorithm within a finite mixture framework [2], the Markov random
field (MRF) K-Wishart algorithm [1], MRF Kummer-Ud clustering [9], and the
MRF EM algorithm [6], to name a few [13]. A unifying aspect of these techniques
is the need to define the number of classes beforehand.

Addressing the challenge of automatically determining the optimal number
of clusters, a new methodology is presented, namely the Bayesian nonparametric (BNP) approach. Unlike finite mixture models with a predetermined number of components, BNP generalizations allow for the simultaneous inference of
both the number of mixture components and their parameters directly from the
data [17]. A prominent example of a BNP model for clustering is the Dirichlet process mixture model (DPMM), which adapts its cluster count to match
the inherent structure of the data [17]. At the heart of the DPMM lies the
Dirichlet process (DP) [10], which generates random probability distributions
upon sampling. The DPMM can be understood through three distinct frameworks: the Chinese restaurant process (CRP), the Stick-breaking construction
[18], and the Polya urn scheme [5]. These frameworks give rise to a range of algorithms, including Markov Chain Monte Carlo (MCMC) methods (like Gibbs and
Metropolis-Hastings), Variational Bayes (VB) methods, Particle filters, and the
Expectation-Maximization (EM) algorithm. The development of BNP models
has been significantly accelerated by MCMC methods. For instance, [19] demonstrates the use of a DPMM with a Gibbs sampler for unsupervised PolSAR image
segmentation. Variational inference offers an alternative to MCMC sampling. In
[7], a variational textured DPMM based on the stick-breaking representation
is used for the unsupervised segmentation of polarimetric SAR images. In [3]
EM is combined with DPMM to segment polarimetric SAR images for change
detection application.

A key contribution of this work is to introduce a novel approach for parameter estimation and segmentation within Dirichlet Process Mixture Models
(DPMMs), serving as an alternative to existing Variational Bayes (VB) and
Markov Chain Monte Carlo (MCMC) methods. This method leverages the
Expectation-Maximization (EM) algorithm to simultaneously estimate both the
number of components in the mixture model and their respective parameters,
guided by Dirichlet process priors. To mitigate the visual artifact of significant
"salt-and-pepper" noise in the segmentation results, we integrate the local spatial
dependencies between neighboring pixels into the DPMM using a Markov Random Field (MRF) model. The resulting unsupervised segmentation technique is
termed EM-DPMM-MRF.

The rest of this paper is structured as follows: Section 2 introduces the fundamental concepts of PolSAR data modeling and the Dirichlet Process Mixture
Model (DPMM). Section 3 presents the Markov Random Field (MRF) framework and the concept of local interaction between adjacent pixels in PolSAR
images, followed by the integration of the MRF and DPMM to construct the
prior distribution for class labels. Section 4 provides a detailed explanation of

the proposed EM-DPMM-MRF method, first outlining the steps of the EM algorithm in this specific application and then deriving the analytical forms for parameter updates. Section 5 presents the experimental evaluation of our method on PolSAR image segmentation. Finally, Sect. 6 offers concluding thoughts to summarize the paper's findings.

2 PolSAR Data Model

PolSAR imagery at high resolution is represented by the multilook polarimetric covariance matrix $\mathbf{C}$. This matrix belongs to the cone $\mathbf{\Omega}_+$ of positive definite complex Hermitian matrices, where $\mathbf{C} \in \mathbf{\Omega}_+ \subset \mathbb{C}^{d \times d}$. Following the multilook polarimetric product model, as described in [6], the covariance matrix $\mathbf{C}$ can be expressed as the product of two components: texture (τ) and speckle ($\mathbf{X}$), such that $\mathbf{C} = \tau \mathbf{X}$. The texture component, τ, is a positive scalar random variable characterized by a probability density function (pdf) $f_\tau(\tau | \lambda)$ depending on a texture parameter λ. The speckle component, $\mathbf{X}$, is a random matrix that adheres to a scaled complex Wishart distribution, denoted as $s\mathcal{W}_d^{\mathbb{C}}(L, \mathbf{\Sigma})$, where L represents the number of looks and $\mathbf{\Sigma}$ is the covariance matrix of the speckle. The pdf of $\mathbf{C}$ is given

$$f_{\mathbf{C}}(\mathbf{C}) = \frac{L^{Ld}|\mathbf{C}|^{L-d}}{\Gamma_d(L)|\mathbf{\Sigma}|^L} \frac{(\lambda-1)^\lambda \Gamma(dL+\lambda)}{\Gamma(\lambda)} \left(L\mathrm{tr}(\mathbf{\Sigma}^{-1}\mathbf{C}) + \lambda - 1 \right)^{-(dL+\lambda)} \quad (1)$$

where tr(.) is the trace operator, |.| is the determinant operator, $\Gamma_d(L)$ is the multivariate gamma function of the complex kind and λ is the texture parameter.

3 Dirichlet Process Mixture Model

The Dirichlet Process Mixture Model (DPMM) framework offers the capability to represent a PolSAR image as a mixture of distributions, with the number of components being inferred directly from the data. This type of model utilizes a Dirichlet process (DP) [10], where each realization of the DP yields a random probability distribution. We consider a collection of independent and identically distributed (iid) random matrices, denoted as $\mathbf{C} = \{\mathbf{C}_s, \forall s \in S\}$, defined over a regular rectangular lattice S. We assume that the probability density function (pdf) of each $\mathbf{C}_s$ can be described by the following infinite mixture of distributions:

$$f_{\mathbf{C}_s}(\mathbf{C}_s|\Theta) = \sum_{m=1}^{+\infty} \pi_m f_{\mathbf{C}_s|X_s}(\mathbf{C}_s|\boldsymbol{\theta}_m), \quad (2)$$

where X denotes a set of latent random labels $\{X_s, \forall s \in S\}$, x_s is a specific realization or value of X_s, π_m represents the weight of each mixture component and corresponds to the prior probabilities, such that $P_{X_s}(x_s = m) = \pi_m$, with $\pi_m \geq 0$ and $\sum_{m=1}^{\infty} \pi_m = 1$. In this work, a prior distribution on the number of mixture components is incorporated through the stick-breaking construction. Specifically, the sequence of mixture weights $\{\pi_m\}$ is defined as follows:

$$\pi_m = v_m \prod_{j=1}^{m-1} (1 - v_j) \tag{3}$$

where $\{v_m\}$ is an infinite sequence of independent and identically distributed (i.i.d.) random variables following a Beta distribution $v_m \sim Beta(1, \alpha) = \alpha(1 - v_m)^{(\alpha-1)}$. The hyperparameter α, known as the concentration parameter of the DP, controls the expected number of significant components, with higher values indicating a greater number. Here, we consider $\alpha > 0$ and estimate it automatically from the data. The pdf $f_{C_s|X_s}(C_s|\theta_m)$ is a distribution describing component m in the form given by (1) with $\theta_m = (L, \Sigma_m, \theta_{\tau,m})$ where $\theta_{\tau,m}$ are the parameters of the texture distribution of partition m.

For practical implementation, a truncated Dirichlet prior is considered, selecting a sufficiently large value M instead of infinity and setting $v_M = 1$. As shown in [12], this truncation has minimal impact when M is reasonably large (e.g., $M = 100$). In this case, the complete set of parameters Θ to be estimated is defined as $\Theta = \{\alpha, v, \theta_1, ..., \theta_M\}$, where $v = (v_1, v_2, ..., v_{M-1})$. The number of looks is assumed to be consistent across all clusters.

$$f_C(C|\Theta) = \prod_{s \in S} \sum_{m=1}^{M} \pi_m f_{C_s|X_s}(C_s|\theta_m) \tag{4}$$

4   DPMM with Markov Random Field

Markov Random Fields (MRFs) have found extensive use in a variety of image processing tasks. Specifically, in image segmentation, the challenge lies in assigning labels to individual pixels, with the dependencies between these labels being modeled by an MRF. MRF models are characterized by energy functions derived from a Gibbs distribution, which capture the local interactions occurring between neighboring pixels. These energy functions are constructed as the sum of potential functions over all possible cliques $\mathcal{C}$. Let $X = \{X_s, \forall s \in S\}$ represent the unobserved label field, a collection of random variables defined on a regular rectangular lattice S. At each spatial location s, the random variable X_s can take a value x_s from the set of possible configurations $E_s = \{1, 2, ..., M\}$. The notation $(X = x)$ denotes the joint realization of the label field. Considering the unobserved class label X as an MRF defined on S with respect to a neighborhood system $\mathcal{N} = \{\mathcal{N}_s, \forall s \in S\}$, where $\mathcal{N}_s$ is the set of sites neighboring s, its joint probability takes the following form:

$$P_X(x) = \frac{1}{Z} \exp\left(- \sum_{\{s,r\} \in \mathcal{C}} V_c(x_s, x_r) \right) \tag{5}$$

where $\mathcal{C}$ represents the set of all cliques. In this paper, we will consider cliques consisting of all pairs of spatially adjacent pixels, either horizontally or vertically.

The normalizing constant Z is the summation over all possible configurations of the random field X, and $V_c(.)$ is the potential function associated with a clique c. This potential function is assumed to have the form:

$$V_c(x_s, x_r) = \beta_c t(x_s, x_r), \quad t(x_s, x_r) = \begin{cases} 0 & \text{if } x_s = x_r \\ 1 & \text{if } x_s \neq x_r \end{cases} \tag{6}$$

where $\beta_c > 0$ is the spatial interaction parameter. The probability $P_{X_s}(x_s|x_r, r \in \mathcal{N}_s)$ at site s, given the labels of its neighbors in $\mathcal{N}_s$, is defined with respect to the clique potentials in that neighborhood as follows:

$$P_{X_s}(x_s|x_{\mathcal{N}_s}, \beta_c) = \frac{\exp\left(-\sum_{c \in \mathcal{C}/s \in c} V_c(x_s)\right)}{\sum_{x_s \in E_s} \exp\left(-\sum_{c \in \mathcal{C}/s \in c} V_c(x_s)\right)}. \tag{7}$$

The integration of the DPMM with the MRF framework results in an unsupervised segmentation method, the proposed DPMM-MRF model, which can automatically determine the number of clusters while also incorporating spatial contextual information present in the image. Specifically, the prior distribution of the class label x_s at a given site s is formulated as a combination of two components: a site-specific term, π_m, representing the global prior probabilities of the classes drawn from a DP, and an interaction term, $P_{X_s}(x_s|x_r, r \in \mathcal{N}_s)$, which captures the spatially varying local prior probabilities. This novel prior distribution, denoted $\Pi_{X_s}(x_s = m|x_{\mathcal{N}_s})$, can be expressed as follows:

$$\Pi_{X_s}(m|x_{\mathcal{N}_s}) = \pi_m P_{X_s}(m|x_{\mathcal{N}_s}) \tag{8}$$

As demonstrated in [16], this model constitutes a valid MRF. Based on this new formulation of the prior distribution for the class label, the joint distribution of independent and identically distributed random matrices is given by:

$$f_{\mathbf{C}}(\mathbf{C}|\Theta) = \prod_{s \in S} \sum_{m=1}^{M} \Pi_{X_s}(m|x_{\mathcal{N}_s}) f_{\mathbf{C}_s|X_s}(\mathbf{C}_s|\boldsymbol{\theta}_m) \tag{9}$$

5   EM for DPMM-MRF

In what follows, we detail the Expectation-Maximization (EM) algorithm adapted for the DPMM-MRF framework, specifically within the context of high-resolution multilook PolSAR images. Our approach builds upon the methodologies presented in [6]. The EM algorithm is an iterative process that can yield an estimated parameter set $\hat{\Theta}$ based on a current parameter estimate Θ'. Given that a prior distribution $f(\boldsymbol{v})$ is defined over the parameters $\boldsymbol{v}$, the EM algorithm can be employed to find Maximum A Posteriori (MAP) solutions.

$$\hat{\Theta} = \arg\max_{\Theta} Q(\Theta, \Theta') + \ln f(\boldsymbol{v}) \tag{10}$$

where $Q(\Theta, \Theta') = E_{X,\tau|\mathbf{C}}\left\{\ln f_{\mathbf{C},X,\tau}(\mathbf{C}, X, \tau|\Theta)\big|\mathbf{C}, \Theta'\right\}$ and $f(\boldsymbol{v}) = \prod_{m=1}^{M-1} \alpha(1 - v_m)^{\alpha-1}$. Consequently, the maximization of the conditional expectation of the complete log-likelihood with respect to the parameters $\Theta = \{\alpha, \boldsymbol{v}, \boldsymbol{\theta}_1, ..., \boldsymbol{\theta}_M, \beta_c\}$, is given by:

$$
\hat{\Theta} = \arg\max_{\Theta} E_{X,\tau|\mathbf{C}}\left\{\sum_{s\in S} \ln f_{\mathbf{C}_s,\tau_s|X_s}(\mathbf{C}_s, \tau_s|X_s, \Theta)\big|\mathbf{C}, \Theta'\right\}
$$
$$
+ E_{X,\tau|\mathbf{C}}\left\{\ln \Pi_X(X|\Theta)\big|\mathbf{C}, \Theta'\right\} + (M-1)\ln\alpha + (\alpha-1)\sum_{m=1}^{M-1} \ln(1 - v_m) \tag{11}
$$

The preceding equation, denoted by (11), comprises two independent parts: one that depends on $\boldsymbol{\theta}_m$ and another that depends on (v_m, β_c, α). These two parts can therefore be maximized independently.

The estimation of speckle and texture parameters are given by the following equations

$$
\hat{\Sigma}_m = \frac{\sum_{s\in S} P_{X_s|\mathbf{C}}(m|\mathbf{C}, \Theta')g_s^{(m)}\mathbf{C}_s}{\sum_{s\in S} P_{X_s|\mathbf{C}}(m|\mathbf{C}, \Theta')} \tag{12}
$$

$$
\ln(\hat{\lambda}_m - 1) - \psi(\hat{\lambda}_m) + \frac{\hat{\lambda}_m}{\hat{\lambda}_m - 1} = \frac{\sum_{s\in S} P_{X_s|\mathbf{C}}(m|\mathbf{C}, \Theta')(g_s^{(m)} + a_s^{(m)})}{\sum_{s\in S} P_{X_s|\mathbf{C}}(m|\mathbf{C}, \Theta')}. \tag{13}
$$

where $a_s^{(m)}$ and $g_s^{(m)}$ are posterior expectations defined by

$$
a_s^{(m)} = E_{\tau_s|\mathbf{C}_s,X_s}\{\ln\tau_s|\mathbf{C}_s, X_s = m, \Theta'\} \tag{14}
$$
$$
g_s^{(m)} = E_{\tau_s|\mathbf{C}_s,X_s}\{\tau_s^{-1}|\mathbf{C}_s, X_s = m, \Theta'\} \tag{15}
$$

The posterior pdf of τ_s given $\mathbf{C}_s$ and $X_s = m$ corresponds to an inverse gamma distribution with parameters $(\alpha_1 = dL + \lambda_m, \beta_1 = Ltr(\Sigma_m^{-1}\mathbf{C}_s) + \lambda_m - 1)$. Then, the posterior expectation expression in terms of these parameters can be deduced. The estimation of (v_m, α) parameters are given by

$$
\hat{v}_m = \frac{\sum_{s\in S} P_{X_s|\mathbf{C}}(m|\mathbf{C}, \Theta')}{\alpha - 1 + \sum_{s\in S}\sum_{j=m}^{M} P_{X_s|\mathbf{C}}(j|\mathbf{C}, \Theta')} \tag{16}
$$

$$
\hat{\alpha} = \frac{1 - M}{\sum_{m=1}^{M} \ln(1 - v_m)} \tag{17}
$$

The estimation of β_c is performed through the following equation

$$
\sum_{s\in S}\sum_{x_s=1}^{M} \left[P_{X_s|\mathbf{C}}(x_s|\mathbf{C}, \Theta') - P_{X_s}(x_s|x_{\mathcal{N}_s}, \beta_c)\right]\sum_{r\in\mathcal{N}_s} t(x_s, x_r) = 0 \tag{18}
$$

This equation is solved numerically to find the solution $\hat{\beta}_c$.

The computation of all expressions for $(\hat{\boldsymbol{\theta}}_m, \hat{\beta}_c, \hat{\alpha}, \hat{v}_m)$ relies on the class posterior probabilities $P_{X_s|\mathbf{C}}(x_s|\mathbf{C}, \Theta')$, which are obtained through the Maximum Posterior Marginal (MPM) algorithm [15]. This method involves utilizing a Gibbs sampler [11] to produce a discrete-time Markov chain. This chain's distribution converges to that of a random field defined by the probability mass function $P_{X|\mathbf{C}}(x|\mathbf{C}, \Theta)$. By using the Bayes' rule, the conditional probability mass function of X given $\mathbf{C}$ to segment the image is given by

$$P_{X|\mathbf{C}}(x|\mathbf{C}, \Theta) \propto f_{\mathbf{C}|X}(\mathbf{C}|x, \Theta)\Pi_X(x) \tag{19}$$
$$= \frac{1}{Z} \exp\left(\sum_{s\in S} \ln f_{\mathbf{C}_s|X_s}(\mathbf{C}_s|x_s, \Theta) + \ln \pi_{x_s} - \sum_{\{s,r\}\in\mathcal{C}} V_c(x_s, x_r) \right).$$

The marginal conditional probability mass functions, $P_{X_s|\mathbf{C}}(m|\mathbf{C}, \Theta)$, which are to be maximized are then approximated as a fraction of time the Markov chain spends in state m at pixel s, for each m and s

$$P_{X_s|\mathbf{C}}(m|\mathbf{C}, \Theta) \approx \frac{1}{T_s} \sum_{t=1}^{T_s} u_{m,s}(t), \quad \forall m, s \tag{20}$$

with $u_{m,s}(t) = 1$ if $X_s(t) = m$, and $u_{m,s}(t) = 0$ if $X_s(t) \neq m$, T_s is the number of visits to pixel s made by the Gibbs sampler.

6 Experimental Results and Analysis

In this section, we experiment the EM-DPMM-MRF algorithm with a pixel-based segmentation algorithm using different PolSAR data.

6.1 Simulated Data

In the first set of experiments, we use simulated images generated according to the product model. The simulated image consists of five different regions and has a resolution of 200 × 200 pixels. This controlled environment allows us to evaluate the performance of the segmentation algorithm. The results are shown in Fig. 1. In fact, Fig. 1(a) shows the simulated PolSAR image, Fig. 1(b) shows the ground truth of the simulated image and Fig. 1(c) shows the result of the DPMM-EM-MRF segmentation. The first observation is that the algorithm successfully detects the exact number of components, with N = 5, thanks to the implementation of the Dirichlet Process Mixture Model (DPMM). This result highlights the effectiveness of the model in accurately identifying the components within the data. Furthermore, upon comparing the algorithm's segmentation with the ground truth, we obtain an accuracy percentage equal to 99,93, so it is clear that the segmentation performance is nearly perfect. The alignment between the detected components and the true segmentation further demonstrates the robustness and precision of our approach in handling the segmentation task.

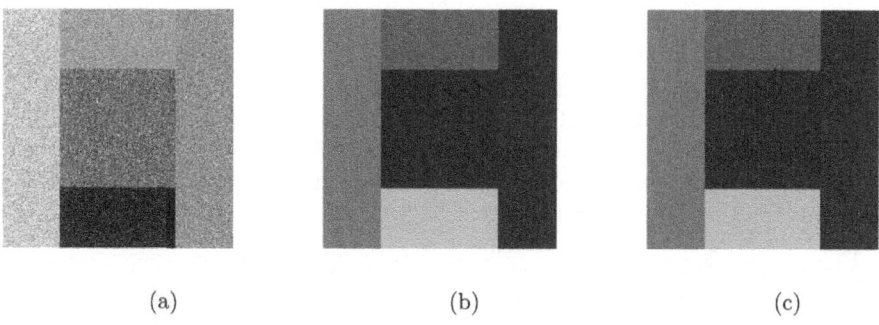

(a)   (b)   (c)

Fig. 1. Simulated data. (a) RGBPauli image of the simulated data. (b) Ground truth of the segmentation. (c) Segmentation result with EM-DPMM-MRF

6.2   Image Constructed from Real Data

In this section we use a real image from the AIRSAR Flevoland dataset, which consists of data from 15 different classes. However, due to the incomplete ground truth associated with this dataset, we have chosen to assemble a PolSAR image based on the available AIRSAR Flevoland (Fig. 2(a)) image and its corresponding ground truth [4]. To achieve this, we selected six different regions from the original image, each representing a different area of interest, and combined them into a single cohesive image. This ensured that the newly created image had a known and complete ground truth, allowing for a more controlled evaluation of our approach. The final image constructed from these six different regions provides a robust basis for testing, as we have precise knowledge of its ground truth. This process allows us to evaluate the performance of our algorithm with greater accuracy and confidence. The new image has a resolution of 80×166 pixels. Figure 2(b) shows the PolSAR image, Fig. 2(c) shows the corresponding ground truth and Fig. 2(d) shows the result of the segmentation using our approach. The result shows that the segmentation algorithm successfully infers the correct number of clusters, which is six, perfectly matching the known ground truth. Furthermore, the algorithm produces a highly accurate segmentation of the classified image, achieving an impressive accuracy of 98.73. The accuracy of each region is detailed in the Table 1 of the confusion matrix. This result demonstrates the effectiveness of the approach in correctly identifying and segmenting the different regions within the image, further validating the reliability and precision of the proposed method.

6.3   Real Data

In this section, we examine actual PolSAR data from the EMISAR Foulum image. We select a specific area of this image and use our approach to test our segmentation system. However, due to the lack of ground truth information, we do not compute segmentation performance metrics in this case, unlike in previous

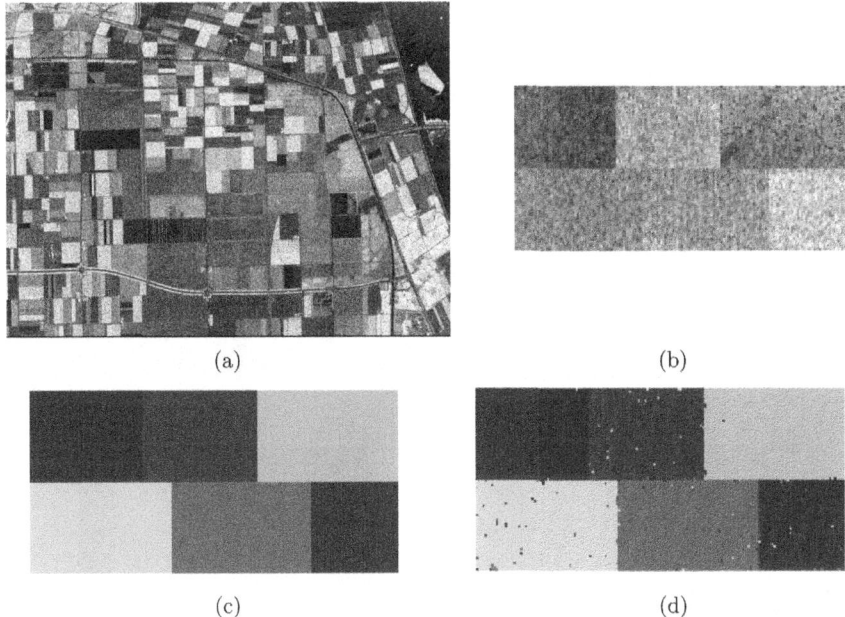

(a)   (b)

(c)   (d)

Fig. 2. Image constructed from airsar Flevoland data. (a) RGBPauli image of the simulated data. (b) Ground truth of the segmentation. (c) Segmentation result with EM-DPMM-MRF.

Table 1. Confusion Matrix (Percentage)

	Bare soil	Rapeseed	Water	Wheat	Lucerne	Wheat 2
bare soil	98.41	0.8209	0.2414	0.2897	0.1449	0.09657
Rapeseed	0	**98.59**	0.09747	0.5361	0.4386	0.3411
Water	0.07921	0.2772	**99.52**	0.0396	0	0.07921
Wheat	0	0.4332	0	**98.98**	0.3939	0.1969
Lucerne	0	0.2759	0.9854	0	**98.27**	0.473
Wheat 2	0	0.964	0.2571	0.3213	0	**98.46**

experiments. Despite this limitation, we can still visually inspect the segmentation results, which allows us to evaluate the effectiveness of the algorithm at a qualitative level. We can gain important insights into the performance of the algorithm on real PolSAR data by examining the output to see how effectively the segmentation captures different structures and patterns within the selected area. Figure 3(a) shows the PolSAR image of EMISAR Foulum. Figure 3(b) shows the selected area and Fig. 3(c) the segmentation result for this image. The segmented image shows seven distinct clusters, clearly demonstrating the effectiveness of the segmentation process.

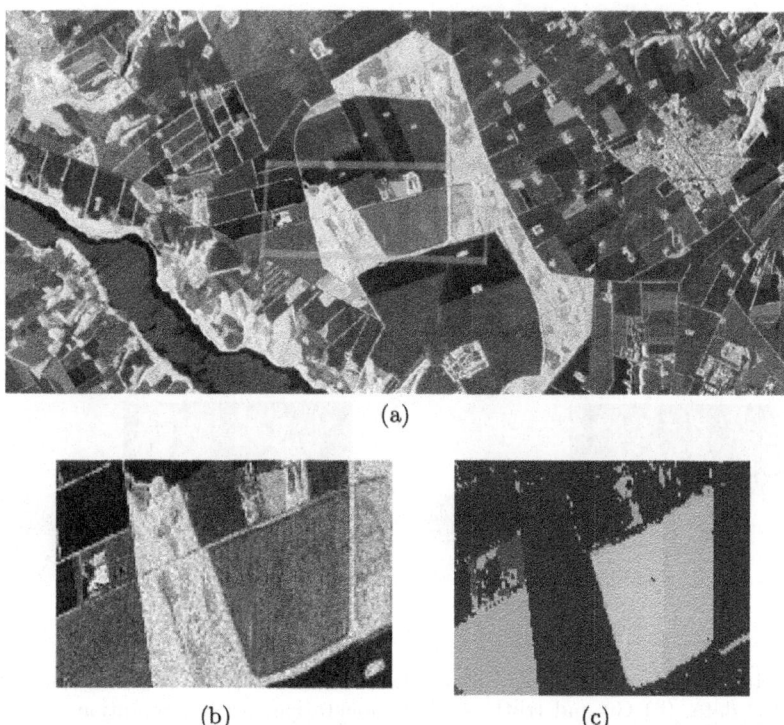

(a)

(b)   (c)

Fig. 3. (a) RGBPauli image of EMISAR Foulum (b) Segmentation result with EM-DPMM-MRF.

7 Conclusion

In this work, we have introduced our method, EM-DPMM-MRF, for PolSAR image segmentation. This approach provides an efficient unsupervised segmentation framework suitable for any PolSAR image. The proposed method effectively estimates the optimal number of clusters with DPMM while ensuring high-quality segmentation with EM and by incorporating spatial information through the Markov Random Field (MRF) model. Experimental results on both simulated and real PolSAR images demonstrate the robustness and accuracy of our approach, highlighting its ability to capture spatial dependencies and improve segmentation performance.

References

1. Akbari, V., Doulgeris, A.P., Moser, G., Eltoft, T., Anfinsen, S.N., Serpico, S.B.: A textural contextual model for unsupervised segmentation of multipolarization synthetic aperture radar images. IEEE Trans. Geosci. Remote Sens. 51(4), 2442–2453 (2013)

2. Anfinsen, S.N., Eltoft, T.: Application of the matrix-variate Mellin transform to analysis of polarimetric radar images. IEEE Trans. Geosci. Remote Sens. **49**(6), 2281–2295 (2011)

3. Bdiri, W., Bouhlel, N., Méric, S., Pottier, E., Kallel, F.: A Bayesian nonparametric model for unsupervised change detection of fully polarimetric SAR images. In: IGARSS 2024, pp. 2789–2794 (2024)

4. Bi, H., Xu, L., Cao, X., Xue, Y., Xu, Z.: Polarimetric SAR image semantic segmentation with 3D discrete wavelet transform and Markov random field. IEEE Trans. Image Process. **29**, 6601–6614 (2020)

5. Blackwell, D., MacQueen, J.B.: Ferguson distributions via polya urn schemes. Ann. Stat. **1**(2), 353–355 (1973)

6. Bouhlel, N., Méric, S.: Unsupervised segmentation of multilook polarimetric synthetic aperture radar images. Trans. Geosci. Remote Sens. **57**(8), 6104–6118 (2019)

7. Chi, L., Heng-Chao, L., Wenzhi, L., Wilfried, P., William, E.: Variational textured dirichlet process mixture model with pairwise constraint for unsupervised classification of polarimetric SAR images. IEEE Trans. Image Proc. **28**(8), 4145–4160 (2019)

8. Cloude, S., Pottier, E.: An entropy based classification scheme for land applications of polarimetric SAR. Trans. Geosci. Remote Sens. **35**(1), 68–78 (1997). https://doi.org/10.1109/36.551935

9. Doulgeris, A.P.: An automatic U-distribution and Markov random field segmentation algorithm for PolSAR images. IEEE Trans. Geosci. Remote Sens. **53**(4), 1819–1827 (2015)

10. Ferguson, T.S.: A Bayesian analysis of some nonparametric problems. Ann. Stat. **1**(2), 209–230 (1973)

11. Geman, S., Geman, D.: Stochastic relaxation, Gibbs distributions, and the Bayesian restoration of images. IEEE Trans. Pattern Anal. Mach. Intell. **PAMI-6**(6), 721–741 (1984)

12. Ishwaran, H., James, L.F.: Gibbs sampling methods for stick-breaking priors. J. Am. Stat. Assoc. **96**(453), 161–173 (2001)

13. Khan, S., Doulgeris, A.P.: Unsupervised clustering of PolSAR data using polarimetric G distribution and Markov random fields. In: EUSAR 2014; 10th European Conference on Synthetic Aperture Radar, pp. 1–4 (2014)

14. Lee, J.S., Grunes, M.R., Ainsworth, T.L., Du, L.J., Schuler, D.L., Cloude, S.R.: Unsupervised classification using polarimetric decomposition and the complex wishart classifier. IEEE Trans. Geosci. Remote Sens. **37**(5), 2249–2258 (1999)

15. Marroquin, J., Mitter, S., Poggio, T.: Probabilistic solution of ill-posed problems in computational vision. J. Am. Stat. Assoc. **82**(397), 76–89 (1987)

16. Orbanz, P., Buhmann, J.M.: Nonparametric Bayesian image segmentation. Int. J. Comput. Vision **77**, 25–45 (2008)

17. Orbanz, P., Teh, Y.W.: Bayesian Nonparametric Models, pp. 81–89. Springer, Boston (2010)

18. Sethuraman, J.: A constructive definition of dirichlet priors. Stat. Sin. **4**(2), 639–650 (1994)

19. Wanying, S., Ming, L., Peng, Z., Yan, W., Lu, J., Lin, A.: Unsupervised PolSAR image classification and segmentation using dirichlet process mixture model and markov random fields with similarity measure. IEEE J. Sel. Topics Appl. Earth Observ. Remote Sens. **10**(8), 3556–3568 (2017)

20. Yueh, S.H., Kong, J.A., Jao, J.K., Shin, R.T., Novak, L.M.: K-distribution and polarimetric terrain radar clutter. J. Electromagn. Waves Appl. **3**(8), 747–768 (1989)

Semantic Segmentation for Coastal Monitoring: Region Extraction and Overtopping Detection

Fernando Sanfiel-Reyes[1] , Jonay Suárez-Ramírez[2] ,
Miguel Alemán-Flores[1] , and Nelson Monzón[1]([⊠])

[1] CTIM, Instituto Universitario de Cibernética, Empresa y Sociedad, University of
Las Palmas de Gran Canaria, 35017 Las Palmas de Gran Canaria, Spain
{fernando.sanfiel,miguel.aleman,nelson.monzon}@ulpgc.es
[2] Qualitas Artificial Intelligence and Science, Las Palmas de Gran Canaria, Spain
jsuarez@qaisc.com
https://iuces.ulpgc.es/, https://qaisc.com/

Abstract. This work presents a method for analyzing coastal areas to extract regions of interest and identify significant events near the shore, using semantic segmentation adapted to these environments. The segmentation approach is applied to label all pixels in an image according to a predefined set of classes. Two additional classes—namely, foam and wet sand—are introduced to the typical categories used in coastal dynamics, allowing for more detailed differentiation of areas that are important for specific purposes. The resulting classifications are then analyzed, either individually or as a sequence of frames in a video, to detect the occurrence of relevant events, such as waves overtopping dikes and reaching pedestrian or vehicle areas, or to extract regions of interest, such as the intertidal zone. In particular, detecting overtopping involves selecting a critical region and monitoring when it is reached by the sea. On the other hand, extracting the intertidal zone implies processing sequences spanning several hours to track the sea's temporal changes. With this approach and the additional classes, the proposed method enables more robust detection of overtopping events and more accurate delineation of the region between high and low tides.

Keywords: Deep learning · Semantic segmentation · Coastal dynamics · Wave overtopping · Intertidal zone

1 Introduction

The analysis of coastal regions is essential for addressing climate change, risk mitigation, and urban planning [3]. Beyond traditional physical measurements, such as sea level, tides, and wave activity [6], visual data from coastal imagery are increasingly being leveraged to improve environmental monitoring. In this

M. Castrillón-Santana et al. (Eds.): CAIP 2025, LNCS 15621, pp. 174–185, 2026.
https://doi.org/10.1007/978-3-032-04968-1_15

context, semantic segmentation has emerged as a key computer vision technique, enabling pixel-wise classification of images and providing detailed spatial information about relevant coastal elements, in contrast to conventional image-level classification.

In this work, we address the analysis of coastal dynamics for applications such as risk detection and environmental monitoring. First, we apply semantic segmentation to identify the different regions in a coastal scene. Typical approaches to the segmentation of coastal areas include natural regions (beach, sea, rock, sky, and mountain), artificial constructions (pier, floor, ship, and building), humans (person) and other small elements (object) [20]. The initial approach is improved by adding two specific classes for foam and wet sand, which make it possible to identify risky scenarios in coastal monitoring and trigger alerts when necessary. Furthermore, some regions which are significant for additional applications can be extracted. For instance, the intertidal zone, which is above water at low tide and underwater at high tide, can be estimated more robustly thanks to the analysis of these regions throughout a long temporal sequence.

2   Related Works

Previous works have dealt with the detection of tidal movements and storm surges in different coastal regions. Guillou et al. [4] addressed the detection of flood events and the estimation of peak sea level in an estuary in Brittany (France) from two different perspectives: multiple regression techniques (linear and polynomial) and artificial neural networks. Rus et al. [12] focused on forecasting sea surface height and storm surges in the northern Adriatic. The authors evaluated their results both globally and in relation to flood events. Sabato et al. [13] used surveillance cameras for the remote measurement of tides and surges in two coastal areas in Italy. Similar to the research conducted on ocean coastlines, some studies, such as that by Zhang et al. [19], have dealt with this problem in estuaries and rivers.

To avoid *ad hoc* solutions, Riazi [11] presented a method for tide estimation which is based on historical data and depends on physical aspects of tides. Instead of extracting a new model for each location, the generalization is achieved with an additional neuron in the input layer.

In order to compare new trends in the segmentation of coastal areas with traditional approaches, Tiggeloven et al. [16] explored the performance of deep learning in predicting surges in coastal areas. The authors concluded that neural network models are able to capture the temporal evolution of surges and outperform large-scale hydrodynamic models.

Different image modalities have been used in prior research. Semantic segmentation by means of deep learning techniques was applied by Scala et al. [14] on aerial and satellite images for coastline detection. O'Sullivan et al. [10] included spectral indices to improve the interpretability of machine learning models. Jeon et al. [7] performed a semantic segmentation of images acquired with drones, and both color and grayscale images were considered.

Particular applications include tracking the erosion of the coastal areas and the recovery process after a storm, as in the work by Kang et al. [8], or detecting obstacles in marine environments, as in Hansen et al. [5]. Regarding the detection of specific elements, algae detection was addressed by Arellano et al. [1], who introduced a mapping methodology for the coverage of a type of algae, and by Valentini et al. [17], who presented a model for the use of low-cost cameras in the segmentation of coastal images and included its use in the detection of algae on beaches and in harbors.

3   Dataset and Segmentation Architecture

In this work, we use semantic segmentation to enable pixel-wise classification of coastal images into meaningful categories, offering fine-grained spatial analysis, which is essential for dynamic coastal monitoring. Our approach adapts this technique to detect static elements (both natural and artificial, such as rock, pier, dike, and building) as well as transient elements (such as foam, wet sand, beach, and sea), which are critical for identifying wave overtopping and delineating the intertidal zone. The following sections describe the dataset and the class definitions used to train the segmentation model.

3.1   Dataset Construction

Given the absence of a dedicated dataset for coastal dynamics analysis, we created a new one using a similar structure of the well-known ADE20K dataset [20]. It consists of images from cameras installed in coastal and port environments in Gran Canaria, Barcelona, and Gijón (Spain), and owned by the company Qualitas Artificial Intelligence and Science[1]. To increase environmental diversity, we also included images from SkyLine Webcams[2], a public platform with cameras located in seaports and coastal zones worldwide.

The images were captured under varying weather conditions and at different times of day, and were automatically collected using a scheduled Python scraper to avoid manual selection bias. In total, the dataset comprises 364 images (292 for training and 72 for validation) from 110 distinct coastal environments, offering broad coverage of diverse geographic and visual conditions. A notable feature of the dataset is that it contains no more than four snapshots per environment, which helps reduce overfitting and prevents the model from memorizing specific scenes. Figure 1 shows representative samples illustrating the different environments, visibility conditions, and times of day.

3.2   Semantic Classes for Coastal Dynamics

Existing datasets from urban-coastal scenes commonly include classes such as sea, sand, person, building, and dike. However, to capture relevant aspects of

---

[1] https://qaisc.com/.

[2] https://www.skylinewebcams.com/es/live-cams-category/seaport-cams.html.

Fig. 1. Sample images from the constructed coastal dataset, encompassing diverse environments (beach, harbor, city promenade, etc.), a variety of weather conditions (with particularly challenging blurred images), and a range of lighting conditions and times of day (from bright daylight to nighttime scenes).

sea behavior, it is beneficial to introduce additional categories. In particular, we classify foam as a distinct class, separate from the sea class, and differentiate between dry and wet sand, which we refer to as beach and sand, respectively. These additions are grounded in prior studies of sea dynamics, which highlight the role of foam in analyzing the sea's impact on coastal and port environments. Including foam as a class allows the trained model to detect wave overtopping on structures such as port dikes and seafront promenades. Similarly, the distinction between beach and sand facilitates the identification of the intertidal zone from the analysis of the involved regions over time. Overall, this approach enables a more detailed study of tidal variability and coastal dynamics [15].

3.3 Net Configuration and Training

In order to select an appropriate segmentation architecture, we first reviewed the top-ranked models on the ADE20K benchmark, as presented on Papers with Code[3]. According to this evaluation, Mask2Former [2] stood out for its consistent performance and proven effectiveness in dense prediction tasks. Based on this analysis, we adopted it as the segmentation head in our system.

Subsequently, several encoders were tested in combination with Mask2Former on the coastal image dataset described in Sect. 3.1. The models were assessed using standard semantic segmentation metrics: mean intersection over union (mIoU), mean pixel accuracy (mPA), and global pixel accuracy (GPA). While mIoU and mPA better reflect performance on underrepresented classes, GPA more directly captures overall pixel-wise accuracy. The results are summarized

---

[3] https://paperswithcode.com/sota/semantic-segmentation-on-ade20k, Papers with Code compiles benchmark results from peer-reviewed publications and serves as a reliable reference for identifying current trends in the field.

in Table 1 and, as observed, Swin-b outperforms the others in all three metrics, providing the best results in both global and class-averaged accuracy.

Table 1. Comparison of the metrics obtained from different backbones/encoders when used in combination with Mask2Former.

	R50 [18]	R101 [18]	Swin-t [9]	Swin-s [9]	Swin-b [9]
mIoU	39.31	39.31	40.28	40.56	**59.25**
mPA	47.01	47.01	52.45	56.51	**70.86**
GPA	77.95	77.95	78.48	78.73	**89.59**

To train the segmentation models, we used the MMSegmentation[4] toolbox. The selected architecture was implemented within this framework, and we adopted the default ADE20K training hyperparameters, running the training process for 20,000 iterations.

4 Results and Discussion

This section presents the main results of evaluating the performance of the proposed method for coastal image segmentation. Subsequently, two specific applications are shown: the detection of overtopping events on critical seafront structures and the extraction of the intertidal zone.

4.1 Segmentation of Coastal Images

Figure 2 illustrates the regions extracted in different scenarios. As observed, two distinct areas are identified in the water (sea and foam) and on partially wet sand (beach and sand). The approach yields reasonably good performance, even under challenging visibility conditions such as fog, low illumination, shadows, and overexposed scenes.

Table 2. Net metrics per class on the created dataset: intersection over union (IoU) and pixel accuracy (PA) obtained for the 15 classes under consideration. Those corresponding to the most relevant classes have been underlined.

Class	beach	foam	sand	sea	building	dike	floor	mountain	object	person	pier	rock	ship	sky	mean
IoU	<u>40.99</u>	<u>78.32</u>	<u>66.04</u>	<u>92.94</u>	76.25	53.34	74.85	82.30	52.16	13.72	26.63	30.13	44.46	97.34	**40.56**
PA	<u>47.59</u>	<u>86.29</u>	<u>82.22</u>	<u>97.08</u>	88.08	60.89	85.23	91.52	77.80	28.30	34.21	61.34	52.16	99.36	**56.51**

---

[4] https://github.com/open-mmlab/mmsegmentation.

Class	Color	Class	Color	Class	Color	Class	Color	Class	Color
beach		floor		object		rock		ship	
building		foam		person		sand		sky	
dike		mountain		pier		sea		void	

Fig. 2. Sample results illustrating the different regions obtained for the 15 classes, including wet sand shown in brown (as in the 1st image) and foam in purple (as in the 2nd image), and testing the performance under challenging weather conditions (3rd image) and in low-light scenes (4th image). The legend at the bottom indicates the color assigned to each region.

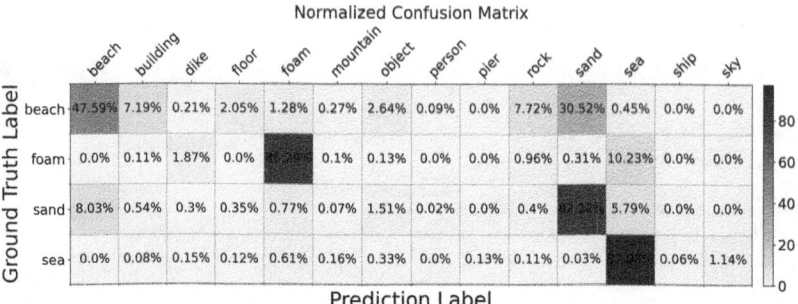

Fig. 3. Confusion matrix (only rows for beach, foam, sand and sea have been selected) indicating the percentage of each actual region (row) which is classified with each label (column).

Table 2 presents the intersection over union (IoU) and pixel accuracy (PA) obtained for each class when the method is evaluated on the constructed dataset. The most relevant classes for the applications described above achieve satisfactory performance (PA exceeds 95% for the sea class and surpasses 80% for the foam and sand classes). The performance on the beach class is comparatively lower, but the reason can be observed in the confusion matrix shown in Fig. 3 (only the rows for the classes beach, sand, sea, and foam are shown to focus attention on the most relevant ones). The moderate values for beach stem from the distinction made between sand and beach, based on whether the sand is dry or wet. Although a portion of the beach area is not labeled as such, approximately 80% of that region is classified as either beach or sand, which represents a reasonably good outcome. Since classifications are performed on individual images, some pixel-level misclassifications occur. However, analyzing full regions and temporal image sequences allows for reliable identification of critical zones and events.

4.2 Wave Overtopping Detection

Wave run-up is the maximum elevation reached by waves, i.e., the height to which waves run up the slopes of beaches or structures. Overtopping occurs when a volume of water passes over the top of a structure due to wave run-up exceeding the structure's freeboard (the vertical distance between the water level and the top of the structure).

At certain points along the coast, it is crucial to detect when overtopping occurs. To automatically identify such events, we propose the following method. Using an image from a static camera monitoring the area, the user selects a set of points to define a polygonal region that should remain uncovered by the sea under normal conditions. If semantic segmentation detects a connected region of pixels labeled as foam or sea within that polygon, and the area of that region exceeds a predefined threshold (30 pixels in our examples), the frame is flagged as containing an overtopping event. Some examples of these events and how they can be captured are shown in Fig. 4. The region outlined in red is the critical area where overtopping is evaluated, whereas the blue and purple regions in the segmentation correspond to sea and foam, respectively. Once a significant part of those regions enters the critical zone, overtopping is signaled. To evaluate this overtopping detection strategy, we tested it on a dataset containing 2, 998 images with and without overtopping. The method achieved an accuracy of **0.86**, a precision of **0.82**, a recall of **0.81**, and an $F1$ score of **0.82**. These results indicate that over 80% of overtopping events were correctly detected and that over 80% of the generated alarms corresponded to actual events. To assess our results, we also evaluated the performance of the model trained with ADE20K, obtaining an accuracy of **0.71**, a precision of **0.82**, a recall of **0.32**, and an $F1$ score of **0.462**. These results indicate that overtopping events were detected significantly more effectively with our proposed model.

4.3 Intertidal Zone

The intertidal zone, or foreshore, is the region that is exposed above sea level at low tide and submerged at high tide, i.e., the part of the seashore within the tidal range. Therefore, this region is visible at certain times of the day and submerged at others. Additionally, wave oscillations cause the shoreline to fluctuate within the same period of time, creating a transition area between the sea and the dry land.

To segment the intertidal zone using semantic segmentation, we analyze a time period that includes both low and high tides. The areas covered by the foam and sea classes indicate where water has been present. The union of these regions across the entire sequence defines the maximum area covered by water, while the intersection of regions from different frames identifies the minimum area covered by water. The difference between the union and the intersection provides the intertidal zone:

$$InterTidal = \bigcup_{i=1}^{n} M_i - \bigcap_{i=1}^{n} M_i, \tag{1}$$

where M_i is the region labeled as sea or foam in frame i, and the intersection and union are calculated throughout the n frames. To avoid outliers in the detection of sea and foam in individual frames due to different artifacts, when these masks are identified inside areas corresponding to certain elements (building, floor and mountain), they are removed from the cumulative intersection and union masks. The process can be performed for a long or short period of time, depending on the intended objective. Figure 5 illustrates the variation between the area that is always covered by the sea within a video sequence and the area that is covered at any time in the sequence (after removing possible outliers). In other words, the region highlighted in red corresponds to the difference between the high and low tide extents over the observed time period.

Class	Color	Class	Color	Class	Color	Class	Color	Class	Color
beach		floor		object		rock		ship	
building		foam		person		sand		sky	
dike		mountain		pier		sea		void	

Fig. 4. Overtopping detection: original frames (left) and corresponding results (right) for two scenes from San Sebastián's promenade (top and middle) and one from Gijón's harbor (bottom). Red lines delineate critical areas; purple and blue regions represent foam and water, respectively. Wave overtopping is identified when these regions appear within the predefined areas.

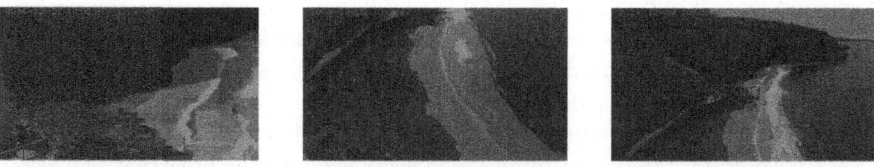

Fig. 5. Intertidal zone (red area) extracted from video sequences captured with different cameras and perspectives, by comparing the regions consistently covered by the sea with those temporarily submerged over several-hour sequences. (Color figure online)

5   Conclusion

Monitoring coastal areas over both short and long terms is crucial for ecological, urban, and economic purposes. In this work, we have presented an application of semantic segmentation specifically tailored for coastal dynamics analysis. The proposed approach incorporates specialized classes (foam and wet sand) that extend beyond conventional categories and provide essential information for detecting critical phenomena, such as wave overtopping of seafronts and port structures. Furthermore, the temporal aggregation of segmentation outputs enables the estimation of the intertidal zone by tracking the spatial evolution of sea-related regions over time.

A robust coastal image dataset was constructed, comprising scenes from over 110 distinct environments, encompassing diverse meteorological and lighting conditions, as well as varying sea states. Sampling was deliberately sparse across locations to reduce overfitting and promote model generalization.

The results demonstrate that the most relevant classes are accurately segmented, even under challenging visibility conditions. These outcomes underscore the effectiveness of the proposed framework for spatial and temporal analysis in coastal monitoring applications. Future work will explore the integration of physical measurements (such as data from tide gauges and radar systems) to validate segmentation outputs and enhance applicability within multimodal coastal monitoring systems. This extension will support the development of more robust, sensor-informed risk detection frameworks.

Acknowledgements. This work stems from the collaboration between the company Qualitas Artificial Intelligence and Science (QAISC) and the Imaging Technology Center (CTIM) at the University of Las Palmas de Gran Canaria, under the research contract C2024/54, signed between the company and the Canarian Science and Technology Park Foundation of the University of Las Palmas de Gran Canaria.

It has also been supported by Vicepresidencia Primera, Consejería de Vicepresidencia Primera y de Obras Públicas, Infraestructuras, Transporte y Movilidad from Cabildo de Gran Canaria through the project "*DETECCIÓN PRECISA IA*".

The PORTS 4.0 equity fund also supports this work through a grant to the precommercial project *SMART COAST AI SOLUTIONS 4.0*, in which the Oceanic Platform of the Canary Islands (PLOCAN, https://plocan.eu/en), QAISC, and CTIM are partners, as part of the Spanish State Port Authorities initiative to promote the transition to Economy 4.0.

References

1. Arellano-Verdejo, J., Santos-Romero, M., Lazcano-Hernandez, H.: Use of semantic segmentation for mapping sargassum on beaches. PeerJ **10**, e13537 (2022). https://doi.org/10.7717/peerj.13537
2. Cheng, B., Misra, I., Schwing, A.G., Kirillov, A., Girdhar, R.: Masked-attention mask transformer for universal image segmentation. CoRR abs/2112.01527 (2021). https://arxiv.org/abs/2112.01527

3. European Environment Agency: European climate risk assessment – Executive summary. Publications Office of the European Union (2024). https://doi.org/10.2800/204249

4. Guillou, N., Chapalain, G.: Machine learning methods applied to sea level predictions in the upper part of a tidal estuary. Oceanologia **63**(4), 531–544 (2021). https://doi.org/10.1016/j.oceano.2021.07.003, https://www.sciencedirect.com/science/article/pii/S0078323421000634

5. Hansen, K.F., Yao, L., Ren, K., Wang, S., Liu, W., Liu, Y.: Image segmentation in marine environments using convolutional LSTM for temporal context. Appl. Ocean Res. **139**, 103709 (2023). https://doi.org/10.1016/j.apor.2023.103709, https://www.sciencedirect.com/science/article/pii/S014111872300250X

6. Intergovernmental Panel on Climate Change (IPCC): Technical Summary, pp. 33–144. Cambridge University Press, Cambridge (2023). https://doi.org/10.1017/9781009157896.002

7. Jeon, E., Kim, S., Park, S., Kwak, J., Choi, I.: Semantic segmentation of seagrass habitat from drone imagery based on deep learning: a comparative study. Ecol. Inform. **66**, 101430 (2021). https://doi.org/10.1016/j.ecoinf.2021.101430, https://www.sciencedirect.com/science/article/pii/S1574954121002211

8. Kang, B., Vinent, O.: The application of cnn-based image segmentation for tracking coastal erosion and post-storm recovery. Remote Sens. **15**, 3485 (2023). https://doi.org/10.3390/rs15143485

9. Liu, Z., et al.: Swin transformer: hierarchical vision transformer using shifted windows. In: Proceedings of the IEEE/CVF International Conference on Computer Vision (ICCV) (2021)

10. O'Sullivan, C., Coveney, S., Monteys, X., Dev, S.: Analyzing water body indicies for coastal semantic segmentation. In: 2023 Photonics & Electromagnetics Research Symposium (PIERS), pp. 1792–1799 (2023). https://doi.org/10.1109/PIERS59004.2023.10221289

11. Riazi, A.: Accurate tide level estimation: a deep learning approach. Ocean Eng. **198** (2020). https://doi.org/10.1016/j.oceaneng.2020.107013

12. Rus, M., Fettich, A., Kristan, M., Ličer, M.: HIDRA2: deep-learning ensemble sea level and storm tide forecasting in the presence of seiches – the case of the northern adriatic. Geosci. Model Dev. **16**(1), 271–288 (2023). https://doi.org/10.5194/gmd-16-271-2023, https://gmd.copernicus.org/articles/16/271/2023/

13. Sabato, G., et al.: Remote measurement of tide and surge using a deep learning system with surveillance camera images. Water **16**(10) (2024). https://doi.org/10.3390/w16101365, https://www.mdpi.com/2073-4441/16/10/1365

14. Scala, P., Manno, G., Ciraolo, G.: Semantic segmentation of coastal aerial/satellite images using deep learning techniques: an application to coastline detection. Comput. Geosci. **192**, 105704 (2024). https://doi.org/10.1016/j.cageo.2024.105704, https://www.sciencedirect.com/science/article/pii/S0098300424001870

15. Stewart, R.: Introduction to Physical Oceanography. University Press of Florida (2009). https://books.google.es/books?id=3dXTRAAACAAJ. ISBN 9781616100452

16. Tiggeloven, T., Couasnon, A.A., van Straaten, C., Muis, S., Ward, P.J.: Exploring deep learning capabilities for surge predictions in coastal areas. Sci. Rep. **11** (2021). https://api.semanticscholar.org/CorpusID:237323782

17. Valentini, N., Balouin, Y.: Assessment of a smartphone-based camera system for coastal image segmentation and sargassum monitoring. J. Marine Sci. Eng. **8**(1) (2020). https://doi.org/10.3390/jmse8010023, https://www.mdpi.com/2077-1312/8/1/23

18. Wightman, R., Touvron, H., Jegou, H.: ResNet strikes back: an improved training procedure in timm. In: NeurIPS 2021 Workshop on ImageNet: Past, Present, and Future (2021). https://openreview.net/forum?id=NG6MJnVl6M5
19. Zhang, Z., Zhang, L., Yue, S., Wu, J., Guo, F.: Correction of nonstationary tidal prediction using deep-learning neural network models in tidal estuaries and rivers. J. Hydrol. **622**, 129686 (2023). https://doi.org/10.1016/j.jhydrol.2023.129686, https://www.sciencedirect.com/science/article/pii/S0022169423006285
20. Zhou, B., et al.: Semantic understanding of scenes through the ADE20K dataset. Int. J. Comput. Vision **127**(3), 302–321 (2018). https://doi.org/10.1007/s11263-018-1140-0

A Deep-Learning-Based Method for Real-Time Barcode Segmentation on Edge CPUs

Enrico Vezzali[1,2], Lorenzo Vorabbi[2], Costantino Grana[1],
and Federico Bolelli[1(✉)]

[1] University of Modena and Reggio Emilia, Modena, Italy
{enrico.vezzali,costantino.grana,federico.bolelli}@unimore.it
[2] Datalogic, S.p.A, Bologna, Italy
{enrico.vezzali,lorenzo.vorabbi}@datalogic.com

Abstract. Barcodes are a critical technology in industrial automation, logistics, and retail, enabling fast and reliable data capture. While deep learning has significantly improved barcode localization accuracy, most modern architectures remain too computationally demanding for real-time deployment on embedded systems without dedicated hardware acceleration. In this work, we present **BaFaLo** (**Ba**rcode **Fa**st **Lo**calizer), an ultra-lightweight segmentation-based neural network for barcode localization. Our model is specifically optimized for real-time performance on low-power CPUs while maintaining high localization accuracy for both 1D and 2D barcodes. It features a two-branch architecture—comprising a local feature extractor and a global context module—and is tailored for low-resolution inputs to improve inference speed further. We benchmark BaFaLo against several lightweight architectures for object detection or segmentation, including YOLO Nano, Fast-SCNN, BiSeNet V2, and ContextNet, using the BarBeR dataset. BaFaLo achieves the fastest inference time among all deep-learning models tested, operating at 57.62 ms per frame on a single CPU core of a Raspberry Pi 3B+. Despite its compact design, it achieves a decoding rate nearly equivalent to YOLO Nano for 1D barcodes and only 3.5% points lower for 2D barcodes while being approximately nine times faster.

Keywords: Barcodes · Embedded Systems · Object Detection

1 Introduction

Barcodes are a pivotal technology for automated data capture and identification, relying on simple visual codes to convey complex information cost-effectively. Their versatility has led to widespread adoption across numerous domains: from inventory tracking and logistics [8,19] to automated warehouse operations [7], manufacturing [19], retail product recognition [9], and even robot navigation [13]. Recognizing the vital role of barcodes, both academia and industry have increasingly focused on enhancing barcode localization techniques [20].

© The Author(s), under exclusive license to Springer Nature Switzerland AG 2026
M. Castrillón-Santana et al. (Eds.): CAIP 2025, LNCS 15621, pp. 186–196, 2026.
https://doi.org/10.1007/978-3-032-04968-1_16

The first attempts at barcode localization in 2D images, dating back to the 1990 s, relied on traditional computer vision techniques such as edge detection, Hough transforms, and texture direction analysis [5, 10, 18]. While computationally efficient, these approaches were sensitive to noise, lighting variations, and perspective distortions. The methods proposed by Sörös *et al.* [14] and Zamberletti *et al.* [23] improved the robustness of traditional techniques by incorporating structural analysis and learning-based refinement, but remained limited by resolution sensitivity and focused primarily on 1D barcodes.

The shift toward deep-learning-based methods began in the mid-2010 s with the application of convolutional neural networks (CNN) [1, 3]. Object detectors such as YOLO and Faster R-CNN were adapted for barcode localization, achieving significant improvements in accuracy and robustness. More recently, dedicated architectures for barcode segmentation have been proposed, such as the method introduced by Zharkov *et al.* [24], to achieve faster inference while remaining quite reliable. Recent work by Vezzali *et al.* [17] introduced the BarBeR benchmarking suite alongside a dataset of 8748 images of barcodes, revealing that deep-learning models significantly outperform hand-crafted methods. However, these models often incur high computational costs, limiting deployment on low-power hardware such as the Raspberry Pi CPU, where even compact networks such as YOLO V8 Nano [6] can exceed one second of inference time [17], making real-time processing impractical.

In this paper, we introduce **BaFaLo** (**Ba**rcode **Fa**st **Lo**calizer), an ultra-lightweight architecture for barcode localization and segmentation. By selectively reducing model complexity while preserving critical performance, BaFaLo achieves real-time inference on embedded CPUs for both 1D and 2D barcodes, with less than a 3.5% drop in decoding rate compared to more computationally demanding models like YOLO Nano. Furthermore, BaFaLo significantly outperforms other lightweight segmentation models, such as the one proposed by Zharkov *et al.* [24], as well as traditional hand-crafted approaches from Yun *et al.* and Sörös *et al.*, while maintaining superior speed. This advance opens the door to cost-efficient solutions for industrial, retail, and robotics applications requiring rapid barcode scanning in the field. To support reproducibility and further research, we release all source code for BaFaLo, along with the training and validation scripts and the trained models: https://github.com/Henvezz95/BarBeR.

2   Proposed Method

In this work, we propose a deep learning framework to localize 1D and 2D barcodes on edge devices in real time (processing time < 60 ms). In particular, we want to show that deep-learning-based localization of barcodes is possible even without accelerators and that an ARM CPU is enough for the task (in our case, a Raspberry PI 3B+). Rather than relying on traditional detection networks (e.g., YOLO), our method employs a segmentation-based architecture that offers several distinct advantages. First, segmentation networks can leverage local texture information and, therefore, function effectively with a smaller receptive field, whereas detection networks require having the entire object in view. Second, because segmentation operates on pixel-level classifications using convolutional blocks, the network more naturally adapts to different input resolutions and handles scale variations without relying on fixed-size anchor boxes. Finally, the segmentation output contains richer, pixel-wise information—such as orientation or region boundaries—potentially benefiting the downstream decoding phase, though we do not explore this in the current paper.

Our design takes inspiration from Fast-SCNN [12], an architecture known for its real-time performance on embedded devices. However, the original Fast-SCNN design remained too slow for the use case targeted in this paper. Our proposed approach is organized into four modules: (*i*) a *Learning to Downsample* module that quickly reduces spatial resolution while preserving low-level features; (*ii*) a *Coarse Feature Extraction* module that uses bottleneck residual blocks to capture a broader context. This component is a streamlined, lightweight adaptation of the Global Feature Extractor from Fast-SCNN; (*iii*) a *Feature Fusion* module to merge the high-resolution details from the downsample path with the global context extracted at lower resolution; and (*iv*) a lightweight *Classifier* that upsamples the fused features to produce the final segmentation map. This pipeline provides a strong balance of local detail and global context.

2.1   Balancing Low-Resolution and High-Resolution Features

To achieve the necessary speed, we reduced the number of layers and channels in the architecture. Vezzali *et al.* [17] demonstrated that even very low-resolution images can provide sufficient information to localize barcodes. For instance, YOLO Nano achieved a high mAP@50 (0.961) when localizing barcodes in the BarBeR dataset [15,16] using a 320×320 resolution. Inspired by these findings, we adopted a similar target resolution, which is much lower than the high-resolution Cityscapes dataset [2] originally used to train Fast-SCNN. Consequently, our first major modification was to significantly streamline the Global Feature Extraction module—reducing its nine linear bottleneck layers to three, lowering the output channels from (64, 64, 64, 96, 96, 96, 128, 128, 128) to (32, 32, 32), and decreasing the expansion ratio of each bottleneck from six to two. We also removed the Pyramid Pooling module at the end, as it offered no tangible benefit in our application. Since this branch now provides a more limited receptive field and captures coarser context, we refer to it as the *Coarse Feature Extractor* rather than a *Global Feature Extractor*. In contrast, the Learning to Downsample module plays a pivotal role in feeding subsequent layers with sufficiently low-level features; therefore, reducing it too aggressively would compromise overall network capacity. We preserved its three convolutional layers but lowered their channel counts from (32, 48, 64) to (12, 24, 32). Furthermore, we opted for regular 3×3 convolutions in the second and third layers rather than depthwise separable convolutions. Our tests showed that the latter would diminish the model capacity too much for a very modest speed improvement. As a result, this new module contains 9828 parameters—compared to 6192 in the original Fast-SCNN—yet still operates more efficiently under our real-time constraints.

2.2   Pixel Shuffle

In the original Fast-SCNN design, the classifier produces a segmentation map at $\frac{1}{8}$ of the input resolution and then uses linear upscaling to match the input's dimensions. However, working with low-resolution images makes it difficult to preserve the fine details of small objects at such a low output resolution. To address this, we replace the linear upscaling step with a pointwise convolution followed by a pixel shuffle operation. First, the pointwise convolution expands the feature maps from 32 to 128 channels. Then, the pixel shuffle layer rearranges these channels spatially by a factor of eight, restoring the original input resolution. This process also reduces the final channel count to two—one channel for 1D barcodes and another for 2D barcodes—thereby enabling clear segmentation of both barcode types in a low-resolution setting.

2.3   Training

Figure 1 illustrates our final proposed architecture. We trained this network on the BarBeR dataset for 300 epochs with a batch size of 16. Each image was resized during preprocessing so that its longest side measured 320 pixels while preserving the original aspect ratio. We then zero-padded the shorter dimension until the input reached 320×320. The Adam optimizer was used with an initial learning rate of 1e-3, which we decayed exponentially to 1e-5 by the final epoch. To increase robustness, we applied random scaling (from 0.6× to 1.4×), random flips (horizontal and vertical), and random adjustments to saturation and contrast. We used a pixel-wise binary cross-entropy as a loss function without adding any additional auxiliary loss.

2.4   Inference

During inference, the proposed neural network processes the input image and generates a two-channel heatmap of the same dimensions as the input. The first channel corresponds to 1D barcode regions, while the second corresponds to 2D barcode regions. Although the network produces pixel-level outputs and can be used for segmentation, in this work, we employ it exclusively for detection. To convert the heatmap into detection boxes, each channel is thresholded independently (we use a fixed threshold of 0.4 in all experiments). We then apply blob detection to the binarized map, and a bounding box is computed for each identified blob.

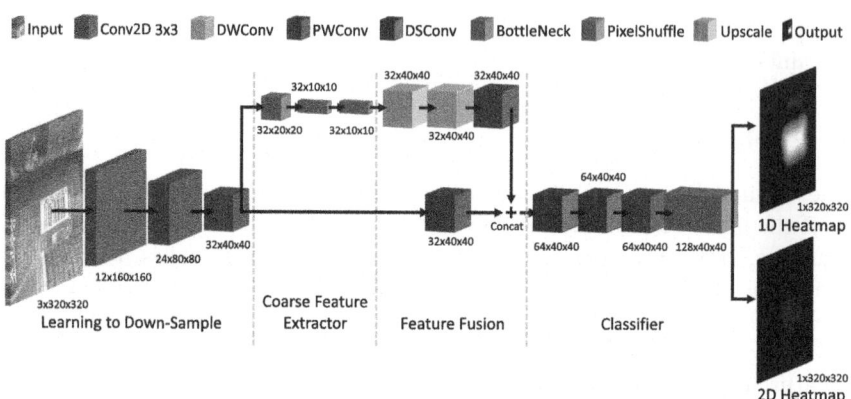

Fig. 1. Architecture of our proposed model. DWConv stands for depthwise convolution, PWConv for pointwise convolution, and DSConv for depthwise separable convolution. We indicate the size of the output tensor for each block, considering an input RGB image of size 320×320.

3   Results

3.1   Experimental Setup

All experiments were conducted on the BarBeR dataset,[1] and the full evaluation pipeline is available on our GitHub repository.[2] We used 5-fold cross-validation to ensure robust evaluation. The dataset was split into five equal subsets; each subset was used once as the test set, with the remaining four used for training. All images were resized so that the longest side measured 320 pixels before being processed in any test.

Localization Accuracy. We evaluated all deep-learning-based localization models, including ours, on the complete dataset of 8748 images. For each method, we computed the average precision scores, AP@0.5 and AP@[0.5:0.95], separately for 1D and 2D barcodes. For segmentation models, region-level confidence scores were computed as the mean of the predicted pixel confidences within each detected region. We also included three traditional methods tailored to 1D barcode localization (Sörös *et al.* [14], Yun *et al.* [22], and Zamberletti *et al.* [23]), which are part of the BarBeR benchmark. Since these methods only support 1D barcodes, we excluded 2D barcodes from this evaluation. Moreover, as the method by Sörös *et al.* detects only one barcode per image, we restricted this evaluation to the 6811 images containing a single 1D barcode.

Time Benchmark. For each localization method, we measured the inference time per image. Each image was processed three times, and the minimum inference time was recorded to minimize the impact of background processes. The

---

[1] https://ditto.ing.unimore.it/barber.
[2] https://github.com/Henvezz95/BarBeR.

Table 1. Performance Comparison of Deep-Learning-Based Barcode Localization Methods. The table reports average precision (AP@0.5 and AP@[0.5:0.95]), decoding rate (% Dec), and single-threaded inference time on two platforms: a high-end PC and a Raspberry Pi 3B. Results are provided separately for 1D and 2D barcodes. All input images were resized such that the longest side measures 320 pixels.

Model	1D barcodes			2D barcodes			Times	
	AP@0.5 ↑	AP@[.5:.95] ↑	% Dec ↑	AP@0.5 ↑	AP@[.5:.95] ↑	% Dec ↑	PC (ms) ↓	Rasp Pi (ms) ↓
YOLO Nano	**0.976**	**0.860**	**59.3%**	**0.947**	**0.872**	**63,5%**	21.55	509.1
Zharkov *et al.*	0.530	0.254	50.9%	0.571	0.382	55.9%	4.521	180.4
ContextNet 0.25x	0.808	0.592	55.5%	0.413	0.295	33.0%	5.359	209.9
BisenetV2 0.25x	0.909	0.725	58.8%	0.834	0.657	60.9%	5.877	193.4
BisenetV2 0.125x	0.880	0.673	58.3%	0.681	0.479	49.2%	2.943	84.73
Fast SCNN 0.5x	0.863	0.629	57.9%	0.702	0.486	53.7%	3.313	120.0
Fast SCNN 0.25x	0.783	0.523	55.7%	0.534	0.354	42.1%	2.235	64.56
Ours	0.898	0.694	58.5%	0.822	0.628	60.0 %	**1.635**	**57.62**

final timing for each method was computed as the average over all images. We conducted benchmarks on both a Raspberry Pi 3B+ (representing low-power embedded devices) and a high-end workstation equipped with an AMD Ryzen Threadripper Pro 5965WX CPU (24 cores) and 128 GB of DDR4 RAM. All deep-learning architectures have been converted to ONNX to accurately measure their maximum speed in possible real-world applications.

Decoding Test. To assess practical usability, we measured the decoding rate using a virtual barcode reader composed of two stages: localization and decoding. In the first stage, the tested localization method outputs bounding boxes. Each box defines a region, which is then passed to the `pyzbar` library for decoding. Before cropping, each box is extended by 20 pixels on each side since barcode readers usually require a quiet zone around the code. Since `pyzbar` supports only a subset of barcode formats (QR Code, EAN-13, EAN-8, UPC-E, Interleaved 2 of 5, Code 39, and Code 128), we restricted this test to those types. The decoding rate is defined as the proportion of images in which at least one barcode is successfully decoded. If decoding fails at the base resolution (320 pixels on the longest side), we attempt decoding again on a 4× larger version of the image (longest side = 1280 pixels) to accommodate barcodes that cover a small area of the overall image.

3.2 Experimental Results

We first evaluated our model against several deep-learning-based localization methods. As baselines, we included YOLO Nano and the architecture proposed by Zharkov *et al.*, both of which are integrated into the BarBeR benchmark. Additionally, we selected three real-time segmentation networks commonly used in lightweight vision tasks: ContextNet [11], BiSeNet V2 [21], and Fast-SCNN [12].

Table 2. Comparison with traditional methods for 1D barcode localization. Each model is evaluated using the Precision (P), Recall (R), and F1 score at an IoU threshold of 0.5, decoding rate (% Dec), and single-threaded inference time on a PC and a Raspberry Pi 3B. All input images were resized such that the longest side measures 320 pixels.

Model	1D barcodes				Times	
	P@0.5 ↑	R@0.5 ↑	F1@0.5 ↑	% Dec ↑	PC (ms) ↓	Rasp Pi (ms) ↓
Sörös *et al.*	0.429	0.429	0.429	51.4%	2.782	92.07
Yun *et al.*	0.918	0.436	0.591	39.9%	2.171	**52.83**
Zamberletti *et al.*	0.103	0.125	0.113	11.0%	17.42	855.7
Ours	**0.976**	**0.947**	**0.962**	**66.0%**	**1.635**	57.62

To ensure a fair comparison on embedded hardware, we tested reduced versions of these models using uniform channel-width scaling. This technique, commonly known as a width multiplier [4], uniformly reduces the number of filters in each convolutional layer by a fixed ratio. For example, a 0.25× configuration reduces all channels to 25% of their original count, significantly lowering computational complexity. Based on this, we evaluated different variants: ContextNet (0.25×), Fast-SCNN (0.5× and 0.25×), and BiSeNet V2 (0.25× and 0.125×).

The performance of all deep-learning models is reported in Table 1, in terms of AP@0.5, AP@[0.5:0.95], processing times on both PC and Raspberry Pi (single-threaded), and decoding rate using `pyzbar`. Our method achieves the fastest inference time across all platforms, requiring only 57.62 ms per image on a single CPU core, making it capable of near real-time operation even on embedded devices. Despite its simplicity, our method ranks third among all models in both AP@0.5 and AP@[0.5:0.95] and in decoding rate.

The second-best decoding rate is achieved by BiSeNet V2 (0.25×), which outperforms our model by just 0.3% points for 1D barcodes (58.8% vs. 58.5%) and 0.9 for 2D barcodes (60.9% vs. 60.0%), while being approximately 3.5 times slower. YOLO Nano achieves the highest accuracy overall, as expected, but its inference time of 509.1 ms per image makes it impractical for many real-time embedded applications. Notably, despite being over 9 times slower than our model, it achieves only a marginal improvement of 0.8% points in 1D barcode decoding and 3.5 points in 2D barcode decoding.

Our method also outperforms several slower models, including Zharkov *et al.*, ContextNet (0.25×), Fast-SCNN (0.5× and 0.25×), and BiSeNet V2 (0.125×), making it the only approach to achieve such high accuracy at this speed. This result underscores that our architectural choices offer a substantial advantage over Fast-SCNN, from which we drew inspiration.

Table 2 compares our method to three classical, hand-crafted approaches: Sörös *et al.*, Yun *et al.*, and Zamberletti *et al.*. Our method is the fastest on PC and the second-fastest on Raspberry Pi 3B+, with only a slight difference from the fastest traditional model (Yun *et al.*). However, it achieves significantly higher Precision, Recall, F1 scores, and decoding rates. Traditional methods

Table 3. Pipeline timing breakdown for barcode detection, comparing localization plus decoding (*Loc + Decoding*) against running `pyzbar` alone (*No Localization*). The table lists per-image processing times on a PC and a Raspberry Pi 3B, with all input images resized so their longest side is 320 pixels. Total time is the sum of localization and decoding.

Model	Times PC (ms)			Times Raspberry Pi (ms)		
	Loc ↓	Decoding ↓	Total ↓	Loc ↓	Decoding ↓	Total ↓
No Localization	-	35.944	35.944	-	299.22	299.22
YOLO Nano	21.558	5.369	27.197	509.16	31.17	540.33
Zharkov	4.521	13.056	17.577	180.41	66.29	246.70
ContextNet 0.25x	5.359	12.223	17.582	209.93	28,02	237.95
BisenetV2 0.25x	5.877	10.685	16.562	193.41	30.54	223.95
BisenetV2 0.125x	2.943	9.118	12.061	84.73	**29.80**	114.53
Fast SCNN 0.5x	3.313	**4.935**	8.248	120.00	30,09	150.09
Fast SCNN 0.25x	2.235	5.190	7.425	64.56	40.39	104.95
Ours	**1.635**	5.431	**7.066**	**57.62**	37.54	**95.16**

are highly sensitive to resolution and require higher pixel density for accurate decoding [17]. Operating at a higher resolution could improve the results a bit but would make these methods around 4× times slower. Additionally, these methods are limited to 1D barcode detection, whereas our model supports both 1D and 2D barcode localization.

3.3 Decoding Time

In addition to measuring localization speed in isolation, we also evaluated the time required to run the entire detection pipeline, including both localization and decoding. This serves two main purposes. First, it reveals how effectively each model limits false positives and unnecessarily large bounding boxes. A model with overly generous detections may achieve high recall but will suffer increased decoding overhead, as the `pyzbar` library would process more (or larger) crops. Second, it clarifies how much overall latency is reduced by speeding up the localization step.

As shown in Table 3, running `pyzbar` directly on the entire image (*No Localization*) at a maximum scale of 1280 pixels for the longest edge, takes 35.94 ms on PC and 299.22 ms on Raspberry Pi. With our localization approach, the total pipeline time drops to 7.07 ms on PC and 95.16 ms on the Pi—the fastest among the methods tested on both platforms. For reference, YOLO Nano requires 27.20 ms on PC and 540.33 ms on the Raspberry Pi, highlighting that a higher-accuracy model can become bottlenecked by slow inference on edge devices. Notably, our method does not inflate decoding times with excess bounding boxes, indicating few false positives.

Table 4. Ablation study on key architectural components of our model. We report AP@0.5, AP@[0.5:0.95], decoding rate (% Dec), and single-threaded inference time on PC and Raspberry Pi 3B. All images were resized to have a longest side of 320 pixels. Variants include the removal of pixel shuffle and the re-introduction of the pyramid pooling module.

Model	1D barcodes			2D barcodes			Times	
	AP@0.5 ↑	AP@[.5:.95] ↑	% Dec ↑	AP@0.5 ↑	AP@[.5:.95] ↑	% Dec ↑	PC (ms) ↓	Rasp PI (ms) ↓
With PPM	0.897	0.691	**58.8%**	0.783	0.568	56.2%	2.518	74.87
No PixelShuffle	0.893	0.681	58.5%	0.817	0.622	59.6%	1.829	57.93
Ours	**0.898**	**0.694**	58.5%	**0.822**	**0.628**	**60.0 %**	**1.635**	**57.62**

Finally, although 95.16 ms per image on the Pi is not yet real-time on a single CPU core, we anticipate that basic parallelization or multithreading could push the pipeline into real-time territory. Anyway, for most applications, a reading rate of 10 FPS is enough. On PC, the decoding phase appears to be more time-consuming relative to localization, suggesting that further optimizations should focus more on the decoding component than the localizer.

3.4   Ablation Studies

We evaluated the impact of two architectural modifications that differentiate our model from the original Fast-SCNN design, beyond the already reduced depth and channel width. Specifically, we assess (i) the use of a pixel shuffle module for upsampling and (ii) the removal of the pyramid pooling module (PPM) from the global feature extractor. Results are shown in Table 4. All tests were conducted at a resolution where the longest image side was set to 320 pixels.

Our full configuration achieves the best overall performance, with the fastest inference time (57.62 ms on Raspberry Pi 3B+) and the highest accuracy across most metrics. The version without pixel shuffle shows slightly lower performance on both AP and decoding rate, particularly on 2D barcodes (60.0% vs. 59.6%), suggesting that pixel shuffle contributes positively to precise localization.

Reintroducing the PPM results in a significant slowdown (74.8 ms vs. 57.6 ms) and mixed accuracy changes: a small improvement in 1D decoding rate (58.8% vs. 58.5%), but noticeably worse performance on 2D barcodes, with lower AP scores (0.783 vs. 0.822) and decoding rate (56.2% vs. 60.0%). These results confirm that removing the PPM and adopting pixel shuffle leads to a better overall balance between speed and performance in our final model.

4   Conclusion and Future Research

In this paper, we introduced **BaFaLo**, an ultra-lightweight neural network architecture for barcode localization and segmentation. Designed for real-time performance on embedded CPUs, BaFaLo achieves real-time speed while resulting in a similar decoding rate to much slower architectures. Unlike many deep-learning models that require powerful GPUs or accelerators, BaFaLo can localize both 1D and 2D barcodes in real time on a Raspberry Pi 3B+, requiring only 57.62 ms per image on a single CPU core.

Looking ahead, now that a high-speed and accurate localization framework is available, future research should explore the optimization of the decoding step. While `pyzbar` provides broad format compatibility, it is not optimized for speed and can become a bottleneck in real-time systems. Developing a fast, lightweight decoder—potentially using deep-learning techniques—could significantly improve end-to-end performance. Moreover, additional tests could evaluate various multithreading strategies that leverage multiple cores on embedded CPUs, and further speed-ups may be achieved through model quantization (e.g., 8-bit or lower).

In parallel, more extensive testing on video streams would offer insights into real-world deployment scenarios. In such contexts, fast but slightly less accurate methods might still yield better results, as higher frame rates provide more decoding opportunities over time. Understanding the trade-off between speed and temporal redundancy will be crucial for robust barcode reading in industrial and retail applications.

Finally, because our network excels at pixel-level texture segmentation, BaFaLo can potentially address other challenges that hinge on subtle textural cues, ranging from surface defect detection in manufacturing to AR marker recognition or fast pattern analysis for robots and drones. This opens the door to a broad spectrum of applications beyond barcode localization.

Acknowledgments. This work was supported by the University of Modena and Reggio Emilia and Fondazione di Modena through the "Fondo di Ateneo per la Ricerca - FAR 2024" (CUP E93C24002080007).

Disclosure of Interests. The authors have no conflicts of interest to declare.

References

1. Chou, T.H., Ho, C.S., Kuo, Y.F.: QR code detection using convolutional neural networks. In: International Conference on Advanced Robotics and Intelligent Systems (ARIS) (2015)
2. Cordts, M., et al.: The cityscapes dataset for semantic urban scene understanding. In: Computer Vision and Pattern Recognition (2016)
3. Hansen, D.K., Nasrollahi, K., Rasmussen, C.B., Moeslund, T.B.: Real-time barcode detection and classification using deep learning. In: International Joint Conference on Computational Intelligence (2017)
4. Howard, A.G., et al.: MobileNets: efficient convolutional neural networks for mobile vision applications. arXiv preprint arXiv:1704.04861 (2017)
5. Hu, H., Xu, W., Huang, Q.: A 2D barcode extraction method based on texture direction analysis. In: International Conference on Image and Graphics (2009)
6. Jocher, G., Chaurasia, A., Qiu, J.: Ultralytics YOLOv8 (2023)
7. Kubáňová, J., Kubasáková, I., Čulík, K., Štítik, L.: Implementation of barcode technology to logistics processes of a company. Sustainability **14**(2) (2022)

8. McCathie, L.: The advantages and disadvantages of barcodes and radio frequency identification in supply chain management. PhD Thesis, School of Information Technology and Computer Science (2004)
9. Melek, C.G., Battini Sönmez, E., Varli, S.: Datasets and methods of product recognition on grocery shelf images using computer vision and machine learning approaches: an exhaustive literature review. Eng. Appl. Artif. Intell. **133** (2024)
10. Muniz, R., Junco, L., Otero, A.: A robust software barcode reader using the Hough transform. In: International Conference on Information Intelligence and Systems (1999)
11. Poudel, R.P., Bonde, U., Liwicki, S., Zach, C.: Contextnet: exploring context and detail for semantic segmentation in real-time. arXiv preprint arXiv:1805.04554 (2018)
12. Poudel, R.P., Liwicki, S., Cipolla, R.: Fast-SCNN: fast semantic segmentation network. arXiv preprint arXiv:1902.04502 (2019)
13. Soliman, A., et al.: AI-based UAV navigation framework with digital twin technology for mobile target visitation. Eng. Appl. Artif. Intell. **123** (2023)
14. Sörös, G., Flörkemeier, C.: Blur-resistant joint 1D and 2D barcode localization for smartphones. In: International Conference on Mobile and Ubiquitous Multimedia (2013)
15. Vezzali, E., Bolelli, F., Santi, S., Grana, C.: Barber: A barcode benchmarking repository. In: International Conference on Pattern Recognition, Springer (2025)
16. Vezzali, E., Bolelli, F., Santi, S., Grana, C.: BarBeR-barcode benchmark repository: implementation and reproducibility notes. In: International Workshop on Reproducible Research in Pattern Recognition (2025)
17. Vezzali, E., Bolelli, F., Santi, S., Grana, C.: State-of-the-art review and benchmarking of barcode localization methods. Eng. Appl. Artif. Intell. (2025)
18. Viard-Gaudin, C., Normand, N., Barba, D.: A bar code location algorithm using a two-dimensional approach. In: International Conference on Document Analysis and Recognition (1993)
19. Weng, D., Yang, L.: Design and implementation of barcode management information system. In: Zhu, R., Ma, Y. (eds.) Information Engineering and Applications. LNEE, vol. 154, pp. 1200–1207. Springer, London (2012). https://doi.org/10.1007/978-1-4471-2386-6_158
20. Wudhikarn, R., Charoenkwan, P., Malang, K.: Deep learning in barcode recognition: a systematic literature review. IEEE Access **10** (2022)
21. Yu, C., Gao, C., Wang, J., Yu, G., Shen, C., Sang, N.: BiSeNet V2: bilateral network with guided aggregation for real-time semantic segmentation. Int. J. Comput. Vision **129**(11), 3051–3068 (2021). https://doi.org/10.1007/s11263-021-01515-2
22. Yun, I., Kim, J.: Vision-based 1D barcode localization method for scale and rotation invariant. In: TENCON - IEEE Region 10 Conference (2017)
23. Zamberletti, A., Gallo, I., Albertini, S.: Robust angle invariant 1D barcode detection. In: 2013 2nd IAPR Asian Conference on Pattern Recognition (2013)
24. Zharkov, A., Zagaynov, I.: Universal barcode detector via semantic segmentation. In: International Conference on Document Analysis and Recognition (2019)

Edge-Aware Camouflaged Object Detection

Patricia L. Suárez[1] and Angel D. Sappa[1,2]

[1] ESPOL Polytechnic University, Guayaquil, Ecuador
{plsuarez,asappa}@espol.edu.ec
[2] Computer Vision Center, 08193 Bellaterra, Barcelona, Spain
asappa@cvc.uab.es

Abstract. This paper presents a novel edge-aware transformer-based framework for camouflaged object detection. By integrating a spatial attention module guided by structural cues extracted from edge information, the model is directed toward visually ambiguous regions that commonly hinder segmentation performance. This design enables more effective global context modeling and improves the delineation of camouflaged object boundaries. Extensive experiments on multiple benchmark datasets validate the effectiveness of the proposed approach, demonstrating consistent performance gains over state-of-the-art methods in all key evaluation metrics.

Keywords: Edge · Attention mechanisms · Camouflaged Objects · Pyramid Vision Transformer

1 Introduction

Camouflaged object detection (COD) poses a significant challenge in computer vision due to the minimal appearance differences between foreground objects and their backgrounds. In such settings, traditional detection strategies often struggle to identify target regions, particularly when objects are intentionally embedded within complex visual contexts. Addressing this problem requires models capable of capturing both high-level semantic features and low-level structural cues.

Recent advances in vision transformers have demonstrated strong potential for dense prediction tasks, thanks to their ability to model long-range dependencies and contextual relationships across the image [6]. Additionally, attention mechanisms enhance feature selectivity by enabling the model to prioritize task-relevant regions [26], making these components particularly well-suited for COD, where visual signals are often sparse and ambiguous.

In scenarios where texture boundaries are weak or indistinct, global semantic features alone may not suffice. Structural information, such as edges, plays a critical role in enhancing the model's sensitivity to object contours. Prior work highlights the benefit of integrating such information into the learning process to improve boundary precision [22].

© The Author(s), under exclusive license to Springer Nature Switzerland AG 2026
M. Castrillón-Santana et al. (Eds.): CAIP 2025, LNCS 15621, pp. 197–208, 2026.
https://doi.org/10.1007/978-3-032-04968-1_17

This work proposes a transformer-based framework for camouflaged object detection, enhanced with a spatial attention mechanism that integrates structural cues from edge information. Building upon the OAFormer backbone, the method introduces key innovations such as multistage edge injection, edge-guided attention, and an edge-aware loss function, which collectively improve boundary sensitivity and detection accuracy in challenging scenarios.

The method is evaluated on four public COD benchmarks, showing consistent improvements in detection quality. These results demonstrate the effectiveness of combining edge-guided attention and structural supervision within a transformer-based architecture. The main contributions of this paper are:

- A transformer-based architecture is enhanced with a spatial attention mechanism guided by edge-derived structural cues.
- An edge-aware loss function is proposed to improve the consistency between predicted masks and the underlying structural layout of the image.

The remainder of this paper is structured as follows, Sect. 2 reviews relevant literature. Section 3 presents the proposed methodology, including the architecture and fusion mechanism. Section 4 discusses results and evaluations, and Sect. 5 concludes the paper with future directions.

2   Related Work

Camouflaged object detection has been extensively studied, and numerous approaches have been proposed to address the challenges posed by the high similarity between foreground objects and background environments. Unlike standard object detection, COD has required models to capture both global context, to differentiate between subtle environmental variations, and local edge cues, to delineate objects from their surroundings. Early COD methods have been based primarily on convolutional neural networks [5], but recent advances have shown that transformer-based architectures can achieve superior performance by exploiting self-attention to model long-range dependencies across the image (e.g., [30,35]).

On the other hand, to enhance the capability of camouflaged object detection in visually ambiguous scenes, structure-aware methods have been developed to enhance spatial discrimination and improve the localization of subtle visual cues (e.g., [3,13]). To further advance contextual understanding, hierarchical graph interaction transformers have been introduced. These models leverage dynamic token clustering to strengthen the detection process and have demonstrated increased robustness in complex environments [33]. In parallel, adaptive guidance learning has been proposed to tailor the detection pipeline to the specific characteristics of camouflaged targets, offering adaptive supervision that improves segmentation accuracy [4]. Additionally, zero-shot learning techniques have been explored to reduce reliance on large-scale annotated datasets, enhancing the scalability and applicability of camouflaged object detection systems [11].

The field has additionally evolved to tackle complex scenarios involving multi-object camouflage and overlapping instances. Multi-object detection pipelines have been proposed that group relevant tokens and refine them using edge-based representations [27]. Collaborative models such as CoCOD [14] have been designed to utilize multi-image contexts for improved inference in challenging environments. These efforts have addressed issues related to boundary ambiguity, occlusion, and limited data availability. Furthermore, minimal-supervision and zero-shot approaches have also been explored, such as ZeroScope [12], which has demonstrated effective detection in data-scarce conditions by leveraging generalized feature representations.

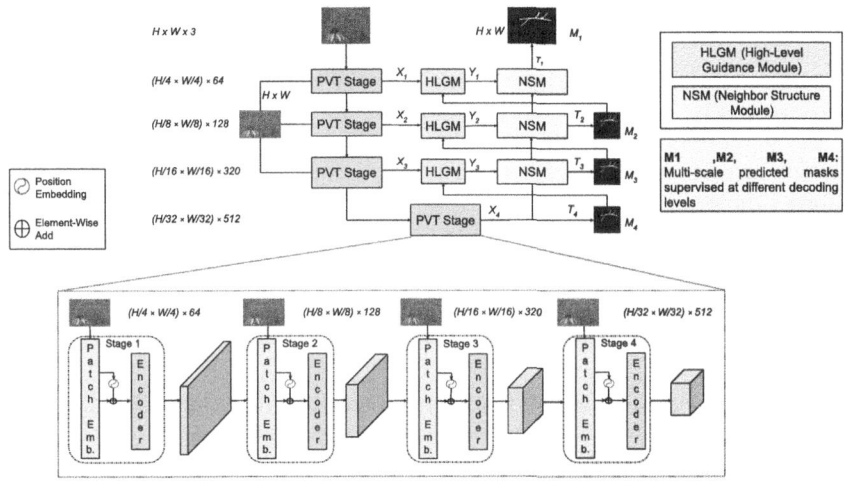

Fig. 1. Overview of the proposed architecture. A four-stage PVT backbone extracts multi-scale features, refined by HLGM and NSM modules to produce predictions at different levels.

Beyond architectural advances, attention mechanisms have been widely adopted in image segmentation to enable models to concentrate on the most relevant features [2], while edge detection has served to improve the localization of object boundaries [22].

3   Proposed Approach

An edge-aware transformer design is introduced to enhance the model's ability to capture spatial structure. This component builds upon the baseline architecture by integrating structural edge information to guide attention in visually ambiguous regions. This design improves the model's ability to capture fine-grained variations between camouflaged objects and their surroundings, thereby enabling more accurate and precise segmentation.

3.1   Edge-Based Attention Mechanism

This paper proposes to integrate edge detection into the model to supply structural information about object edges. The Canny edge detector [1] has been applied to the RGB input images to extract edge maps, which are subsequently normalized and used to implement the attention map. This configuration has allowed the network to better localize object contours that are otherwise difficult to distinguish in camouflaged environments. The attention map A has been computed through a convolutional operation over the edge map E followed by a sigmoid activation:

$$A = \sigma\left(\mathrm{Conv}(E)\right), \tag{1}$$

where $\sigma(\cdot)$ denotes the sigmoid function. The feature map X from the RGB image has been modulated as follows:

$$X_{\mathrm{att}} = X \odot (1 + A), \tag{2}$$

where $\odot$ denotes element-wise multiplication. The term X_{att} represents the spatially enhanced feature map where structural information from the edge map A reinforces the original feature X.

3.2   Model Architecture

The proposed approach extends the OAFormer framework [32], originally developed for camouflaged object detection, by incorporating an edge-guided attention mechanism into its architecture (see Fig. 1). In this design, the PVT backbone is composed of four hierarchical stages, each consisting of a patch embedding layer and a transformer encoder. At each stage, edge-salience maps are injected before the encoder to enrich spatial encoding. The outputs are progressively refined through HLGM and NSM modules. This guidance is integrated into the transformer encoder alongside the RGB input, allowing the model to capture both contextual semantics and spatial patterns. The resulting features are subsequently refined through convolutional layers to produce the final segmentation output.

3.3   Loss Function

The original OAFormer loss [32] is adopted and extended to improve structural accuracy. It combines three components: weighted binary cross-entropy $\mathcal{L}_{\mathrm{bce}}^{\omega}$, weighted IoU $\mathcal{L}_{\mathrm{iou}}^{\omega}$ [29], and uncertainty-aware loss $\mathcal{L}_{\mathrm{ual}}$ [8]. These are computed at four decoder stages M_i, with outputs resized to the input resolution. The OAFormer base loss is defined as:

$$\mathcal{L}_{\mathrm{OAFormer}} = \sum_{i=1}^{4} \frac{1}{2^{i-1}} \left(\mathcal{L}_{\mathrm{bce}}^{\omega}(M_i, G) + \mathcal{L}_{\mathrm{iou}}^{\omega}(M_i, G) + \mathcal{L}_{\mathrm{ual}}(M_i)\right), \tag{3}$$

where G denotes the ground-truth mask, and M_i represents the predicted mask produced at the i-th decoding stage. To enhance contour alignment, an additional edge loss is introduced:

$$\mathcal{L}_{\text{edge}} = \frac{1}{N} \sum_{i,j} |\nabla PM_{ij} - E_{ij}|, \tag{4}$$

where PM is the final predicted mask, ∇PM its gradient, E edge map, and N the number of pixels.

The final training objective is formulated as:

$$\mathcal{L}_{\text{final}} = \mathcal{L}_{\text{OAFormer}} + \lambda_{\text{edge}} \cdot \mathcal{L}_{\text{edge}}, \tag{5}$$

the weight $\lambda_{\text{edge}} = 0.3$ has been empirically selected to balance the impact of each component, enhancing edge quality without disrupting the original training dynamics of OAFormer.

4 Experimental Results

This section presents the experimental results obtained with the proposed architecture. The model's object detection performance is evaluated and compared with state-of-the-art approaches.

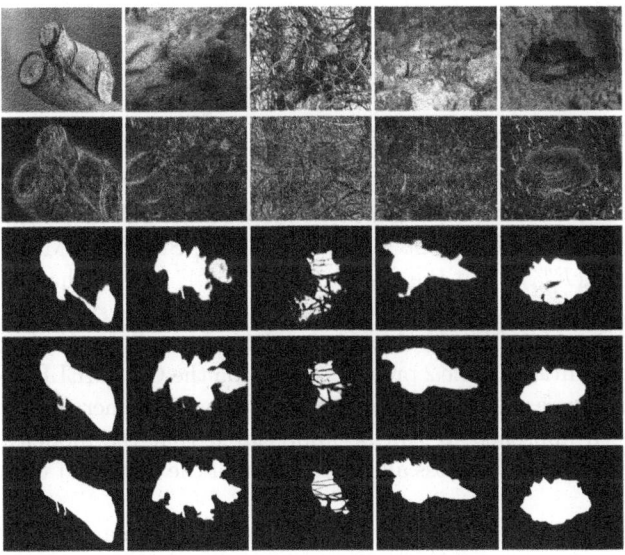

Fig. 2. Results from COD10K dataset: (*1st. row*) Input RGB images; (*2nd. row*) Edge salience images; (*3rd. row*) Camouflaged masks detected by [32]; (*4th. row*) Camouflaged masks from the proposed approach; (*5th. row*) Ground truth masks.

4.1   Datasets

The model has been trained and evaluated using four benchmark datasets for camouflaged object detection: COD10K [8], CAMO [10], CHAMELEON [21], and NC4K [16]. Images from COD10K and CAMO datasets have been jointly considered for training and validation, while CHAMELEON and NC4K were used just for evaluating the generalization capability of the proposed framework. COD10K contains 10,000 high-resolution images, of which 6,066 have been used for training and validation; and 2,026 images have been used for testing. CAMO comprises 1,250 images, with 1,000 used for training and validation; and 250 images have been used for testing. In other words, the model has been trained and validated with 7,066 images and tested with 2,276 images. On the other hand, CHAMELEON includes 76 challenging images and NC4K consists of 4,121 images, collected from various sources. The model was trained and evaluated using two NVIDIA A100 GPUs, each with 24 GB of VRAM.

4.2   Pre-proccesing

To ensure consistency in preprocessing across all datasets, each image and its corresponding ground-truth mask has been resized to a fixed resolution of 416 × 416 pixels before being input into the model. To enhance robustness and generalization during training, several data augmentation techniques have been applied. These have included random horizontal flipping, random cropping, and resizing, while color jittering has been introduced in selected experiments to simulate lighting variations.

4.3   Results and Comparisons

Results from the proposed approach have been compared with respect to several state-of-the-art models using four widely adopted metrics: Mean Absolute Error (MAE, $\downarrow$), Enhanced-alignment Measure (E_m, $\uparrow$), Structure Measure (S_m, $\uparrow$), and F-measure (F_β^w, $\uparrow$). In the results table, "$\uparrow$"/"$\downarrow$" means that larger/smaller is better. As shown in Table 1, the approach demonstrates superior performance on the CAMO and COD10K datasets, which are used for both training and validation. In addition, Table 2 presents results on the CHAMELEON and NC4K datasets, which were used exclusively for evaluating the generalization capability of the proposed approach. It can be appreciated that the model also achieves consistent improvements across all metrics—in all the metrics it obtains the best result.

To contextualize the evaluated models, Table 3 summarizes the publication venue, year, and number of parameters for each approach considered in the comparison. This information provides a reference for understanding differences in model complexity and development timelines.

Complementing the quantitative results, qualitative comparisons are illustrated in Figs. 2, 3, 4 and 5. The masks generated by the proposed model exhibit

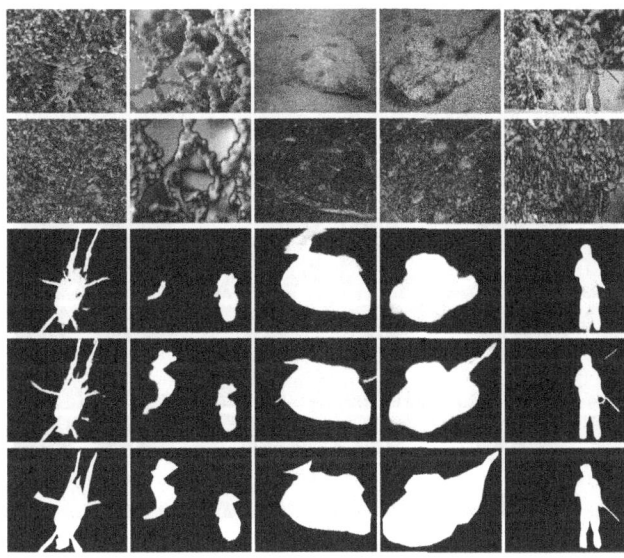

Fig. 3. Results from CAMO dataset: (*1st. row*) Input RGB images; (*2nd. row*) Edge salience images; (*3rd. row*) Camouflaged mask detected by [32]; (*4th. row*) Camouflaged mask from the proposed approach; (*5th. row*) Ground truth masks.

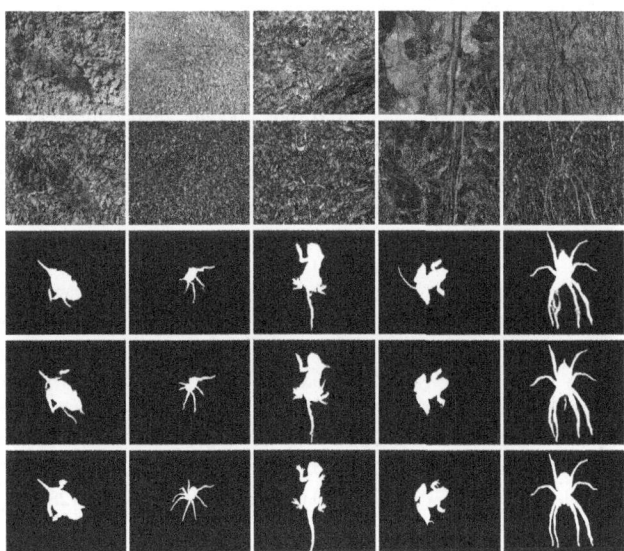

Fig. 4. Results from CHAMELEON dataset: (*1st. row*) Input RGB images; (*2nd. row*) Edge salience images; (*3rd. row*) Camouflaged masks detected by [32]; (*4th. row*) Camouflaged masks from proposed approach; (*5th. row*) Ground truth masks.

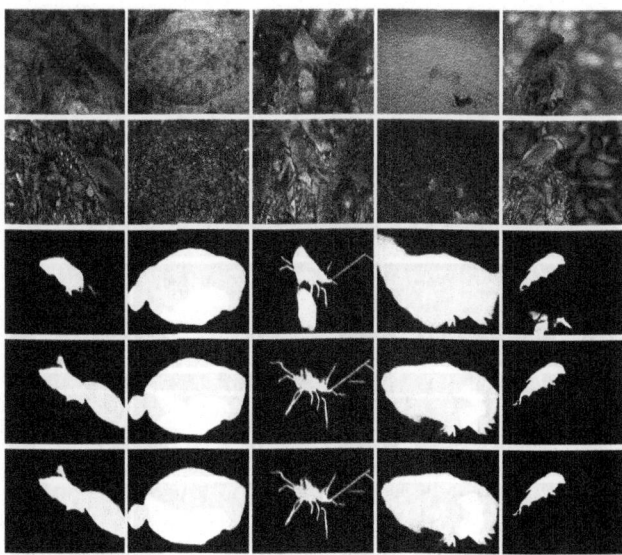

Fig. 5. Results from NC4K dataset: (*1st. row*) Input RGB images; (*2nd. row*) Edge salience images; (*3rd. row*) Camouflaged masks detected by [32]; (*4th. row*) Camouflaged masks from the proposed approach; (*5th. row*) Ground truth masks.

Table 1. Comparison of the proposed method with state-of-the-art methods on CAMO and COD10K datasets. "↑"/"↓" means that larger/smaller is better (best values in bold black; second-best values in bold blue).

Method	CAMO [10]				COD10K [8]			
	$S_m \uparrow$	$E_m \uparrow$	$F_\beta^w \uparrow$	$MAE \downarrow$	$S_m \uparrow$	$E_m \uparrow$	$F_\beta^w \uparrow$	$MAE \downarrow$
EGNet [36]	0.732	0.7996	0.6036	0.1095	0.7365	0.8097	0.5174	0.0607
PraNet [9]	0.7692	0.8245	0.6625	0.0942	0.7894	0.8606	0.6294	0.0451
F2Net [28]	0.7113	0.7407	0.5636	0.1087	0.7386	0.7951	0.5438	0.0513
MINet [19]	0.748	0.7911	0.637	0.0903	0.7697	0.832	0.6085	0.0417
SINet [8]	0.7454	0.8035	0.6443	0.0915	0.7794	0.8642	0.631	0.0426
C2FNet [24]	0.7961	0.8537	0.7187	0.0799	0.813	0.8902	0.6862	0.036
PFNet [17]	0.7823	0.841	0.6952	0.0849	0.7998	0.8772	0.6599	0.0396
MGL [34]	0.7755	0.816	0.6728	0.0884	0.8139	0.8513	0.66566	0.0353
UGTR [31]	0.7839	0.8215	0.6836	0.0863	0.8171	0.8519	0.6656	0.0356
SINet [7]	0.8201	0.8817	0.7426	0.0705	0.8151	0.887	0.6796	0.0368
BSANet [20]	0.7943	0.8511	0.7174	0.0786	0.8176	0.8905	0.699	0.0342
OCENet [15]	0.8019	0.8518	0.7234	0.0804	0.8272	0.8935	0.7071	0.0327
BGNet [25]	0.8116	0.8698	0.7485	0.0734	0.8307	0.9007	0.7219	0.0326
ZoomNet [18]	0.8197	0.877	0.7522	0.0659	0.8384	0.8876	0.7288	0.0289
OAFormer [32]	0.8659	0.9238	0.8263	0.048	0.8599	0.9274	0.7732	0.0245
FSENet [23]	**0.885**	**0.942**	**0.85**	**0.04**	**0.873**	**0.928**	**0.8**	**0.021**
Our Approach	**0.8912**	**0.9574**	**0.8699**	**0.042**	**0.8674**	**0.9324**	**0.7798**	**0.0221**

†FSENet omits CHAMELEON dataset, limiting fair comparison and generalization assessment.

Table 2. Comparison of the proposed method with state-of-the-art methods on CHAMELEON and NC4K datasets. "↑"/"↓" means that larger/smaller is better (best values in bold black; second-best values in bold blue).

Method	CHAMELEON [21]				NC4K [16]			
	S_m ↑	E_m ↑	F_β^w ↑	MAE ↓	S_m ↑	E_m ↑	F_β^w ↑	MAE ↓
EGNet [36]	0.7975	0.8599	0.6486	0.0648	0.7771	0.8408	0.6386	0.0751
PraNet [9]	0.86	0.9073	0.7633	0.0437	0.8222	0.8761	0.7245	0.0588
F^3Net [28]	0.848	0.8943	0.7436	0.0467	0.78	0.8244	0.6556	0.0695
MINet [19]	0.8548	0.9139	0.7713	0.0358	0.8122	0.8618	0.7195	0.0555
SINet [8]	0.872	0.9363	0.8056	0.0341	0.808	0.8173	0.7227	0.0576
C^2FNet [24]	0.8881	0.9353	0.8284	0.0316	0.8383	0.8974	0.7624	0.049
PFNet [17]	0.8819	0.9306	0.8099	0.0325	0.829	0.8874	0.7453	0.0527
MGL [34]	0.8932	0.9171	0.8123	0.0305	0.8326	0.8666	0.7392	0.0526
UGTR [31]	0.8857	0.9097	0.7939	0.0314	0.8394	0.8744	0.7463	0.0519
SINet-v2 [7]	0.8882	0.9417	0.816	0.0297	0.8472	0.9027	0.7698	0.0476
BSANet [20]	0.8954	0.9458	0.841	0.0272	0.8414	0.8968	0.7708	0.0479
OCENet [15]	0.8972	0.9402	0.8333	0.0269	0.8533	0.9025	0.7846	0.045
BGNet [25]	0.9012	0.943	0.8505	0.0268	0.851	0.9067	0.7884	0.0444
ZoomNet [18]	0.9017	0.9427	0.8451	0.0229	0.8528	0.8957	0.7844	0.0434
OAFormer [32]	**0.9035**	**0.9608**	**0.8583**	**0.0227**	0.8833	0.9343	0.8372	0.0325
FSENet† [23]	-	-	-	-	0.892	0.941	0.853	0.03
Our Approach	**0.9067**	**0.9714**	**0.8620**	**0.0223**	**0.8986**	**0.9476**	**0.8418**	**0.0297**

†FSENet is not evaluated on CHAMELEON dataset, limiting generalization assessment.

Table 3. Model size and venue details by approach.

Approach	Venue'Year	Parameters (M)
EGNet [36]	ICCV'2019	56.02
PraNet [9]	MICCAI'2020	32.55
F^3Net [28]	AAAI'2020	25.54
MINet [19]	CVPR'2020	162.38
SINet [8]	CVPR'2020	48.95
C^2FNet [24]	IJCAI'2021	28.41
PFNet [17]	CVPR'2021	46.5
MGL [34]	ICCV'2021	63.6
UGTR [31]	ICCV'2021	48.87
SINet-v2 [7]	TPAMI'2022	26.98
BSANet [20]	AAAI'2022	32.58
OCENet [15]	WACV'2022	58.17
BGNet [25]	IJCAI'2022	79.85
ZoomNet [18]	CVPR'2022	32.38
OAFormer [32]	ICME'2023	39.86
Our Approach		39.86

sharper contours and more complete object coverage than those of other evaluated approaches. These visual results confirm that the integration of edge-guided attention enables the model to distinguish camouflaged objects more effectively from their background, even in scenarios with minimal visual contrast.

5 Conclusions

This work presents a transformer-based camouflaged object detection framework enhanced with edge-guided attention, extending the OAFormer backbone to explicitly model structural cues. The integration of edge features and an edge-aware loss function enables more accurate contour localization and improves mask consistency. Extensive experiments across four benchmark datasets confirm consistent improvements over state-of-the-art methods in both quantitative and qualitative evaluations. Future research will explore the incorporation of additional domain modalities, such as depth or thermal imagery, to further improve segmentation under challenging camouflage scenarios.

Acknowledgements. This work was supported in part by the ESPOL project CIDIS-003-2024-T, in part by Grant PID2021-128945NB-I00 funded by MCIN/AEI/ 10.13039/5011000 11033 and by "ERDF A way of making Europe", in part by the Air Force Office of Scientific Research Under Award FA9550-24-1-0206. The second author acknowledges the support of the Generalitat de Catalunya CERCA Program to CVC's general activities, and the Departament de Recerca i Universitats from Generalitat de Catalunya with reference 2021 SGR01499.

References

1. Canny, J.: A computational approach to edge detection. IEEE Trans. Pattern Anal. Mach. Intell. **8**(6), 679–698 (1986)
2. Chen, L., Wang, M., Zhang, K.: Attention-based models for image segmentation: a survey. IEEE Trans. Neural Networks Learn. Syst. **32**(6), 1234–1245 (2021)
3. Chen, M.: Non-local attention mechanisms for COD. CVPR (2023)
4. Chen, Z., Zhang, X., Xiang, T.Z., Tai, Y.: Adaptive guidance learning for camouflaged object detection. preprint arXiv:2401.12345 (2024)
5. Cui, L.: Contrastive learning for camouflaged object detection. TPAMI (2023)
6. Dosovitskiy, A., Beyer, L., Kolesnikov, A.: An image is worth 16x16 words: Transformers for image recognition at scale. In: International Conference on Learning Representations (ICLR) (2021)
7. Fan, D.P., Ji, G.P., Cheng, M.M., Shao, L.: Concealed object detection. IEEE Trans. Pattern Anal. Mach. Intell. **44**(10), 6024–6042 (2021)
8. Fan, D.P., Ji, G.P., Sun, G., Cheng, M.M., Shen, J., Shao, L.: Camouflaged object detection. In: Proceedings of the IEEE/CVF Conference on Computer Vision and Pattern Recognition, pp. 2777–2787 (2020)
9. Fan, D.P., et al.: Pranet: parallel reverse attention network for polyp segmentation. In: International Conference on Medical Image Computing and Computer-assisted Intervention, pp. 263–273. Springer (2020)
10. Le, T.N., Nguyen, T.V., Nie, Z., Tran, M.T., Sugimoto, A.: Anabranch network for camouflaged object segmentation. Comput. Vis. Image Underst. **184**, 45–56 (2019)
11. Lei, C., Fan, J., Li, X., Xiang, T., Li, A., Zhu, C., Zhang, L.: Towards real zero-shot camouflaged object segmentation without camouflaged annotations. arXiv preprint arXiv:2402.12345 (2024)
12. Lei, X.: Zeroscope: Minimal supervision techniques for cod tasks. CVPR (2024)

13. Li, X.: Edge-resnet: Boundary-aware networks for camouflaged object detection. ECCV (2022)
14. Liu, F.e.a.: Collaborative camouflaged object detection: A large-scale dataset and benchmark. Preprint arXiv:2205.11333 (2022)
15. Liu, J., Zhang, J., Barnes, N.: Modeling aleatoric uncertainty for camouflaged object detection. In: Proceedings of the IEEE/CVF Winter Conference on Applications of Computer Vision, pp. 1445–1454 (2022)
16. Lv, Y., et al.: Simultaneously localize, segment and rank the camouflaged objects. In: Proceedings of the IEEE/CVF Conference on Computer Vision and Pattern Recognition, pp. 11591–11601 (2021)
17. Mei, H., Ji, G.P., Wei, Z., Yang, X., Wei, X., Fan, D.P.: Camouflaged object segmentation with distraction mining. In: Proceedings of the IEEE/CVF conference on Computer Vision and Pattern Recognition, pp. 8772–8781 (2021)
18. Pang, Y., Zhao, X., Xiang, T.Z., Zhang, L., Lu, H.: Zoom in and out: a mixed-scale triplet network for camouflaged object detection. In: Proceedings of the IEEE/CVF Conference on Computer Vision and Pattern Recognition, pp. 2160–2170 (2022)
19. Pang, Y., Zhao, X., Zhang, L., Lu, H.: Multi-scale interactive network for salient object detection. In: Proceedings of the IEEE/CVF Conference on Computer Vision and Pattern Recognition, pp. 9413–9422 (2020)
20. Qin, X., Zhang, Z., Huang, C., Gao, C., Dehghan, M., Jagersand, M.: Basnet: boundary-aware salient object detection. In: Proceedings of the IEEE/CVF Conference on Computer Vision and Pattern Recognition, pp. 7479–7489 (2019)
21. Skurowski, P., Abdulameer, H., Błaszczyk, J., Depta, T., Kornacki, A., Kozieł, P.: Animal camouflage analysis: chameleon database. https://www.polsl.pl/rau6/chameleon-database-animal-camouflage-analysis/ (2018)
22. Smith, J., Doe, J.: Edge detection techniques: a comprehensive review. J. Comput. Vision $45(3)$, 567–589 (2021)
23. Sun, Y., Xu, C., Yang, J., Xuan, H., Luo, L.: Frequency-spatial entanglement learning for camouflaged object detection. In: European Conference on Computer Vision, pp. 343–360. Springer (2024)
24. Sun, Y., Chen, G., Zhou, T., Zhang, Y., Liu, N.: Context-aware cross-level fusion network for camouflaged object detection. arxiv 2021. preprint arXiv:2105.12555 (2021)
25. Sun, Y., Wang, S., Chen, C., Xiang, T.Z.: Boundary-guided camouflaged object detection. arXiv preprint arXiv:2207.00794 (2022)
26. Vaswani, A., Shazeer, N., Parmar, N., et al.: Attention is all you need. In: Advances in Neural Information Processing System, vol. 30, pp. 5998–6008 (2017)
27. Wang, H.: Multi-scale transformer networks for camouflaged object detection. ICCV (2021)
28. Wei, J., Wang, S.: F^3net: Fusion, feedback and focus for salient object detection. In: Proceedings of the AAAI Conference on Artificial Intelligence, vol. 34, pp. 12321–12328 (2020). https://doi.org/10.1609/aaai.v34i07.6916
29. Wei, J., Wang, S., Huang, Q.: F^3net: fusion, feedback and focus for salient object detection. In: Proceedings of the AAAI Conference on Artificial Intelligence, vol. 34, pp. 12321–12328 (2020)
30. Wu, J.: Lightweight models for camouflaged object detection. NeurIPS (2022)
31. Yang, F., et al.: Uncertainty-guided transformer reasoning for camouflaged object detection. In: Proceedings of the IEEE/CVF International Conference on Computer Vision, pp. 4146–4155 (2021)

32. Yang, X., Zhu, H., Mao, G., Xing, S.: Oaformer: occlusion aware transformer for camouflaged object detection. In: 2023 IEEE International Conference on Multimedia and Expo (ICME), pp. 1421–1426 (2023)
33. Yao, S., Sun, H., Xiang, T.Z., Wang, X., Cao, X.: Hierarchical graph interaction transformer with dynamic token clustering for camouflaged object detection. IEEE Trans. Image Process. **33**, 1234–1245 (2024)
34. Zhai, Q., Li, X., Yang, F., Chen, C., Cheng, H., Fan, D.P.: Mutual graph learning for camouflaged object detection. In: Proceedings of the IEEE/CVF conference on Computer Vision and Pattern Recognition, pp. 12997–13007 (2021)
35. Zhang, J.: Transformer-based object detection. CVPR (2020)
36. Zhao, J.X., Liu, J.J., Fan, D.P., Cao, Y., Yang, J., Cheng, M.M.: EGNET: edge guidance network for salient object detection. In: Proceedings of the IEEE/CVF International Conference on Computer Vision, pp. 8779–8788 (2019)

DVS-StereoInsect: An Event-Based Stereo Dataset for Foreground-Background Insect Segmentation

Colin Gebler[1,2]($^{(\boxtimes)}$)(iD), Jonas Funk[3](iD), Regina Pohle-Fröhlich[1,2](iD), and Andreas Wagner[3,4](iD)

[1] Institute for Pattern Recognition, Niederrhein University of Applied Sciences, Krefeld, Germany
colin.gebler@hs-niederrhein.de
[2] Department CDS, Graduate School for Applied Research in North Rhine-Westphalia, Bochum, Germany
[3] Karlsruhe University of Applied Sciences, Karlsruhe, Germany
[4] Fraunhofer Institute for Industrial Mathematics ITWM, Kaiserslautern, Germany

Abstract. Insect monitoring is a field of growing importance as the need to evaluate the effects of various stressors on insect populations rises. Existing methods for insect monitoring are often unsuitable for continuous monitoring of larger areas. This work presents an approach that aims to separate insect trajectories from background information in dynamic vision sensor (DVS) recordings on a new dataset. The dataset consists of approximately one hour of training data and six one-minute-long test sets of varying difficulty. A method to synthetically generate foreground-background segmentation-labeled data for this task given appropriately created source recordings is presented. To demonstrate the performance of the approach, an existing method for insect tracking in DVS data and a U-Net-based method are evaluated. The U-Net method achieves a Matthews correlation coefficient (MCC) of 0.955 detecting wasps and 0.850 detecting varied insects in front of a complex natural background. The evaluation of the existing method on the new dataset shows that it is not applicable in all use cases.

Keywords: Dataset · Dynamic Vision Sensor · Insect Monitoring · Deep Learning · Segmentation

1 Introduction

The insect biodiversity crisis has been a central topic for multiple years. 75% of our agriculture is dependent on pollination provided by insects [13]. Consequently, a reduction in insect-provided pollination will lead to a shift of plant biodiversity towards wind-pollinated species. In addition, insects are themselves an important food source for many other species.

In 2017, the "Krefeld Study" [8] revealed a concerning reduction in insect biodiversity by utilizing malaise traps. These operate by capturing some of the

© The Author(s), under exclusive license to Springer Nature Switzerland AG 2026
M. Castrillón-Santana et al. (Eds.): CAIP 2025, LNCS 15621, pp. 209–219, 2026.
https://doi.org/10.1007/978-3-032-04968-1_18

insects occurring in the area in a jar, allowing for species level identification. However, since it necessitates the capturing and killing of insects, it is an invasive method which must be deployed in a controlled manner. In addition, the evaluation of the trap contents is highly labor-intensive.

Due to these shortfalls, the development of non-invasive monitoring methods suitable for wide-scale deployment is necessary. Dynamic Vision Sensors (DVS), also called event-cameras [1], have much potential in this area due to their capability to capture high-resolution temporal information at comparatively low data-rates. The DVS only captures per-pixel changes in illumination rather than entire images, eliminating redundant information which would usually be captured by a frame-camera. Events at each pixel are triggered independently and asynchronously instead of being tied to a set frame rate. A DVS generates events when illumination changes occur which can be described as tuples (x, y, p, t) where (x, y) are the position of the event, the polarity p indicates whether illumination increased or decreased and t is a microsecond resolution timestamp.

In this work, we present a dataset designed to accomplish the first necessary step towards DVS-based insect monitoring: separating insect-caused activity from background activity in DVS data. Section 2 outlines related work in the field of insect monitoring, Sect. 3 outlines the methodology and structure of the dataset, Sect. 4 describes foreground-background segmentation methods applied to the dataset and Sect. 5 evaluates the results and provides an outlook to further work.

2   Related Work

There are several approaches being used in the area of automated insect monitoring. The approach most similar to the usage of DVS is using frame cameras, which is generally divided into the usage in traps [5] and recording of insects in the wild [3]. The former allows for imaging of the specimens in a predictable situation and in front of a homogeneous background, enhancing system accuracy. The latter is less invasive and capable of capturing information about the behavior of observed insects. However, the more complex backgrounds make localization and classification of the observed insects more challenging. One approach attracts insects onto an illustrated platform without leading them into a trap [16], achieving a less invasive method of capturing images of insects in homogeneous scenarios.

To capture the motion of insects, video-based camera traps are also used. However, storage limitations generally prohibit the continuous recording of videos in a sufficient resolution, especially at high frame-rates [17]. This makes video-based approaches challenging to use for long-term monitoring.

In addition, acoustic monitoring [14] and radar-based monitoring [9] are used to track insects in the field. These methods have been shown to be capable of differentiating between different species in the surrounding area. However, they can only obtain limited information about the behavior and count of the observed insects.

Gebauer et al. [7] presented an asynchronous insect detection and tracking algorithm that identifies insect presence in the event stream based on the density of triggered events. Foreground-background segmentation on DVS data has also been investigated, albeit on simple backgrounds [11]. In addition, the classification of manually segmented event-based insect trajectories has been investigated [12].

3   Dataset

3.1   Reasons for Segmentation

DVS-based insect monitoring is currently challenging in particular due to the low availability of labeled data. Especially for the classification of insect-trajectories based on the spatio-temporal (x, y, t) point cloud, an event-level separation between events resulting from insect activity and events resulting from background activity is neccessary. Background activity most commonly consists of flora occurring where insect activity is expected. However, existing approaches to DVS-based insect monitoring rely on bounding-box detection and image-based classification [7]. The segmentation of insect-caused events in DVS data has been implemented before, though in front of simpler backgrounds which were easier to segment both manually and automatically [11]. Consequently, additional datasets and approaches for the task of extracting insect trajectories from DVS-Data are needed to develop a system that can segment reliably in real-world scenarios. One likely cause for the lack of such datasets is the labor-intensive nature of creating event-level segmentation labels. While the manual creation of bounding boxes is also a labor intensive task, it permits the inclusion of some background activity in the marked areas. This makes the process of creating ground truth bounding boxes easier than creating ground truth segmentation labels. Unfortunately, extracting insect trajectories from bounding-box-based approaches also leads to background events being included in the trajectories for the same reason, presenting an obstacle to classification accuracy. In addition, especially dense backgrounds would necessitate aggressive subsampling when working with point cloud classification approaches which expect a fixed input size. If the number of background events compared to the number of insect events in a given bounding box is especially high, the necessary subsampling would cause most of the insect events to be lost.

Semantic segmentation on DVS data has been previously implemented. Due to the difficulty of manually annotating events for segmentation, ground-truth segmentation labels are often transferred from grayscale or color frame recordings, accepting some degree of error [4]. Applying such label generation methods to this use case is challenging, as one of the main reasons to use DVS for insect monitoring is their ability to capture high-speed insects with more clarity than a frame-based camera can. A label transfer method would, therefore, only be able to reliably annotate insects in situations in which they can already be tracked using existing solutions.

In addition, obtaining depth information is necessary in the field of insect monitoring. Depth information can support the tracking of insect behavior and classification. It can also be used in combination with the high-resolution times-tamps provided by the DVS to obtain additional features, such as physical size and speed [12]. Therefore, stereo pairs of the recordings in this dataset are provided as they are expected to be useful in further work.

3.2   Methodology

The raw recordings were created using pairs of SENSING SE1-S4-USB DVS, which use the SONY IMX646 event-based image sensor. This results in a spatial resolution of 1280×720 pixels. Each DVS is fitted with a Tamron M117FM06-RG objective lens with a focal length of $6mm$.

To bypass the labor-intensive nature of labelling each event, the dataset presented here is generated synthetically by combining two separate recordings, a *foreground* recording and a *background* recording, per camera.

The *foreground* recording contains insects and no background activity. Since recordings which do not contain background activity are required, recording meadows or flowering plants, where one can expect to find many pollinators, is not feasible for this purpose, as even slight winds or the insects themselves cause the plants to move. In the presented dataset, three scenes were used for this purpose.

An insect hotel An insect hotel is a man-made structure containing holes for various insect species to use as a nest. Unlike plants, which also attract insects, it is itself static, meaning that it will not cause background activity to be recorded by the DVS.

A wasp nest on a balcony A wasp's nest on a balcony provides another location frequented by insects, which is itself static. Compared to the insect hotel, the wasps recorded here are generally larger, adding variety to the data.

An open area next to a flowerbed Recording next to a flowerbed allows for the recording of insects that are approaching it without capturing the plants themselves, providing additional varied examples.

The *background* recordings contain differing levels of plant activity. They were recorded while most insects were out of season or in the early morning to minimize the amount of insect activity. They were then manually screened to extract sections of the recordings without any insect activity. One of three levels of difficulty, easy, moderate and hard, is assigned to each background based on the complexity of the background movements. The backgrounds are depicted in Fig. 1.

Shrubby cinquefoil in weak wind (easy) (Fig.1 a) Flowering bushes are backgrounds that can be expected to be encountered in insect monitoring appli-cations. Since their branches are rigid, they exhibit less movement than a thin-stilted plant would in the wind.

Meadow with umbellifers (moderate) (Fig.1 b) The individual plants and wind conditions lead to more complex movement patterns than exhibited by the shrubby cinquefoil.

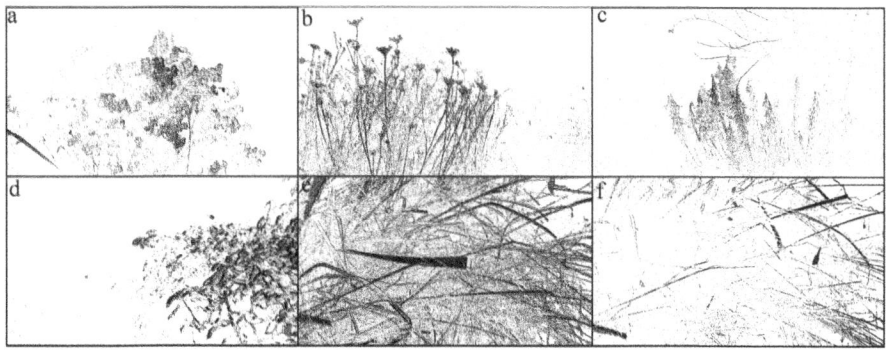

Fig. 1. Images of the background scenes used in the dataset. Each image depicts events captured over a period of $60000\mu s$. Events with a polarity of 1 (an increase in brightness) are drawn in blue, while events with a polarity of 0 (a decrease in brightness) are drawn in black on a white background. Images a and c show easy backgrounds, images b and d show moderate difficulty backgrounds and image e shows a hard difficulty background. In image f, the mixed-in events caused by 5 wasps are marked in red. (Color figure online)

Camphor mugwort (easy) (Fig.1 c) Due to the relatively slow winds in the scene, the movements are minor, making the plants easier to separate.

Blackberry bush (moderate) (Fig. 1 d) The bush was shaken from off-screen to create additional movement, simulating stronger winds.

Dense grass in strong wind (hard) (Fig. 1 e) The long grass in strong wind provides a challenging background. The fast movements of individual blades can easily be mistaken for insect movements by a detector which is overly reliant on speed.

To eliminate electrical noise caused by the DVS, the recordings are filtered using the *ActivityNoiseFilter* provided by Metavision SDK[1]. This filter discards events that occur isolated, without another event occurring in its neighborhood in a given time span. To ensure that all false positives are eliminated, the filter is applied 5 times consecutively with a time span of $10000\mu s$. For use as ground truth data, falsely discarding some insect-caused events is less of an issue than falsely including noise and labeling it as insect-caused events. The streams are merged, sorted by their timestamps. Events from the *foreground* recordings are assigned the label 1 while events from the *background* recordings are assigned the label 0. Occlusion effects are ignored. Since the insects are generally fast and small, this has a similar effect to assuming that the insects are always in the foreground.

Combining the recordings, seven components of the dataset are created. The largest component is the *training set*. In the foreground, it consists of 30 min each of the *insect hotel* and *wasp nest* scenes. This provides a mix of large and small

---

[1] Metavision SDK by Prophesee (Version 4.6.2) https://docs.prophesee.ai/4.6.2/index.html.

insects. The background consists of short sequences. 90 s of shrubby cinquefoil (see Fig. 1a), 70 s of grass and umbellifers (see Fig. 1b) and 180 s of complex grass (see Fig. 1e) are included.

As the foreground events and background events are merged, the backgrounds are repeated until the length of the foreground data is reached, increasing the timestamps accordingly. Manually screening longer background sequences would increase the risk of missed insects, leading to incorrect labels. Due to the repetitive nature of plant movements in the wind, this repetition is not expected to harm segmentation performance. This is confirmed by using separate data in the test sets. An example image generated from the mixed events streams is given in Fig. 1f.

The six remaining components are test sets. They are generated from the combination of the *wasp nest* and *open area next to a meadow* foreground recordings and the backgrounds depicted in Figs. 1c, 1d and 1e, each 60 s long. For the *wasp* foreground and the hard difficulty scene, which also occur in training, separate sections of the recordings unseen in training are used. This results in a total of six test sets of varying difficulty, combining one foreground scene and one background scene each. The *wasp* foreground is considered to be easier to segment than the *open area* foreground due to the occurrence of smaller insects and the higher variance in speed in the latter. To ensure the quality of the validation, backgrounds are not repeated in individual test sets.[2]

4   Experiments

4.1   Clustering-Based Approach

In this section, the effectiveness of the approach presented by Gebauer et al. [7] on this dataset is investigated. The algorithm comprises three distinct steps for the detection of flying insects. Initially, non-target events are identified and disregarded, avoiding further processing. Subsequently, motion patterns resembling insect trajectories are detected within the remaining event stream. Finally, a process of disambiguation occurs, distinguishing between clusters representing flying insects and those comprising moving background elements, such as plants and other objects.

Given that the code provided was not utilized for the purpose of labeling the events, a function was incorporated that outputs a list comprising the events (x, y, p, and t) and the assigned cluster ID. This function was implemented to facilitate the calculation of scores and the visualization of results.

Preliminary Analysis. To assess the basic segmentation functionality of the algorithm, we applied it to a simple recording of a honeybee and evaluated the events classified as flying insects. The results indicated that the algorithm could sufficiently remove the background events, which proved its basic functionality.

---

[2] The dataset is available at https://github.com/Event-Based-Insects/DVS-StereoInsect.

To see how the algorithm performs on a larger, more complex recording, we applied it to our dataset. We tested each parameter set provided by the authors with our data. Our experiments have shown that these parameter sets are not suitable for the scenarios in our dataset, with a maximum MCC of 0.03358.

Parameter Optimization. To improve performance in complex scenarios, Bayesian optimization using the Scikit-Optimize package[3] was applied to all parameters. A multiprocessing approach enabled the parallel evaluation of 128 parameter combinations per iteration, resulting in a total of 25,600 evaluations across 200 iterations. Each output file, containing cluster assignments for over 400 million events, was processed in chunks to reduce memory usage. The bounds of the parameter space were chosen based on the parameters provided by Gebauer et al. [7]. The Matthews correlation coefficient is employed as the scoring metric. It provides a more informative score in binary classification with imbalanced classes [6]. Due to memory limitations, the events were evaluated in chunks of 10 million. Only the best-scoring configurations were retained to manage storage constraints.

The first optimization runs yielded MCC values below 0.1. Many of the best-performing parameters were located at the bounds of the defined search space, suggesting suboptimal limits. Therefore, we iteratively relaxed the parameter boundaries. After three relaxation steps, further expansion was unnecessary as bounds were no longer reached. The new parameter limits indicated the necessity of a significantly lower time difference between events to be assigned together, suggesting that in complex datasets, temporal coherence must be stricter to avoid linking unrelated or noisy events. This prevents the inclusion of background noise into insect clusters and favors temporally compact trajectories. Due to the denser background activity, the activity threshold for a cluster to be considered insect-caused had to be increased substantially. Additionally, the initial observation area during cluster initialization had to be increased. This expansion improved the algorithm's ability to handle denser or overlapping trajectories, likely due to insects flying close to the camera or exhibiting more complex flight behavior.

For these computations, we used 2 AMD EPYC 7742 with 64 cores, 2.25 GHz and 256 MB cache each in combination with 768 GB of memory. The computation time for the runs averaged in 15 d, 12 h and 24 min per run.

Optimization Results. After parameter optimization, we observed substantial improvements in several performance metrics. The F1 score increased from 0.061 to 0.51, and the MCC rose from 0.03 to 0.55, indicating a notable enhancement in the algorithm's overall classification quality. Precision improved markedly from 0.03 to 0.83, indicating that the algorithm became more selective and precise in identifying true positive instances. However, this improvement came at the cost of a reduction in recall, which dropped from 0.67 to 0.38. This trade-off reflects

---

[3] Scikit-Optimize (Version 0.9.0) https://scikit-optimize.readthedocs.io.

an increased precision at the cost of sensitivity. The overall improvements–particularly in F1 and MCC–suggest a more balanced and robust classification performance, despite reduced recall.

Since the goal is to develop a generalizable solution, we refrained from creating separate optimized parameter sets for the different validation scenarios. Therefore, the best parameter combination after optimization was used to predict clusters for each scenario.

4.2   U-Net Based Approach

To achieve a higher quality segmentation utilizing the available data, a baseline deep learning approach is tested. Segmenting the (x, y, t) point cloud directly [4] is made challenging due to the high event rate in the dataset. This is caused by a combination of the comparatively high spatial resolution of 1280×720 pixels, which is desirable for observing small insects and the complex backgrounds, particularly in the more difficult scenarios. To avoid aggressive subsampling, an image-based approach is chosen for this task. After the background events are removed, significantly lowering the event rate, point cloud-based approaches can be utilized in further steps, such as instance segmentation and classification [12].

To obtain an image from the event data without discarding the valuable temporal information, linear time surfaces [2] are generated. Each pixel's value indicates the relative timestamp of the last event that occurred at the corresponding position within a given *accumulation time*. In addition, the image is divided into 2 channels based on event polarity. An accumulation time of $40000\mu s$ was chosen.

In image-based segmentation, U-Net-based [15] approaches are popular and have already been used for the present task [11]. A U-Net is divided into an *encoder* portion to extract multi-scale features in lowering resolution steps and a *decoder* portion that generates the segmentation map from those features. *Skip connections* are inserted between corresponding *encoder* and *decoder* layers to utilize higher resolution information while generating the segmentation map.

As the target objects are small, feeding the entire HD image into the network might lead to poor segmentation results on the presented dataset. This is caused by the low amount of insect events being drowned out by the comparatively high number of background events in each training sample. This effect worsens as the density of background events increases. Therefore, the time surfaces are divided into tiles of 80×80 pixels each before being input into the U-Net. This results in 91195 time surfaces divided into 8949564 tiles containing events in the training set. Tiles not containing events are discarded outright. After inference, the time surfaces are reassembled from the individual segmentation maps, and predicted labels are assigned to the events within the current accumulation window based on the value of the segmentation map at the corresponding position. For training data augmentation, the tiles are randomly flipped along the x and y axes with a probability of 0.5 for each axis and randomly translated along each axis by up to 80 pixels wrapping around the image border.

An implementation of U-Net [10] in TensorFlow[4] is used. The segmentation task considers 3 classes: 0 if no event occurred, 1 if a background event occurred, and 2 if an insect event occurred at that position. The "no event" class is treated separately from "background" to guide the network towards more detailed discrimination and reduce the imbalance between "background" and "insect" pixels. The *encoder* and *decoder* portions consist of four layers each. The learning rate is set to 0.001, the loss function is sparse categorical cross-entropy and the Adam optimizer is used. The *hard wasps* dataset is used as validation data, and training is stopped when the F1-score no longer improves, after 11 epochs. In inference, each event is assigned the label assigned to the corresponding pixel on the time surface.

Table 1. Resulting scores for the test datasets using both approaches

Dataset	U-Net based (ours)				Clustering [7]			
	Precision	Recall	F1	MCC	Precision	Recall	F1	MCC
EASY WASPS	0.995	0.996	0.995	0.993	0.990	0.784	0.875	0.836
EASY MIXED	0.993	0.946	0.969	0.968	0.989	0.541	0.699	0.728
MOD. WASPS	0.996	0.997	0.997	0.995	0.996	0.836	0.909	0.888
MOD. MIXED	0.962	0.941	0.951	0.951	0.496	0.298	0.372	0.382
HARD WASPS	0.951	0.962	0.956	0.955	0.315	0.549	0.361	0.375
HARD MIXED	0.857	0.844	0.850	0.850	0.030	0.099	0.043	0.049

The results in table 1 show that the U-Net approach outperforms the clustering approach in all test sets. While satisfactory results are achieved by both methods on the easy sets and the moderate difficulty wasps set, the clustering approach achieves low scores on the medium mixed set and the high difficulty sets. On the most difficult set, which features small insects in combination with a very high activity background, the U-Net also shows a notable decline in performance. To compare the U-Net results using tiling, we also ran the U-Net on full time surfaces for insect trail segmentation, as presented by Pohle-Fröhlich and Bolten [11]. For this purpose, the approach was retrained on the more varied dataset presented here. The approach showed worse results especially for the more challenging tasks.

5 Conclusions and Future Work

In this work, a foreground-background segmentation approach for insect trajectories in DVS data has been presented. Due to the prohibitive difficulty of manually segmenting the insect tracks in the event stream, the dataset was generated synthetically by combining separate event recordings containing insect

---

[4] TensorFlow (Version 2.7.2) https://tensorflow.org.

trajectories and background activity. A training set with a total length of 60 min and six test sets of varying difficulty were generated. Experimental evaluation of a U-Net on the dataset showed promising results in the central task. In addition, the suitability of an existing method for the localization of insect trajectories in DVS Data which does not rely on training [7] was tested, though the results were not satisfactory in all testing scenarios.

In future work, it is worth investigating more computationally efficient methods of segmentation based on the dataset presented here. This would be especially valuable in providing the possibility of real-time monitoring in the field instead of relying on off-site computation. In addition, the further steps to build a DVS-based insect monitoring solution should be investigated. The next steps should be the separation of individual trajectories and subsequent classification. Obtaining meaningful counts may also require some action recognition, such as recognizing an insect landing on a specified plant in the scene.

Acknowledgments. We would like to thank Leland Gehlen, Michael Glück and Kirsten Traynor of the State Institute of Bee-Research at the University of Hohenheim for their advice in entomological matters and their aid in data collection. This work was funded by the Carl-Zeiss-Foundation as part of Project BeeVision.

Disclosure of Interests. The authors have no competing interests to declare that are relevant to the content of this article.

References

1. AliAkbarpour, H., et al.: Emerging trends and applications of neuromorphic dynamic vision sensors: a survey. IEEE Sens. Rev. pp. 1–53 (2024). https://doi.org/10.1109/SR.2024.3513952
2. Benosman, R., et al.: Event-based visual flow. IEEE Trans. Neural Netw. Learn. Syst. **25**(2), 407–417 (2014). https://doi.org/10.1109/TNNLS.2013.2273537
3. Bjerge, K., et al.: Accurate detection and identification of insects from camera trap images with deep learning. PLOS Sustain. Trans. **2**(3), e0000051 (2023). https://doi.org/10.1371/journal.pstr.0000051
4. Bolten, T., et al.: DVS-OUTLAB: a neuromorphic event-based long time monitoring dataset for real-world outdoor scenarios. In: CVPRW 2021, pp. 1348–1357. IEEE, Nashville, TN, USA (2021). https://doi.org/10.1109/CVPRW53098.2021.00149
5. Chiavassa, J.A., et al.: The field automatic insect recognition-device–a non-lethal semi-automatic malaise trap for insect biodiversity monitoring: proof of concept. Ecol. Evol. **14**(12), e70642 (2024). https://doi.org/10.1002/ece3.70642
6. Chicco, D., et al.: The matthews correlation coefficient (mcc) is more informative than cohen's kappa and brier score in binary classification assessment. IEEE Access **9**, 78368–78381 (2021). https://doi.org/10.1109/ACCESS.2021.3084050
7. Gebauer, E., et al.: Towards a dynamic vision sensor-based insect camera trap. In: Proceedings of the IEEE/CVF Winter Conference on Applications of Computer Vision, pp. 7142–7151 (2024). https://doi.org/10.1109/WACV57701.2024.00700

8. Hallmann, C.A., et al.: More than 75 percent decline over 27 years in total flying insect biomass in protected areas. PLoS ONE **12**(10), e0185809 (2017). https://doi.org/10.1371/journal.pone.0185809

9. Noskov, A., et al.: A review of insect monitoring approaches with special reference to radar techniques. Sensors **21**(4), 1474 (2021). https://doi.org/10.3390/s21041474

10. Planche, B., Andres, E.: Hands-On Computer Vision with TensorFlow 2. Packt Publishing Ltd (2019)

11. Pohle-Fröhlich, R., et al.: Stereo-event-camera-technique for insect monitoring. In: VISAPP 2024, pp. 375–384. INSTICC, SciTePress (2024).https://doi.org/10.5220/0012326500003660

12. Pohle-Fröhlich, R., et al.: Features for classifying insect trajectories in event camera recordings. In: VISAPP 2025, pp. 355–364. INSTICC, SciTePress (2025).https://doi.org/10.5220/0013140100003912

13. Potts, S.G., et al.: The assessment report of the Intergovernmental Science-Policy platform on biodiversity and ecosystem services on pollinators, pollination and food production (2016). https://www.ipbes.net/sites/default/files/downloads/pdf/individual_chapters_pollination_20170305.pdf

14. Rodríguez Ballesteros, A., et al.: Towards acoustic monitoring of bees: wingbeat sounds are related to species and individual traits. Philos. Trans. R. Soc. B: Biol. Sci. **379**(1904), 20230111 (2024). https://doi.org/10.1098/rstb.2023.0111

15. Ronneberger, O., Fischer, P., Brox, T.: U-Net: Convolutional networks for biomedical image segmentation. In: Navab, N., Hornegger, J., Wells, W.M., Frangi, A.F. (eds.) MICCAI 2015. LNCS, vol. 9351, pp. 234–241. Springer, Cham (2015). https://doi.org/10.1007/978-3-319-24574-4_28

16. Sittinger, M., et al.: Insect detect: an open-source DIY camera trap for automated insect monitoring. PLoS ONE **19**(4), e0295474 (2024). https://doi.org/10.1371/journal.pone.0295474

17. Wallace, J.R.A., et al.: Camera-based automated monitoring of flying insects (Camfi). I. Field and computational methods. Front. Insect Sci. **3** (2023). https://doi.org/10.3389/finsc.2023.1240400

ColorEM-Net: Automated Segmentation of Structures in Large-Scale Electron Microscopy Using Element-Derived Ground Truth

Anusha Aswath[1,2]([✉]) [iD], Ahmad M.J. Alsahaf[2] [iD], B. H. Peter Duinkerken[2] [iD], Jacob P. Hoogenboom[3] [iD], Ben N.G. Giepmans[2] [iD], and George Azzopardi[1] [iD]

[1] Bernoulli Institute for Mathematics, Computer Science and Artificial Intelligence, University of Groningen,Groningen, The Netherlands
[2] Department of Biomedical Sciences, University Medical Center Groningen, Groningen, The Netherlands
anusha.aswath@gmail.com
[3] Department of Imaging Physics, Delft University of Technology, Delft, The Netherlands

Abstract. Electron microscopy (EM) combined with energy dispersive x-ray (EDX) imaging (or 'ColorEM') of cells and tissues provides ultrastructural insight complemented with elemental context. The resulting hyperspectral datasets can be used to map the relative abundance of specific elements or subjected to more data-driven approaches such as spectral mixture analysis or clustering to highlight the ultrastructural components of interest. Despite the benefits of automatic segmentation over manual annotation, EDX imaging is two orders of magnitude slower than EM imaging precluding its routine use for segmentation. Large-scale ColorEM, however, does generate sufficient annotated labels, which we use as ground truth to train U-Net models, and thus enables the transfer of these labels to conventional EM data. Here, we present ColorEM-Net, a label-free segmentation technique based on features obtained from unsupervised clustering of ColorEM data. ColorEM-Net achieves label-free identification with over 95% accuracy for nuclei, lysosomes and exocrine granules. However, with an accuracy of 79%, the recognition of endocrine granules needs further effort in training for reliable segmentation. By reusing open-access ColorEM datasets, this approach facilitates automated segmentation of EM data, while eliminating the need for manual annotation and achieving scalability for tissue-scale segmentation.

Keywords: Segmentation · electron microscopy · analytical pixel labels

1 Introduction

Understanding how the building blocks of life are arranged is crucial for biomedical research. The study of biomolecules, organelles and cells at the highest

M. Castrillón-Santana et al. (Eds.): CAIP 2025, LNCS 15621, pp. 220–231, 2026.
https://doi.org/10.1007/978-3-032-04968-1_19

resolution can be done by electron microscopy (EM), which enables cellular ultrastructure to be inspected with nanometer resolution. Visualization of the cellular ultrastructure with relevant biological labels is a powerful approach to understand health and disease. Unbiased imaging of large two-dimensional sections at tissue scale (Nano-anatomy or nanotomy) is nowadays routine, providing datasets similar to zoomable Google Earth images [7]. Moreover, with recent technological developments in faster acquisition and higher throughput [10,12,17,26] a data avalanche is foreseen. Hence, automation of data analysis is crucial to maximize the biological insight gathered from such gigapixel datasets.

Advanced features learned from labeled images via deep learning (DL) are being explored and implemented to advance EM analysis [3], building on traditional methods. However, identifying subcellular structures require DL models to capture diverse morphologies and sizes, typically demanding manual labeling of many grayscale images. Therefore, complementary methods such as ColorEM [22,25] and correlated microscopy, i.e. using fluorescent labels projected on EM maps [6] are used to gain information of biostructures in the grayscale data. The former yields hyperspectral images (HSIs) based on the elemental composition, allowing data-driven and more extensive analysis methods which have been widely used in material sciences and remote sensing for classification, segmentation and anomaly detection [2]. ColorEM includes analytical methods in EM, such as energy dispersive X-ray (EDX; [22]) imaging, that bring the benefits of HSI analysis to cellular biology, combining spectral information with spatial context. Recently, large-scale ColorEM was introduced to aid in the unbiased identification of sub-cellular content through spectral unmixing [11] as well as user-guided iterative clustering approach [4]. Like the endmembers in spectral analysis [11], the resulting cluster masks also correspond to specific biological structures.

EDX acquisition is $\sim$100 times slower than conventional EM, which makes it impractical for its routine implementation for the analysis of large-scale EM datasets. Here, we present an automated segmentation method that learns structures in large-scale EM datasets using supervised models trained on labels derived from ColorEM datasets.

2 Methods

Figure 1 shows the ColorEM-Net pipeline, outlining the key stages from hyperspectral data processing to EM segmentation, using examples such as heterochromatin and two types of endocrine granules: insulin and glucagon. In our main experiments, we focus on four representative pancreatic structures: heterochromatin in the nucleus, lysosomes, and both endocrine and exocrine granules. Cluster masks derived from hyperspectral ColorEM data serve as pseudo-labels to enable label-free segmentation of large-scale EM datasets, following the CLEM-Net-inspired methodology [1].

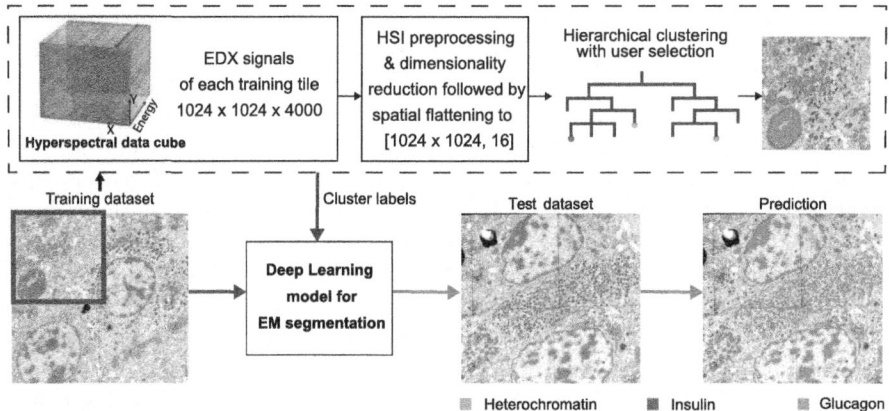

Fig. 1. Overview of the ColorEM-Net pipeline for label-free EM segmentation using EDX-derived pseudo-labels. A hyperspectral data cube ($1024 \times 1024 \times 4000$), shown for a single tile highlighted by a red box in the training dataset, is acquired via energy-dispersive X-ray (EDX) imaging capturing elemental composition at nanometer resolution. After preprocessing and dimensionality reduction, the data is flattened and clustered in a hierarchical way. The resulting cluster masks, each corresponding to a specific biological structure (e.g., insulin, glucagon, heterochromatin), are manually selected and used as pseudo-labels. These masks supervise the training of a deep learning segmentation model using a corresponding grayscale EM image (red path). Once trained, the model is applied to EM-only images without labels (green path), enabling automated segmentation across large-scale datasets. The hyperspectral cube is reproduced from [4]. (Color figure online)

2.1   Segmentation Model

ColorEM-Net uses a U-Net architecture with a ResNet-50 backbone (pre-trained on ImageNet) for enhanced feature extraction [23]. To adapt the pre-trained ResNet model to grayscale images, the first convolutional layer is modified for accepting a single channel by averaging the original multi-channel weights. The input training masks are encoded with zero for the background and one for the foreground in case of binary masks, and values greater than zero (up to one) for the foreground enabling the model to learn confidence-weighted scores for segmenting the target structure from the EM image. Training on large EM images is performed by dividing the image into smaller patches and using data augmentation to increase image diversity for training (Fig. 2).

2.2   Loss Functions

In binary segmentation tasks, addressing class imbalance is a critical challenge, particularly when foreground and background pixels differ significantly in the number of training samples. The Dice loss, which is one minus the Dice similarity coefficient, is widely used to mitigate this issue, especially in EM segmentation tasks [21]. To improve robustness, we employ a combined loss function incorporating both binary cross entropy (BCE) and Dice loss:

$$\text{BCE} = -\frac{1}{N} \sum_{i=1}^{N} y_i \log(\hat{y}_i) + (1 - y_i) \log(1 - \hat{y}_i), \tag{1}$$

$$\text{Dice loss} = 1 - \frac{2 \sum_{i=1}^{N} y_i \hat{y}_i + \epsilon}{\sum_{i=1}^{N} y_i + \sum_{i=1}^{N} \hat{y}_i + \epsilon}, \tag{2}$$

$$\text{Combined loss} = \gamma \cdot \text{BCE} + (1 - \gamma) \cdot \text{Dice loss}, \tag{3}$$

where y_i is the ground truth label, $\hat{y}_i$ is the predicted probability for pixel i, and N is the total number of pixels. A small constant ϵ ensures numerical stability; in our experiments, we set $\gamma = 0.5$ and $\epsilon = 10^{-5}$.

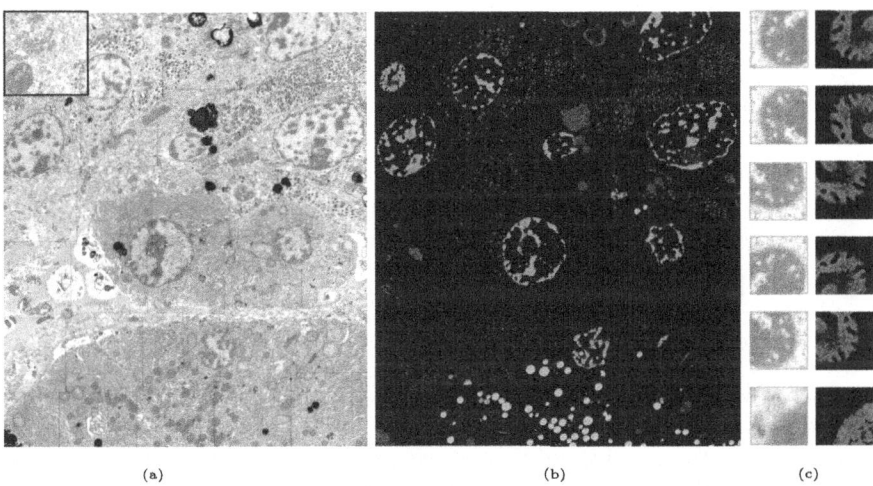

(a)   (b)   (c)

Fig. 2. ColorEM-Net using EDX-derived pseudo-labels for training binary segmentation models enable prediction on EM-only images. (a) The EM dataset used for training is shown as a stitched grayscale image arranged in a 5-column by 6-row grid following preprocessing and inversion. (b) The cluster masks from [4] serve as pseudo-labels for heterochromatin (green), lysosomes (pink), exocrine granules (yellow), and endocrine granules (red), corresponding to the grayscale dataset. The training set is derived from selected grayscale image tiles containing the target structure (e.g., heterochromatin, shown in the black box of the first tile). (c) Each tile is divided into smaller patches for efficient processing. An example patch, highlighted in green, undergoes various rigid transformations along with its cluster mask for five augmented variations. (Color figure online)

While effective against class imbalance, this combined loss is susceptible to noisy pseudo-labels: mislabeled samples contribute equally to the loss, potentially degrading performance. To address this, we incorporate pixel-level confidence scores derived from clustering results. Each pixel's confidence C_i is computed as $C_i = 1 - (d_i/mx)$, where d_i is the distance to its cluster centroid and mx is the maximum distance in the cluster. This gives higher weight to samples closer to the centroid. To incorporate these confidence scores, we use them as weights in the weighted BCE loss [9]:

$$\text{weighted BCE} = -\frac{1}{N}\sum_{i=1}^{N} C_i \left[y_i \log(\hat{y}_i) + (1 - y_i) \log(1 - \hat{y}_i) \right] \qquad (4)$$

Finally, this confidence-weighted BCE loss is incorporated into the previously defined combined loss function (Eq. 3). We refer to the combined version using this weighted BCE as the *confidence loss*.

To improve the robustness of pseudo-labels for confidence scoring, we apply hierarchical clustering (Fig. 1). This method captures structural variability by identifying nested substructures, and filtering out high-dimensional noise or outliers at higher cluster levels. Hence, the interactive clustering algorithm is suitable for segmentation tasks with ambiguous boundaries or varying object sizes [27], even when spatial information is flattened.

Since clustering is inherently ill-posed, both model choice and uncertainty characterization are crucial [5]. For instance, [20] shows how confidence calibration via Monte Carlo dropout or deep ensembles can support robust loss weighting or using uncertainty from variance or entropy of predictions [13].

3   Experiments and Results

3.1   Datasets and Image Processing

The pseudo-labels are derived from cluster masks, reusing an open access ColorEM dataset [11] (Fig. 1). The transfer was performed on a separate dataset from the same tissue, but a different section. This dataset consists solely of grayscale EM images and is four times larger than the training dataset, comprising 120 tiles, each of 2000×2000 pixels at a resolution of 4nm. The predictions are included only for the first four rows (12 columns each) and the bottom four rows (also 12 columns each). The remaining 24 tiles were excluded due to section folds and four out of focus tiles.

The EM images are acquired as high-angle annular dark-field (HAADF) images that contain overlapping areas for stitching of the images into a single large dataset using TrakEM2 plugin in ImageJ [24]. Although the granule content is relatively uniform and theoretically possible to annotate, manual quantification remains difficult due to discrepancies between expert annotations leading to incomplete data. From the clustering method that was applied on the ColorEM dataset [4], the obtained cluster masks enable the automatic identification of 198 endocrine granules, 54 exocrine granules, and 10 heterochromatin regions (classified as nucleus), which are used as pseudo-labels for learning EM image segmentation models. The spectrally distinct lysosomes were additionally included as training masks.

First, pre-processing is performed by subjecting the datasets to 2×2 binning [4] followed by global normalization within each dataset using min-max scaling. Here the 1st and 99th percentiles are used as the lower and upper bounds, respectively. Values below the 1st percentile and those above the 99th percentile are clipped to remove artifacts, effectively scaling the data in the range [0, 1]. By minimizing the impact of outliers, this step improves the uniformity of intensity distributions across each dataset. Following normalization, all images were inverted to align with the conventional grayscale appearance of EM images, where darker regions correspond to denser materials. Second, image distribution shift is addressed between the datasets (ColorEM and EM-only datasets) by mean-centering to avoid variations in image brightness.

3.2   Training and Inference

To ensure robust model performance, we divided the 30 tiles from the ColorEM dataset into three distinct sets: training (20 tiles), validation (5 tiles), and test (5 tiles). Each set was chosen in such a way that there is sufficient representation of the target structure. Firstly, the labeled tiles for each structure are assigned as 67% for training, 13% for validation, and 13% for testing. Then, the remaining tiles (unlabeled) are allocated randomly to each of the sets to fill the remaining tiles needed for the train-validation-test split.

To facilitate efficient processing of large images, each individual tile was subdivided into patches of 256×256 pixels. Constant padding (with a value of zero) was applied to the right and bottom edges to ensure that no partial patches were formed at the image borders. A sliding window with an overlap of 50% was used to extract patches. To enhance image diversity, five types of data augmentations were separately applied; horizontal and vertical flips, random rotations, transpositions, and crops [19].

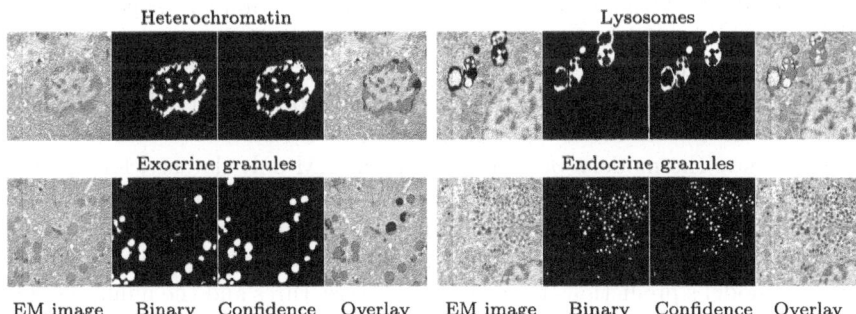

Fig. 3. Qualitative results on EM-only images showing the effective transfer of EDX-derived cluster masks as ground truth for training. Each EM image (1000×1000 pixels) shows predictions from two methods: (i) binary segmentation masks trained with the combined loss, and (ii) masks obtained using confidence-weighted BCE loss in the combined loss, as an overlay indicating differences between the two methods. Predictions from the confidence-weighted method (green) show improved localization of heterochromatin, lysosomes, and exocrine granules, while binary masks (red) perform better for endocrine granules but exhibit a higher rate of false positives. (Color figure online)

Training was conducted for up to a maximum of 100 epochs, with each epoch iterating through all training patches in batches of size 32. For the encoder, a pre-trained ResNet backbone initialized with ImageNet weights was used, which leveraged transfer learning to provide a strong starting point for feature extraction. The decoder, on the other hand, was trained from scratch. Optimization was performed using the Adam optimizer with an initial learning rate of 10^{-3}. To prevent overfitting and to refine learning, the learning rate was reduced by a factor of 10 if validation performance did not improve for 20 consecutive epochs. Additionally, model weights were saved whenever validation performance improved by more than 10^{-3}. The best model based on validation performance was retained, which could be restored after training for testing or deployment.

For inference, the input image was divided into overlapping patches (50%). Each patch was processed by a neural network that generated a confidence map representing the likelihood of each pixel belonging to the target structure. Overlapping regions were averaged to produce a smooth, artifact-free prediction. The final output was obtained by thresholding pixels with confidence above zero as foreground; others were background.

3.3   Evaluation Protocol

The EM-only images lack labels of ground truth, as cluster masks are only available for the EDX dataset. The evaluation aims to address two key aspects: first, the comparative performance of two training approaches – binary and confidence mask-based on segmentation metrics, and second the transferability of labels from ColorEM to the EM-only dataset.

Models trained with cluster masks are evaluated for each target structure in their respective test sets, which were set aside during training. While the cluster masks were used as-is during training, for evaluation they were manually corrected by removing evident false positives for a fair assessment. The evaluation metrics include precision, recall and region-based scores, namely Dice Similarity Coefficient (DSC) and Intersection over Union (IoU).

To evaluate the model's performance on the EM-only dataset, for which no EDX-based ground-truth is available, a proxy of ground-truth was synthesized by three separate field experts using ImageJ's multi-point tool through which the individual structures were identified. Subsequently, a separate evaluator compared the model's predictions to the identified structures and the number of true positives is reported upon.

3.4   Results

ColorEM-Net achieves image segmentation on the EM-only dataset by training binary segmentation models using cluster masks derived from EDX-based features for four selected target structures: heterochromatin, lysosomes, exocrine granules, and endocrine granules.

Segmentation performance on the EM-only dataset was evaluated using the pseudo-ground truth labels and model predictions based on the binary and

Table 1. Segmentation performance metrics comparing model predictions using binary-trained and confidence-weighted models. Evaluation metrics include Precision (P), Recall (R), Dice Similarity Coefficient (DSC), and Intersection over Union (IoU) across four cellular structures. The results are obtained from the ColorEM dataset where pseudo-ground truth is available.

Structure	Method	P	R	DSC	IoU
Heterochromatin	Binary/Confidence	0.78/0.82	0.80/0.82	0.79/0.81	0.65/0.69
Lysosomes	Binary/Confidence	0.71/0.75	0.86/0.85	0.78/0.80	0.64/0.67
Exocrine granules	Binary/Confidence	0.84/0.90	0.92/0.87	0.86/0.88	0.78/0.79
Endocrine granules	Binary/Confidence	0.73/0.73	0.98/0.94	0.83/0.82	0.71/0.70

confidence-weighted loss (Fig. 3). The application of the confidence-weighted model yielded substantial improvements in precision and recall for most target structures. Specifically, heterochromatin segmentation shows consistent enhancements, with reduced false positives and false negatives, highlighting the model's ability to refine predictions under confidence-based constraints (Table 1).

Predictions on the unlabeled EM-only dataset for the four structures are shown as colored overlays on the 120-tile dataset in Fig. 4. The displayed predictions demonstrate the model's performance across diverse unlabeled EM tiles, highlighting the spatial distribution of all four cellular structures.

Counts of subcellular structures are derived from 10% of the images in the EM-only dataset (12 tiles). Heterochromatin (nucleus), lysosomes, and zymogen were predicted with over 95% accuracy, closely matching annotator counts (Fig. 5). Predictions for endocrine granules, particularly glucagon and polypeptide were highly concentrated in regions corresponding to these granules, consistent with results from the ColorEM dataset. Insulin was excluded due to a severe distribution shift in image intensity or contrast, making it non-transferable. Among the considered granules, severe class imbalance from an uneven distribution of samples or pixels between granules and the background led to a reduced count, accounting for only 79% of the segmentation results. Also, quantitative analysis using biological labels identified 19 nuclei, 49 lysosomes, 169 exocrine granules and 706 endocrine granules. Note that the endocrine counts here only reflect glucagon and polypeptide granules. Overall, the confidence-weighted model excelled in image quality but not in count accuracy for the endocrine granules.

4   Discussion

Label-free segmentation of large-scale EM images was achieved using EDX-derived cluster labels as pseudo-ground truth. Integrating analytical features from additional EM modalities (e.g. EDX) offers a promising approach for achieving precise pixel-level segmentation of critical image features. By adopting cross-

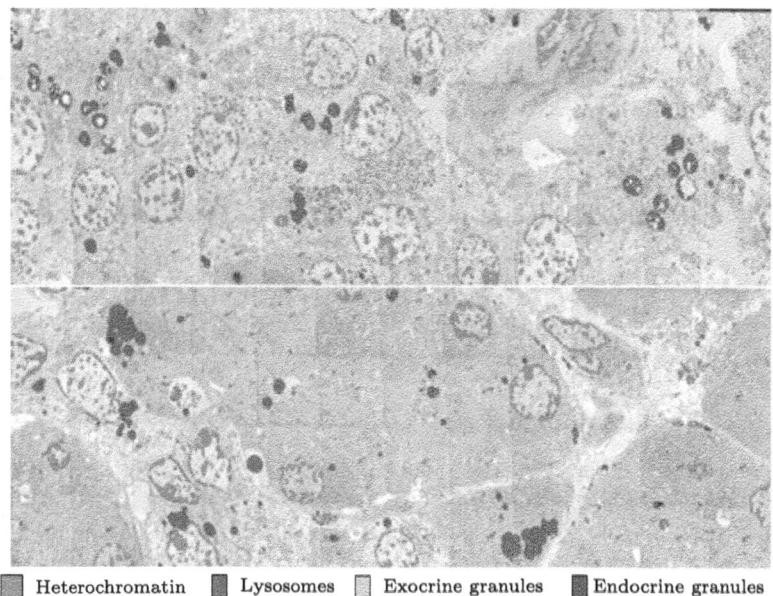

Heterochromatin ■ Lysosomes ▢ Exocrine granules ■ Endocrine granules

Fig. 4. Predictions of the trained binary models on the EM-only dataset.
The EM-only dataset consists of 120 grayscale tiles without EDX signals, in contrast to the 30-tile ColorEM dataset which includes EDX. The white line indicates regions omitted due to artifacts like folds or out-of-focus tiles.

modal insights, similar distributions of image regions were extracted, which are useful to segment grayscale EM in a discriminative manner.

The performance of CNNs depends on the choice of patch size, which must be optimized for each specific structure [18]. For example, the small size of endocrine granules may require smaller patch dimensions to avoid severe class imbalances.

One limitation we observed was the insufficient transferability of insulin-related structures, likely due to strong contrast differences between training and target datasets. Future work could explore contrast enhancement through targeted augmentation. Additionally, while mean subtraction was used to correct brightness variations, more advanced normalization methods, such as histogram matching or class-specific corrections, may further improve robustness.

While our method focuses on spectral features derived from elemental composition, an important direction for future work involves integrating morphological context. The shift from object-based pseudo-labels to biological measurements presents opportunities for improved transferability. However, different ultrastructural components are not always differentiable based on their elemental composition. Integrating morphology- or shape-based encodings particularly for structures like mitochondria, the Golgi apparatus, or the nuclear envelope could improve segmentation, as these organelles exhibit greater variability in appearance and shape, often appearing as larger-scale patterns in images. Using

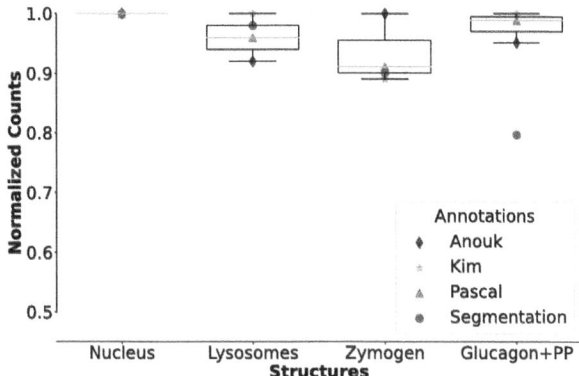

Fig. 5. Quantification of selected biostructures using expert annotations as ground truth and ColorEM-Net outputs as predictions show similar structural counts. The plot shows the variability among expert annotation and normalized counts (scaled by dividing by the maximum of all four counts) for each structure. The accuracy for heterochromatin (nucleus), lysosomes and zymogen (exocrine granules) is very close to all annotators, while that of endocrine granules for glucagon and polypeptide (PP) requires further improvement.

such morphological cues may enable more accurate segmentation of additional sub-cellular structures.

Additional EM and EDX data processed by the methods like those presented in [4,11] could be used to generate pseudo labels that are refined and corrected by experts to produce proper groundtruth labels. This will allow the application of fully and semi-supervised methods, against which label-free method like ours could be consistently benchmarked.

The current confidence score based on pixel distance to cluster centroids showed important improvements in segmentation quality. To build on these results, future work could explore incorporating cluster-level features such as size or density to further capture structural complexity and biological variability.

Finally, a promising direction for improving unsupervised segmentation is the integration of contrastive and self-supervised learning. Including non-target ultrastructural components, even as unlabeled regions, may support contrastive frameworks that help the model distinguish between subtle classes [15]. Self-supervised methods like SimCLR [8], MoCo [16], and BYOL [14] have shown that robust, discriminative visual representations can be learned from unlabeled data. Applying these techniques to EM images could complement our clustering-based confidence scoring by better capturing fine-grained structural similarities and differences. Moreover, contrastive pretraining may improve generalization in settings with sparse or noisy pseudo-labels.

5 Conclusions

Label-free segmentation in large-scale EM images using EDX-derived cluster labels demonstrates a promising approach for precise pixel-level segmentation without requiring manual annotations. The method allows for the extraction of biologically relevant features (using pixel-level information) that traditional object-based labeling approaches often overlook. The integration of biological measurements and advanced staining techniques offers a promising direction for improving feature discrimination and enabling the segmentation of more complex organelles, leading to improved biological interpretation of EM datasets.

Acknowledgments. We thank the University of Groningen for providing funding through the Center for Data Science and Systems Complexity, and the Center for Information Technology for access to the Hábrók high-performance computing cluster. Part of this work was performed at the UMCG Microscopy and Imaging Center (UMIC), sponsored by the NWO National Roadmap for Large-Scale Research Infrastructure (NEMI; NWO 184.034.014, NL-BioImaging: NWO 184.036.012); NWO-ENPPS.LIFT.019.030; The sample shown involves reused data from tissue provided by the Network for Pancreatic Organ donors with Diabetes (nPOD; RRID:SCR 014641). We also thank Jeroen Kuipers, Kim Kats, Anouk Wolters and Pascal de Boer for their expert assistance and feedback.

References

1. Lane et al., R.: Organelle segmentation facilitated by correlative light microscopy data. Microsc. Microanal. (2022)
2. Amigo, J.M., Babamoradi, H., Elcoroaristizabal, S.: Hyperspectral image analysis. a tutorial. Analytica chimica acta **896**, pp.34–51 (2015)
3. Aswath, A., Alsahaf, A.M.J., Giepmans, B.N.G., Azzopardi, G.: Segmentation in large-scale cellular electron microscopy with deep learning: a literature survey. Med. image anal. p. 102920 (2023)
4. Aswath, A., Duinkerken, B.H.P., Giepmans, B.N.G., Azzopardi, G., Alsahaf, A.M.J.: Interactive segmentation of biostructures through hyperspectral electron microscopy. 14th WHISPERS Workshop, IEEE (2024)
5. Binder, D.A.: Bayesian cluster analysis. Biometrika **65**(1), 31–38 (1978)
6. de Boer, P., Hoogenboom, J.P., Giepmans, B.N.G.: Correlated light and electron microscopy: ultrastructure lights up! Nat. Methods **12**(6), 503–513 (2015)
7. de Boer, P., et al.: Large-scale electron microscopy database for human type 1 diabetes. Nat. Commun. **11**(1), 2475 (2020)
8. Chen, T., Kornblith, S., Norouzi, M., Hinton, G.: A simple framework for contrastive learning of visual representations. In: International conference on machine learning, pp. 1597–1607. PmLR (2020)
9. Chen, X., Yuan, Y., Zeng, G., Wang, J.: Semi-supervised semantic segmentation with cross pseudo supervision. In: Proceedings of the IEEE/CVF conference on computer vision and pattern recognition, pp. 2613–2622 (2021)
10. Collinson, L.M., et al.: Volume em: a quiet revolution takes shape. Nat. Methods **20**(6), 777–782 (2023)

11. Duinkerken, B.H.P., Alsahaf, A.M.J., Hoogenboom, J.P., Giepmans, B.N.G.: Automated analysis of ultrastructure through large-scale hyperspectral electron microscopy. npj Imaging **2**(1),pp. 1–9 (2024)
12. Duinkerken, B.P., et al.: Sample processing and benchmarking for multibeam optical scanning transmission electron microscopy. Microsc. Microanal. **31**(2), ozaf024 (2025)
13. Gomez, C., Drost, A., Roger, J.M.: Analysis of the uncertainties affecting predictions of clay contents from VNIR/SWIR hyperspectral data. Remote Sens. Environ. **156**, 58–70 (2015)
14. Grill, J.B., et al.: Bootstrap your own latent-a new approach to self-supervised learning. Adv. Neural. Inf. Process. Syst. **33**, 21271–21284 (2020)
15. Guo, L.Z., Zhang, Y.G., Wu, Z.F., Shao, J.J., Li, Y.F.: Robust semi-supervised learning when not all classes have labels. Adv. Neural. Inf. Process. Syst. **35**, 3305–3317 (2022)
16. He, K., Fan, H., Wu, Y., Xie, S., Girshick, R.: Momentum contrast for unsupervised visual representation learning. In: Proceedings of the IEEE/CVF conference on computer vision and pattern recognition, pp. 9729–9738 (2020)
17. Kievits, A.J., Lane, R., Carroll, E.C., Hoogenboom, J.P.: How innovations in methodology offer new prospects for volume electron microscopy. J. Microsc. **287**(3), 114–137 (2022)
18. Litjens, G., et al.: A survey on deep learning in medical image analysis. Med. Image Anal. **42**, 60–88 (2017)
19. Liu, J., Xu, D., Yang, W., Fan, M., Huang, H.: Benchmarking low-light image enhancement and beyond. Int. J. Comput. Vision **129**, 1153–1184 (2021)
20. Mehrtash, A., Wells, W.M., Tempany, C.M., Abolmaesumi, P., Kapur, T.: Confidence calibration and predictive uncertainty estimation for deep medical image segmentation. IEEE Trans. Med. Imaging **39**(12), 3868–3878 (2020)
21. Milletari, F., Navab, N., Ahmadi, S.A.: V-net: fully convolutional neural networks for volumetric medical image segmentation. In: Fourth international conference on 3D vision (3DV), pp. 565–571 (2016)
22. Pirozzi, N.M., Hoogenboom, J.P., Giepmans, B.N.G.: Colorem: analytical electron microscopy for element-guided identification and imaging of the building blocks of life. Histochem. Cell Biol. **150**(5), 509–520 (2018)
23. Ronneberger, O., Fischer, P., Brox, T.: U-net: convolutional networks for biomedical image segmentation. MICCAI, pp. 234–241 (2015)
24. Schindelin, J., et al.: Fiji: an open-source platform for biological-image analysis. Nat. Methods **9**(7), 676–682 (2012)
25. Scotuzzi, M., et al.: Multi-color electron microscopy by element-guided identification of cells, organelles and molecules. Sci. Rep. **7**(1), 45970 (2017)
26. Shapson-Coe, A., et al.: A petavoxel fragment of human cerebral cortex reconstructed at nanoscale resolution. Science **384** (2024)
27. Yang, S.T., Lu, J.C., Tsao, Y.C.: Clustering and representative selection for high-dimensional data with human-in-the-loop. INFORMS J. Data Sci. (2025)

Object Detection and Applications

LTBoost: Boosting Recall Uniformity in Long-Tailed Learning

Morteza Mohammady Gharasuie[1]([✉]), Fengjio Wang[2], Ravi Mukkamala[1],
and Jiangwen Sun[1]

[1] Department of Computer Science, Old Dominion University, Norfolk, VA 23529,
USA
mmoha014@odu.edu, {mukka,jsun}@cs.odu.edu
[2] School of Computing, University of Utah, Salt Lake City, UT 84112, USA
u6053554@utah.edu

Abstract. Long-tail distribution is widespread in many practical applications, where most categories contain only a small number of samples. This imbalance poses significant challenges for recognizing underrepresented classes, since they can be easily biased towards dominant classes and perform poorly on tail classes. To address these issues, we propose LTBoost, a two-phase approach: (1) confusion-matrix-guided mixup augmentation to improve rare-class representation and (2) logit adjustment via meta-data optimization to reduce majority-class bias. LTBoost enhances recall uniformity, illustrated by balanced accuracy and geometric mean metrics. We also introduce the coefficient of variation to better assess the uniformity of the recall distribution across all classes when the geometric mean can be zero. Experiments confirm LTBoost's effectiveness in improving minority-class recognition and achieving balanced performance in real-world long-tailed datasets.

1 Introduction

Classifying camera trap images is vital for ecological research, supporting biodiversity monitoring, species identification, and conservation. Deep learning has enabled automated classification, reducing manual work and speeding up analysis [12]. However, real-world camera trap classification datasets are highly imbalanced, with a few common species (classes) dominating and many rare ones underrepresented [23]. This long-tailed distribution affects overall accuracy and especially reduces recall for rare species, highlighting the need for methods that ensure balanced and accurate performance across all classes.

Recent advances in long-tailed classification have shown success on benchmark datasets using techniques like re-sampling and re-weighting. Re-sampling techniques address data imbalance by either over-sampling rare classes or under-sampling common ones [16,22,24], whereas re-weighting methods incorporate class frequency-based weights into the loss function [3,18,20]. However, recent research highlights the limitations of these strategies on large-scale datasets

M. Castrillón-Santana et al. (Eds.): CAIP 2025, LNCS 15621, pp. 235–247, 2026.
https://doi.org/10.1007/978-3-032-04968-1_20

[9,11], as their effectiveness depends heavily on the specific dataset and context. This emphasizes the persistent challenge of developing reliable solutions for real-world applications.

To overcome these limitations we propose LTBoost, Long-Tail Boosting (LTBoost). Our method introduces confusion-matrix-guided mixup in the first phase, which selectively augments hard-to-classify class pairs based on empirical misclassification patterns, improving representation learning and distinguishability. In the second phase, we apply logit adjustment via meta-data optimization (MetaLA) as a post-training calibration step that shifts decision boundaries using a small balanced validation set to enhance performance on underrepresented classes. Additionally, we introduce the Coefficient of Variation (CV) to quantify recall dispersion across all classes. LTBoost achieves a more balanced and fair classification across all classes by minimizing CV while maximizing balanced accuracy and geometric mean.

Through extensive experiments on camera trap datasets, we demonstrate that LTBoost not only excels in rare class detection but also provides a robust solution for real-world classification tasks with improved recall uniformity. Our contributions are as follows:

- We propose LTBoost, a method that improves tail class prediction and uniformity of prediction among all classes, measured through geometric mean and balanced accuracy.
- We use the Coefficient of Variation (CV) as an additional metric for measuring the uniformity of class-wise recall values.
- We validate our approach on multiple camera trap datasets, showing consistent improvements in the accuracy and uniformity of LTBoost over state-of-the-art techniques.

By addressing the real-world challenges in long-tailed classification, our work contributes to more effective and reliable species identification in ecological research, ultimately aiding biodiversity conservation efforts.

2   Related Work

The classification of long-tailed datasets and camera trap images are two intersecting research areas that present unique challenges. This section reviews prior work in both domains, highlighting existing methods and their limitations.

2.1   Long-Tailed Classification

Long-tailed classification tackles class imbalance, where a few dominant classes contain most data while minority classes are underrepresented. Solutions fall into three broad categories:

Data-centric strategies include resampling methods like class-balanced [16,22,24] or progressive balancing [3], which adjust class distributions by oversampling rare or undersampling frequent classes, albeit risking overfitting or loss of majority-class information. Reweighting methods [3, 18, 20] complement these by modifying loss functions to prioritize minority classes through higher error costs. One the other hand, **Model-centric adjustments** involve logit adaptation [10, 19], which recalibrates pre-softmax scores using class priors or sample features to mitigate majority-class bias, and feature-space regularization [13], which preserves meaningful representations for rare classes via prototype-based learning or decoupled training strategies. Moreover, **Loss function innovations** address imbalance through specialized designs: focal loss [8] modulates class-specific penalties, asymmetric loss [14] separates positive/negative sample focus, and geometric mean loss [4] prioritizes worst-case classes to avoid majority-class dominance.

Although the aforementioned methods are effective on benchmarks, many struggle in real-world settings with severe class imbalance and rare but essential classes. In this regard, we propose a new technique that improves rare class classification and achieves more uniform class-wise recall on long-tailed datasets.

2.2 Camera Trap Image Classification

Camera trap images present classification challenges due to variable quality and extreme class imbalance. Traditional machine learning approaches initially relied on handcrafted features with classifiers like SVMs [25], but these lacked scalability for large datasets. Deep learning methods have been dominant in the field. Combining CNNs with active learning [12] has enhanced model performance. Additionally, transfer learning on small, high-quality datasets from citizen scientists [23] has improved accuracy while reducing extensive annotation needs. On the other hand, Edge/IoT solutions address real-time needs through lightweight models optimized for embedded systems [26], enabling on-device processing. However, Hybrid approaches leverage transfer learning and meta-learning to generalize across datasets, though platform comparisons [21] reveal persistent limitations in existing AI tools for unseen environmental conditions.

Despite progress in camera trap classification, real-world datasets remain imbalanced, with rare species poorly classified. Our method, LTBoost, addresses this by improving balanced accuracy and geometric mean, enhancing performance on challenging ecological datasets.

3 Methodology

We introduce *LTBoost*, a two-phase framework for camera trap animal classification. Phase 1 focuses on model training with augmentation strategies, and Phase 2 applies logit adjustment optimization via a small meta set (see Figs. 1 and 2).

3.1 Phase 1: Network Training

We train a ResNet-50 model on long-tailed camera trap datasets using two augmentation strategies:

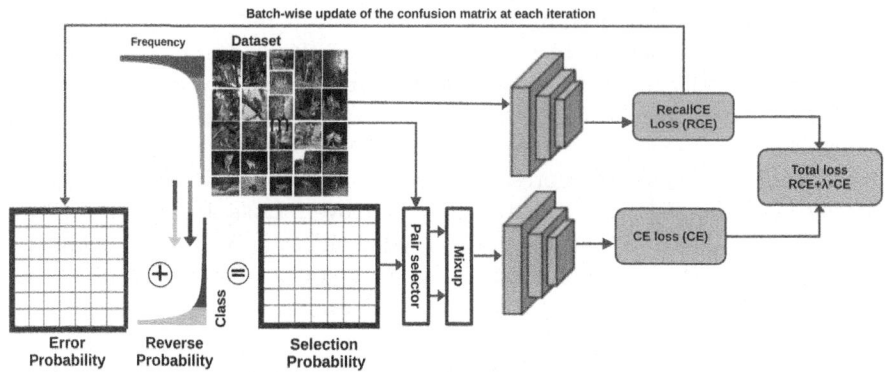

Fig. 1. LTBoost Phase 1: Confusion-matrix-guided mixup and loss functions. RP, EP and SP are defined in Eqs. 1, 2 and 3 respectively.

Default Augmentation and Loss Function: Standard augmentations—random cropping, flipping, and color jittering—are used to improve feature learning. However, uniform sampling overexposes the majority classes. To mitigate this, we apply recall cross-entropy loss [20], which penalizes errors on minority classes more heavily, promoting better recall despite residual class confusion.

Confusion-Matrix-Guided Sample Selection for Augmentation: This selection strategy is used alongside mixup augmentation to further improve representation learning under severe class imbalance. We compute class frequencies (CF), sort them in descending order, and derive reverse probabilities (RP) by inverting the normalized frequencies—assigning higher selection chances to rarer classes (Eq. 1). Unlike ECS-SC [6], which relies on predefined semantic hierarchies, our method leverages the model's confusion matrix to dynamically identify hard-to-classify class pairs during training, enabling adaptive and targeted mixup augmentation.

$$\text{RP}_i = \frac{\text{CF}_{C-i+1}}{\sum_{k=1}^{C} \text{CF}_k} \quad \text{for} \quad i = 1, 2, \ldots, C. \tag{1}$$

Relying solely on reverse probability overlooks inter-class confusion caused by label noise or visual similarity. To address this, we use the model's confusion matrix (CM), where $CM_{i,j}$ counts misclassifications of class i as j (with $CM_{i,j} = 0$), to extract misclassification frequencies and identify frequently confused class pairs. Pair classes with higher error rates receive greater sampling weights. The error probability (EP) is then computed by normalizing CM with class frequencies (CF), as shown in Eq. 2.

$$EP_{i,j} = \begin{cases} \frac{CM_{i,j}}{CF_i}, & i \neq j \\ 0, & i = j \end{cases} \tag{2}$$

We then merge the pairwise confusion probabilities (EP) with inverse class frequencies (RP) (by summing and normalizing) to guide a more informed, targeted sampling process. We compute the selection probability (SP) matrix by combining EP and RP:

$$SP_{i,j} = \frac{EP_{i,j} + RP_j}{\sum_r \sum_k EP_{r,k} + RP_k} \tag{3}$$

Finally, we use selection probabilities to prioritize rare and frequently misclassified classes. Based on these, we sample confusing and hard-to-classify class pairs and apply mixup augmentation in the input space to generate diverse, challenging examples. Training on these interpolated samples helps the model better distinguish difficult cases. We use standard cross-entropy loss weighted by a hyperparameter λ (Fig. 1), as the selection probabilities already account for bias and confusion. Since the effect of confusion-matrix-guided sampling varies across datasets due to differences in class distributions and sample difficulty, we empirically selected λ by evaluating performance (bAcc and GM) on the validation set.

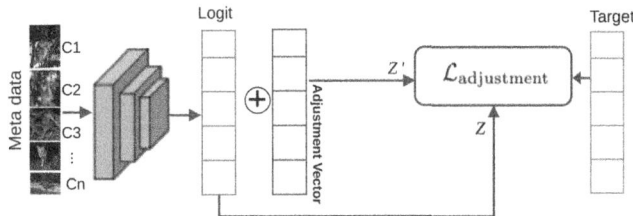

Fig. 2. *LTBoost* Phase 2: Logit adjustment using meta-data and a custom loss.

3.2 Phase 2: Logit Adjustment via Optimization

After training, we refine the model output using logit adjustment based on a balanced meta-data set—collected from the validation set—to improve calibration and performance on rare or confusing classes (see Fig. 2).

A meta-data set X_m, containing at least one sample per class with labels Y_m, is used to calibrate logits:

$$Z_m = f_\theta(X_m) \tag{4}$$

where f_θ is the trained model (not optimized in this phase), and Z_m are the presoftmax outputs. We then introduce a learnable adjustment vector R, modifying logits as:

$$Z'_m = Z_m + R \tag{5}$$

To guide optimization, we compute two cross-entropy losses:

$$L_1 = CE(Z_m, T_m), \quad L_2 = CE(Z'_m, T_m) \tag{6}$$

where T_m represents on-hot encoding of the true class labels (Y_m). The final adjustment loss is:

$$\mathcal{L}_{\text{adjustment}} = L_2 + \beta \max(0, L_2 - L_1) \tag{7}$$

where λ is a hyperparameter and is set to 1 in our experiments. It ensures that the adjustment process does not degrade the original classification performance while calibrating logits. The vector R is optimized via gradient descent for 500–1000 iterations:

$$R^{(t+1)} = R^{(t)} - \eta \nabla_R \mathcal{L}_{\text{adjustment}} \tag{8}$$

Once optimized, R is applied to test logits:

$$Z'_{\text{test}} = Z_{\text{test}} + R \tag{9}$$

This post-training adjustment leads to better-calibrated predictions and more balanced performance, especially for underrepresented classes.

4   Experimental Study

This section presents the results of our approach on various camera trap datasets and compares it with recent long-tailed image classification methods [3, 4, 10, 20]. It also includes a detailed analysis of individual components within our method, as well as the baseline approaches.

4.1   Datasets

We evaluate our method on four diverse camera trap datasets: Island Conservation [2], Snapshot Mountain Zebra [17], Orinoquia [21], and Ohio Small Animals [1]. These datasets span various ecosystems, regions, and species, capturing a wide range of wildlife and environmental contexts for robust evaluation. Table 1 summarizes each dataset's size and number of classes.

We evaluate long-tailed classification using *geometric mean, balanced accuracy*, and *Coefficient of Variation*. After filtering classes with fewer than three samples, datasets were split into train/validation/test sets using stratified sampling to ensure class presence across all splits.

Table 1. Summary of Camera Trap Datasets

Dataset	# Images	# Classes
Island Conservation [2]	123,000	49
Snapshot Mountain Zebra [17]	73,034	54
Orinoquia [21]	104,782	51
Ohio Small Animals [1]	118,554	45

Table 2. Dataset-Specific Hyperparameters and Training Settings for LTBoost

Dataset	λ	β	Meta Samples per Class	MetaLA Iterations	Phase 1 Learning Rate	Phase 2 Learning Rate	Optimizer (Both Phases)
Ohio Small Animals	0.5	1	1	1000	0.001	0.0005	Adam
Orinoquia	1.3	1	1	1000	0.001	0.0005	Adam
Island Conservation	1.3	1	1	500	0.001	0.0005	Adam
Snapshot Mountain Zebra	0.5	1	1	1000	0.001	0.0005	Adam

4.2 Implementation Details

We trained a ResNet-50 model with a batch size of 16 and applied early stopping across all datasets. Table 2 outlines dataset-specific hyperparameters, including loss weight (λ and β), meta-data configuration for MetaLA, and phase-specific learning rates. A balanced meta set with one sample per class was used, and all experiments employed the Adam optimizer.

The results are presented as the mean and standard deviation over four independent runs, using random seeds 42, 82, 220, and 100 to account for variability. We include CDMAD [7], which originally applies logit adjustment during training iterations using outputs of model from a white image. Because the original approach harms the learning process, we adapt CDMAD as a post-hoc method, applying adjustment via white image after training only at test time. This yields more stable and comparable results. We also evaluate post-hoc LA [10] and test GML [4] on a pre-trained ResNet-50 using the same experimental setup. Instead of training from scratch, we fine-tune ImageNet-pretrained weights from the torchvision library with each method.

4.3 Evaluation Metrics

In long-tailed classification, traditional metrics can be misleading as they favor majority classes. To address this, balanced Accuracy (bAcc) and Geometric Mean (GM) of per-class recalls are used. bAcc reduces majority bias by averaging recall values across all classes, while the GM highlights rare class performance by emphasizing low-recall values. While the GM is sensitive to zero values and may collapse when some classes receive no predictions, the CV provides a stable, interpretable, and fair measure of recall uniformity, especially when the GM becomes uninformative. The equation for the CV is given below.

$$CV = \frac{\sigma}{\mu} \tag{10}$$

where σ and μ are the standard deviation and mean of class-wise recall values, respectively. Note that a lower CV signifies a more uniform recall value across all classes.

5   Results

This section presents experimental results for LTBoost, comparing it to standard baselines and some state-of-the-art methods using GM, bAcc, and CV. We also analyze the effects of confusion-matrix-guided mixup and MetaLA via per-class performance.

5.1   Overall Performance Comparison

By monitoring bAcc, GM and CV along with overall performance metrics (macro-averaged Precision, Recall and F1 score), we can better understand the impact of enhancing uniformity on minority class detection and prediction consistency across all classes. Tables 3, 4, 5, and 6 group bAcc, GM, and CV in one column to demonstrate how these metrics complement overall performance metrics. The tables also categorize rows to distinguish our approach (R50TL+CMG for phase 1 and LTBoost for both phases) from existing methods, underscoring the trade-offs and improvements achieved by different optimization strategies. From the Tables, we also observe that the GM is 0.0 for particular datasets, indicating it does not provide sufficient information for comparison; CV offers a more informative perspective in these cases.

Table 3 and Table 4 present the performance of different methods on the Ohio Small Animal and Orinoquia datasets, respectively. R50TL+CMG leads in precision, recall, and F1-score in both datasets highlighting the benefit of the Confusion-Matrix-Guided augmentation. On the other hand, LTBoost's post-training logit calibration secures its dominance in long-tail metrics (bAcc, GM, CV). Baseline methods—R50TL and its RecallCE, class-balanced loss, GML, CDMAD, and LA variants—show some improvements but struggle with recall uniformity, often yielding lower GM and higher CV. Overall, R50TL+CMG excels on standard performance measures, while LTBoost delivers more balanced and stable predictions across all classes.

As shown in Tables 5 and 6, the transfer-learning baseline (R50TL) attains the highest macro precision, recall, and F1, yet completely fails to recover certain minority classes. In contrast, R50TL+CMG and LTBoost accept a slight reduction in those aggregate scores to deliver far stronger long-tail performance. In particular, LTBoost boosts balanced accuracy, while simultaneously reducing the coefficient of variation. These findings underscore the inherent trade-off between maximizing overall accuracy and ensuring equitable, stable recognition across all classes.

Table 3. Performance Comparison for Ohio Small Animal Dataset

Method	Precision	Recall	F1	bAcc	GM	CV
ResNet50 (R50) [5]	0.9473 ± 0.00004	0.9467 ± 0.00004	0.9462 ± 0.00004	0.8676 ± 0.00003	0.5658 ± .2402	0.1858 ± 0.00119
R50+Transfer Learning (R50TL) [15]	0.9627 ± 0.00001	0.9624 ± 0.00001	0.9621 ± 0.00002	0.8798 ± .00279	0.6004 ± 0.27066	0.1891 ± 0.0053
R50TL+RecallCE [20]	0.9624 ± 0.00002	0.9605 ± 0.00004	0.9609 ± 0.00003	0.9127 ± 0.00087	0.602 ± 0.27344	0.1511 ± 0.00424
R50TL+Class-Balanced [3]	0.9395 ± 0.00019	0.9375 ± 0.00017	0.9377 ± 0.00019	0.8983 ± 0.001187	0.8725 ± 0.00185	0.1452 ± 0.0017
R50TL+GML [4]	0.9139 ± 0.00007	0.9355 ± 0.00003	0.921 ± 0.00023	0.9206 ± 0.00014	0.0 ± 0.0	0.2343 ± 0.00231
R50TL+CDMAD [7]	0.9474 ± 0.00004	0.9467 ± 0.00004	0.9462 ± 0.00004	0.8676 ± 0.00003	0.5658 ± 0.24022	0.1858 ± 0.00119
R50TL+LA [10]	0.959 ± 0.00001	0.9581 ± 0.00001	0.9571 ± 0.00001	0.8735 ± 0.00142	0.5894 ± 0.26062	0.1 ± 0.05438
R50TL+CMG	**0.9697 ± 0.00001**	**0.9696 ± 0.00001**	**0.9694 ± 0.00001**	0.9333 ± 0.00021	0.9259 ± 0.00033	0.1069 ± 0.00046
LTBoost	0.9689 ± 0.00002	0.9678 ± 0.00003	0.968 ± 0.00003	**0.9466 ± 0.00017**	**0.9397 ± 0.00031**	**0.0863 ± 0.00055**

Table 4. Performance Comparison for Orinoquia Dataset

Method	Precision	Recall	F1	bAcc	GM	CV
ResNet50 (R50) [5]	0.87330 ± 0.00010	0.8668 ± 0.00019	0.8664 ± 0.00019	0.7358 ± 0.00088	0.0 ± 0.0	0.3233 ± 0.00078
R50+Transfer Learning (R50TL) [15]	0.8909 ± 0.00013	0.8827 ± 0.00027	0.8838 ± 0.00022	0.7781 ± 0.00009	0.0 ± 0.0	0.2753 ± 0.00039
R50TL+RecallCE [20]	0.8848 ± 0.00019	0.9114 ± 0.00191	0.8801 ± 0.00023	0.8199 ± 0.00017	0.0 ± 0.0	0.268 ± 0.0026
R50TL+Class-Balanced [3]	0.8341 ± 0.00034	0.830 ± 0.00029	0.8308 ± 0.00033	0.8226 ± 0.00192	0.2802 ± 0.23548	0.2327 ± 0.00377
R50TL+GML [4]	0.8312 ± 0.00044	0.7548 ± 0.00361	0.7726 ± 0.00264	0.802 ± 0.00363	0.499 ± 0.1915	0.2575 ± 0.0089
R50TL+CDMAD [7]	0.8939 ± 0.00074	0.8921 ± 0.00093	0.8911 ± 0.00088	0.7171 ± 0.0017	0.0 ± 0.0	0.3252 ± 0.00072
R50TL+LA [10]	0.886 ± 0.00008	0.8316 ± 0.00225	0.8521 ± 0.00107	0.8176 ± 0.00107	0.5352 ± 0.0004	0.2133 ± 0.00023
R50TL+CMG	**0.8998 ± 0.00001**	**0.8951 ± 0.00001**	**0.8948 ± 0.00003**	0.8431 ± 0.00122	0.5572 ± 0.2339	0.2077 ± 0.09195
LTBoost	0.8804 ± 0.00032	0.8510 ± 0.00074	0.8569 ± 0.00073	**0.8689 ± 0.00062**	**0.7455 ± 0.03776**	**0.1817 ± 0.0022**

Moreover, in the second phase across all datasets, MetaLA consistently outperforms the frequency-based adjustment (R50TL+LA) [10] and other minority-class enhancement techniques (GML and CDMAD). Its higher bAcc, higher non-zero GM, and lower coefficient of variation (CV) demonstrate effectiveness of adjustment via optimization in improving the performance on tail classes.

Table 5. Performance Comparison for Island Conservation Dataset

Method	Precision	Recall	F1	bAcc	GM	CV
ResNet50 (R50) [5]	0.9251 ± 0.00000	0.9279 ± 0.00000	0.9257 ± 0.00000	0.6932 ± 0.00047	0.0 ± 0.0	0.4488 ± 0.0007
R50+Transfer Learning (R50TL) [15]	**0.9305 ± 0.00001**	**0.9339 ± 0.00002**	**0.9310 ± 0.00001**	0.6796 ± 0.00008	0.0 ± 0.0	0.4696 ± 0.0011
R50TL+RecallCE [20]	0.9162 ± 0.00001	0.9043 ± 0.00002	0.9072 ± 0.00002	0.7809 ± 0.00008	0.0 ± 0.0	0.3599 ± 0.00267
R50TL+Class-Balanced	0.9002 ± 0.0006	0.8763 ± 0.0027	0.8847 ± 0.0018	0.783 ± 0.0001	0.0 ± 0.0	0.339 ± 0.00022
R50TL+GML [4]	0.8988 ± 0.0001	0.8046 ± 0.0003	0.8347 ± 0.0018	0.8229 ± 0.0003	0.0 ± 0.0	0.2529 ± 0.00689
R50TL+CDMAD [7]	0.9251 ± 0.00001	0.9279 ± 0.00001	0.9257 ± 0.0002	0.6932 ± 0.00047	0.0 ± 0.0	0.4488 ± 0.0007
R50TL+LA [10]	0.9291 ± 0.00008	0.9289 ± 0.00003	0.9280 ± 0.00003	0.7905 ± 0.00069	0.0 ± 0.0	0.331 ± 0.00146
R50TL+CMG	0.9166 ± 0.00002	0.9100 ± 0.00006	0.9189 ± 0.00011	0.7642 ± 0.00031	0.0 ± 0.0	0.3724 ± 0.00029
LTBoost	0.9107 ± 0.00003	0.8623 ± 0.00222	0.8875 ± 0.00041	**0.8293 ± 0.00046**	0.0 ± 0.0	**0.2349 ± 0.00016**

5.2   Ablation Study

We conduct an ablation study to analyzing class-wise performance and the contribution of each component in our two-phase design, validating their impact on long-tailed distribution challenges.

Class-Wise Comparison: MetaLA, a core part of LTBoost in the second phase, improves model performance on underrepresented classes. Figure 3 shows that MetaLA improves recall mostly among non-head classes. This improvement leads to a more uniform recall distribution and demonstrates MetaLA's effectiveness in reducing majority-class bias, enhancing rare species recognition in real-world camera trap applications.

Table 6. Performance Comparison for Snapshot Mountain Zebra Dataset

Method	Precision	Recall	F1	bAcc	GM	CV
ResNet50 (R50) [5]	0.9452 ± 0.000003	0.9497 ± 0.00003	0.94603 ± 0.000002	0.2734 ± 0.00013	0.0 ± 0.0	0.9961 ± 0.00212
R50+Transfer Learning (R50TL) [15]	**0.9550 ± 00001**	**0.9543 ± 0.0000001**	0.9529 ± 0.000001	0.2936 ± 0.0007	0.0 ± 0.0	0.9445 ± 0.00043
R50TL+RecallICE [20]	0.9455 ± 0.000004	0.9261 ± 0.00001	0.9336 ± 0.00001	0.3215 ± 0.00030	0.0 ± 0.0	0.9112 ± 0.00005
R50TL+Class-Balanced [3]	0.876 ± 0.00265	0.8741 ± 0.00073	0.8736 ± 0.00143	0.1933 ± 0.04902	0.0 ± 0.0	2.5882 ± 3.1415
R50TL+GML [4]	0.9386 ± 0.000003	0.8005 ± 0.00086	0.8567 ± 0.00034	0.3306 ± 0.00083	0.0 ± 0.0	0.8828 ± 0.00197
R50TL+CDMAD [7]	0.953 ± 0.00002	0.951 ± 0.00004	0.9529 ± 0.00003	0.2935 ± 0.0007	0.0 ± 0.0	0.9445 ± 0.00043
R50TL+LA [10]	0.9487 ± 0.00002	0.8494 ± 0.00476	0.892 ± 0.00164	0.3587 ± 0.00148	0.0 ± 0.0	0.7858 ± 0.0179
R50TL+CMG	0.9419 ± 0.000001	0.9257 ± 0.00008	0.9323 ± 0.00002	0.3181 ± 0.00008	0.0 ± 0.0	0.9371 ± 0.0073
LTBoost	0.9484 ± 0.00002	0.791 ± 0.00005	0.8607 ± 0.000012	**0.3737 ± 0.00007**	0.0 ± 0.0	**0.7774 ± 0.0007**

Table 7. Balanced accuracy on head, middle, tail, and overall subsets for each method (seed=42).

Dataset	Subset	R50TL	R50TL+CMG	LTBoost
Ohio Small Animals	Head	0.9305	0.9525	0.9581
	Middle	0.8391	0.9465	0.9514
	Tail	0.8907	0.9353	0.9449
	Overall	0.8854	0.9453	0.9505
orinoquia	Head	0.8152	0.8682	0.8444
	Middle	0.7649	0.7799	0.8001
	Tail	0.7611	0.8861	0.89
	Overall	0.7797	0.84202	0.8452
Island Conservation	Head	0.8231	0.8198	0.8453
	Middle	0.6312	0.7233	0.8149
	Tail	0.6167	0.75	0.8214
	Overall	0.6889	0.7634	0.8263
Snapshot Mountain Zebra	Head	0.5591	0.5573	0.4362
	Middle	0.2608	0.3279	0.4882
	Tail	0.0256	0.0833	0.1218
	Overall	0.2808	0.32312	0.3572

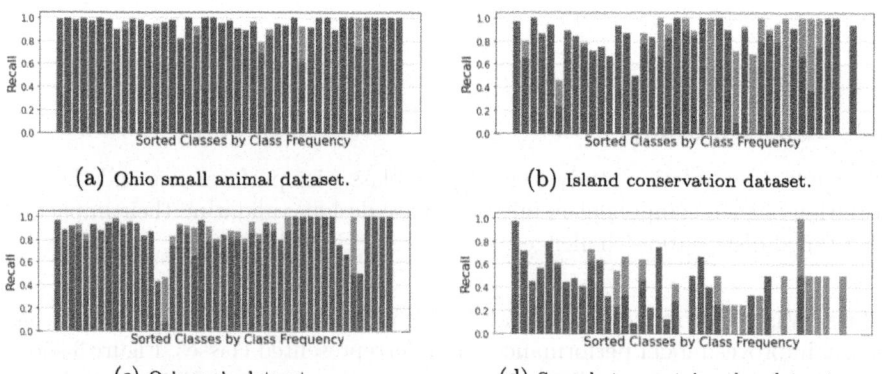

(a) Ohio small animal dataset.

(b) Island conservation dataset.

(c) Orinoquia dataset.

(d) Snapshot mountain zebra dataset.

Fig. 3. Class-wise recall before and after MetaLA adjustment. Blue and orange bars show pre- and post-adjustment values; red indicates their overlap. (Color figure online)

Impact of Components in Two Phases: Table 7 presents an ablation study illustrating the improvement achieved by adding CMG and then MetaLA, boosting on top of the base R50TL model. The classes in each dataset are divided into head, middle, and tail groups by first ranking them in descending order of sample counts and splitting them into three equal-sized bins (top third as head, middle third as medium, bottom third as tail), offering a closer look at how each method tackles the long-tail challenge. As seen across all four datasets, R50TL+CMG consistently provides improved bAcc over R50TL, particularly in the middle and tail subsets, indicating that this additional component effectively enhances performance on underrepresented classes. LTBoost further extends these improvements, showing the strongest performance overall, suggesting its additional boosting mechanisms more robustly handle the imbalance across three groups, with the most impact on the middle and tail groups.

6   Conclusion

In this work, we proposed LTBoost, a novel approach to improve recall uniformity in long-tailed camera trap datasets. LTBoost operates through two key phases: Confusion-Matrix Guided Mixup (CMG) to enhance representation learning and Logit Adjustment via Optimization on meta-data set (MetaLA) to refine decision boundaries, particularly for underrepresented classes. Our experimental results demonstrate that LTBoost effectively mitigates class imbalance, improving balanced accuracy (bAcc) and geometric mean (GM). However, when the geometric mean is zero, it becomes uninterpretable and fails to provide meaningful comparisons across different methods. To address this limitation, we introduced the Coefficient of Variation (CV) as a complementary metric, demonstrating its usefulness in such cases by capturing recall variability and ensuring more reliable evaluations of recall class balance. Our findings highlight LTBoost's ability to produce more stable and equitable predictions across class distributions.

References

1. Balasubramaniam, S.: Optimized classification in camera trap images: an approach with smart camera traps, machine learning, and human inference. Master's thesis, The Ohio State University (2024). https://lila.science/datasets/ohio-small-animals/
2. Conservation, I.: Island conservation camera trap dataset. https://lila.science/datasets/island-conservation-camera-traps/. Accessed 15 Feb 2024
3. Cui, Y., Jia, M., Lin, T.Y., Song, Y., Belongie, S.: Class-balanced loss based on effective number of samples. In: Proceedings of the IEEE/CVF Conference on Computer Vision and Pattern Recognition, pp. 9268–9277 (2019)
4. Du, Y., Wu, J.: No one left behind: improving the worst categories in long-tailed learning. In: Proceedings of the IEEE/CVF Conference on Computer Vision and Pattern Recognition, pp. 15804–15813 (2023)

5. He, K., Zhang, X., Ren, S., Sun, J.: Deep residual learning for image recognition. In: Proceedings of the IEEE Conference on Computer Vision and Pattern Recognition (CVPR), pp. 770–778 (2016). https://doi.org/10.1109/CVPR.2016.90

6. He, W., Xu, J., Shi, J., Zhao, H.: ECS-SC: long-tailed classification via data augmentation based on easily confused sample selection and combination. Expert Syst. Appl. **246**, 123138 (2024)

7. Lee, H., Kim, H.: CDMAD: class-distribution-mismatch-aware debiasing for class-imbalanced semi-supervised learning. In: Proceedings of the IEEE/CVF Conference on Computer Vision and Pattern Recognition, pp. 23891–23900 (2024)

8. Li, B., et al.: Equalized focal loss for dense long-tailed object detection. In: Proceedings of the IEEE/CVF Conference on Computer Vision and Pattern Recognition, pp. 6990–6999 (2022)

9. Mahajan, D., et al.: Exploring the limits of weakly supervised pretraining. In: Proceedings of the European Conference on Computer Vision (ECCV), pp. 181–196 (2018)

10. Menon, A.K., Jayasumana, S., Singh Rawat, A., Jain, H., Veit, A., Kumar, S.: Long-tail learning via logit adjustment. In: International Conference on Learning Representations (2020)

11. Mikolov, T., Sutskever, I., Chen, K., Corrado, G.S., Dean, J.: Distributed representations of words and phrases and their compositionality. In: Advances in Neural Information Processing Systems, vol. 26 (2013)

12. Norouzzadeh, M.S., Morris, D., Beery, S., Joshi, N., Jojic, N., Clune, J.: A deep active learning system for species identification and counting in camera trap images. Methods Ecol. Evol. **12**(1), 150–161 (2021)

13. Parisot, S., Esperança, P.M., McDonagh, S., Madarasz, T.J., Yang, Y., Li, Z.: Long-tail recognition via compositional knowledge transfer. In: Proceedings of the IEEE/CVF Conference on Computer Vision and Pattern Recognition, pp. 6939–6948 (2022)

14. Park, W., Park, I., Kim, S., Ryu, J.: Robust asymmetric loss for multi-label long-tailed learning. In: Proceedings of the IEEE/CVF International Conference on Computer Vision, pp. 2711–2720 (2023)

15. Paszke, A., et al.: PyTorch: an imperative style, high-performance deep learning library. In: Advances in Neural Information Processing Systems (NeurIPS), pp. 8024–8035 (2019)

16. Shi, J.X., Wei, T., Xiang, Y., Li, Y.F.: How re-sampling helps for long-tail learning? In: Advances in Neural Information Processing Systems, vol. 36, pp. 75669–75687 (2023)

17. Snapshot Mountain Zebra: Snapshot Mountain Zebra (2018). https://lila.science/datasets/snapshot-mountain-zebra

18. Tan, J., et al.: Equalization loss for long-tailed object recognition. In: Proceedings of the IEEE/CVF Conference on Computer Vision and Pattern Recognition, pp. 11662–11671 (2020)

19. Tao, Y., et al.: Local and global logit adjustments for long-tailed learning. In: Proceedings of the IEEE/CVF International Conference on Computer Vision, pp. 11783–11792 (2023)

20. Tian, J., Mithun, N.C., Seymour, Z., Chiu, H.P., Kira, Z.: Recall loss for imbalanced image classification and semantic segmentation (2021)

21. Vélez, J., et al.: An evaluation of platforms for processing camera-trap data using artificial intelligence. Methods Ecol. Evol. **14**(2), 459–477 (2023). https://lila.science/datasets/orinoquia-camera-traps/

22. Wang, Y.X., Ramanan, D., Hebert, M.: Learning to model the tail. In: Advances in Neural Information Processing Systems, vol. 30 (2017)

23. Willi, M., et al.: Identifying animal species in camera trap images using deep learning and citizen science. Methods Ecol. Evol. **10**(1), 80–91 (2019)

24. Wu, T., Huang, Q., Liu, Z., Wang, Yu., Lin, D.: Distribution-balanced loss for multi-label classification in long-tailed datasets. In: Vedaldi, A., Bischof, H., Brox, T., Frahm, J.-M. (eds.) ECCV 2020. LNCS, vol. 12349, pp. 162–178. Springer, Cham (2020). https://doi.org/10.1007/978-3-030-58548-8_10

25. Yu, X., Wang, J., Kays, R., Jansen, P.A., Wang, T., Huang, T.: Automated identification of animal species in camera trap images. EURASIP J. Image Video Process. **2013**(1), 1–10 (2013). https://doi.org/10.1186/1687-5281-2013-52

26. Zualkernan, I., Dhou, S., Judas, J., Sajun, A.R., Gomez, B.R., Hussain, L.A.: An IoT system using deep learning to classify camera trap images on the edge. Computers **11**(1), 13 (2022)

Enhancing Agricultural Disease Diagnosis: YOLO-Based Detection of Root Rot in Beans

Renato Cristiano Torres[1] and Díbio Leandro Borges[2(✉)]

[1] Brazilian Agricultural Research Corporation, Brasília 6041, Brazil
renato.torres@embrapa.br
[2] Department of Computer Science, University of Brasília, Brasília, DF, Brazil
dibio@unb.br

Abstract. Fungi are living organisms that inhabit the soil and can cause severe root diseases in various crops. These diseases result from the interaction between the pathogen, the host, and the biotic and abiotic components of the soil. These fungi are generally resilient and remain inactive in the absence of the host plant. However, in the presence of a vulnerable host in the rhizosphere or the absence of adequate nutrients, these resilient structures infect the plant. The fungi in the soil can spread to other plants, in some cases without inducing symptoms that may be visible, and it can survive in crop residues. Therefore, this set of characteristics of biology, ecology, and resilience in the soil results in a complex management situation in cases of root diseases caused by fungi, which mainly cause root rot. Analysis methods are usually based on manually scoring the severity of the disease. The method proposed in this work was to create a system capable of detecting the presence of fungi that cause root rot leveraging a state-of-the-art convolutional neural network model, a YOLO-based object detection framework, and to evaluate its reliability and accuracy by comparing it with other recent work on predicting the state of plant health. The proposed network was trained with 125 samples from 15 bean genotype lines. We obtained a confidence level of 85% in detecting roots and crowns attacked by fungi, and it is a promising tool for root plant health analysis regarding its performance and the real possibility of automating such diagnostics.

Keywords: Fungi root diseases · YOLO-based detection · Phytopathology · Image analysis · Deep learning applications

1 Introduction

Root diseases are common in many crops and are caused by a series of phytopathogens in the soil. These diseases mainly affect the roots of plants. However, this type of disease does not only affect the roots; in contrast, it affects the entire plant, and, in some cases, its aerial part is affected before the roots. The

M. Castrillón-Santana et al. (Eds.): CAIP 2025, LNCS 15621, pp. 248–258, 2026.
https://doi.org/10.1007/978-3-032-04968-1_21

fungi that cause root diseases survive in the soil for a long time, up to several years, in the absence of a host, be it in seeds, grains, or even in plant remains. Their survival method and the difficulty in eliminating them from the soil are some of the most significant problems faced in the management of these diseases [14].

Pathogenic fungi associated with the root system can be ecologically classified into non-specialized and specialized. Non-specialized pathogenic species persist in these conditions not only because of their competitive saprophytic capacity, which allows them to live on organic matter remains in the absence of their specific hosts. Saprophytism is the primary form of life of these fungi, and parasitism is a condition enhanced by suitable environmental conditions. Non-specialized pathogenic fungi include the causal agents of seed rot, seedling damping, stem cancer, and root rot, such as *Rhizoctonia solani*, *Sclerotium rolfsii*, *Macrophomina phaseolina*, and *Fusarium solani*. In turn, specialized fungi are characterized by the short period they remain in the soil, which coincides with their association with the host. Specialized pathogenic fungi include those that cause vascular wall damage, such as *Fusarium oxysporum*, *Verticillium albo-atrum*, and *Verticillium dahliae* [12]. Manual analysis can be performed by extracting a plant sample from a given area and using a photo or the naked eye; the severity and presence of fungal diseases can be classified.

Object detection is an important field in the domain of computer vision and various machine learning (ML) and deep learning (DL) models are employed to improve the effectiveness of the object detection process and related tasks.

The evolution of deep learning has enabled the automatic learning of characteristic details in objects through neural networks, which has led to advances in computer vision. The field has moved towards two main detection paradigms: 2-stage methods with single-stage prediction mechanisms and unified prediction frameworks through a pipeline that first identifies regions of interest (through RPNs or selective search) before conducting region-specific classification and localization refinement [19].

The R-CNN framework developed by Girshick et al. [17] established a new paradigm in object detection using region-based convolutional feature extraction. To address some computational inefficiency of R-CNN in processing numerous overlapping proposals, He et al. [8] developed SPP-Net by implementing spatial pyramid pooling between convolutional and FC layers to reduce processing overhead. Girshick's Fast R-CNN [16] made contributions to object detection by addressing limitations in R-CNN and SPP-Net, setting new standards for real-time performance. Ren et al. Faster R-CNN [20] generated new benchmarks in detection efficiency by integrating RPN into the convolutional backbone, enabling overall optimization.

By adding instance segmentation to the detection pipeline, He et al. [20] promoted an evolution of Faster R-CNN to the Mask R-CNN model to make it more comprehensive. Cai and Vasconcelos [25] proposed Cascade R-CNN to overcome the limitations of fixed IoU thresholds in R-CNN models by using a series of detection heads with increasing IoU values for fine-grained object recognition.

The YOLO-based network, which has come to be used in several applications for object detection and recognition in various contexts, has performed very well compared to its previous detector counterparts [2,7].

The structure of a YOLOv8 is simple, and it can directly generate the position and category of the bounding box through the neural network. The speed of YOLOv8 is fast compared to most of its previous detector counterparts because it only needs to put the image into the network to get the final detection result without any preprocessing, and it has a strong generalization ability since it can learn highly generalized features to be transferred to other fields [23,24], converting the target detection problem into a regression one, and the detection accuracy is improved.

Due to the loss function, the positioning error is the main reason for improving detection efficiency. Especially, the handling of large and small objects needs to be reinforced, so establishing the bounding boxes to specify which classes they belong to generates input data, which is a fundamental step to obtain a good learning response and subsequent detection [7].

Advances in the automation of phenotyping have been achieved with the works [1,3,22], and the works of [19] and [4] build specific tools for root treatment and analysis using a Bayesian approach. However, we note that there is operational complexity, and although several advances have been made, the process is costly. Pierz et al. [14] developed a Python-based analysis pipeline called RootDS to improve disease severity phenotyping in the pipeline in common bean inoculated with Fusarium root rot. They show quantitative disease scores and root area generated by this pipeline had a strong correlation with manually curated values. Furthermore, it provided a broader capture of variation than manual disease scores.

In this way, we developed a system that automatically assesses its phytosanitary status using images of the root system without the need for a trained classifier during the use of the system, and which can be easily embedded in portable devices, such as RaspBerry, Jetson, among others, to be used in the field. Highlights of our method include:

- An enhanced pipeline for diagnosing root rot in bean plants;
- A score to classify the severity of infestations in images;
- A real-time approach that could be deployed for use in practical settings.

A flowchart of the approach is shown in Fig. 1.

2   Materials and Methods

2.1   Image Dataset

The image set from the work of Pierz et al. [14] was used, where the authors stated that the genotypes used in this study were chosen from a previous common bean population screened for the fungus Fusarium brasiliense (known to cause

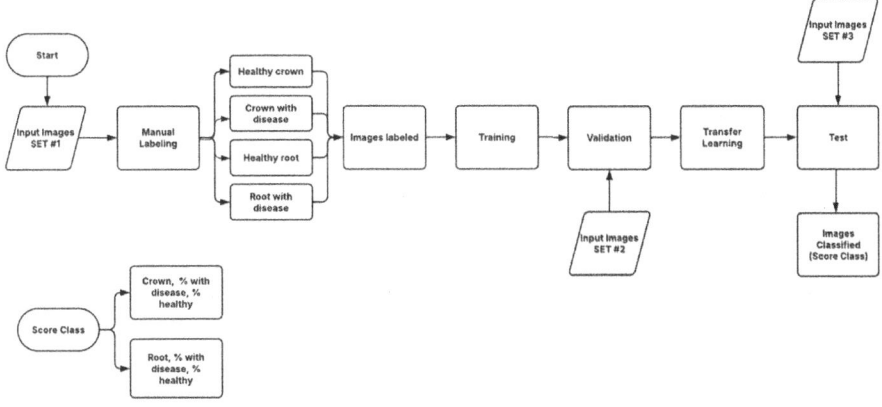

Fig. 1. A flowchart of the proposed approach in this work.

the disease known as Root Rot) to represent diversity in disease severity. From the Cal96 × MLB49-89A population, 13 diverse recombinant inbred lines and the two parental lines were included for a total of 15 lines used, shown in Fig. 2.

According to Pierz et al. [14], seven days after inoculation, plants were harvested, and roots were manually scored for disease severity on a scale of 1 to 9, where "1" indicates no disease symptoms or discoloration and "9" indicates severe discoloration and extreme necrosis. Disease scores were averaged across scorers. Root images were collected using an Epson Perfection V550 photo scanner (Epson, Tokyo, Japan) to preserve quality, lighting, and consistency. Roots were separated from the stem, and the stem color changed from white to purple or green. Roots were dried in an oven for three days and then weighed for biomass.

2.2   A YOLO-Based Detection Model for the Disease Diagnosis

YOLOv8 is an anchorless model that directly predicts an object's center instead of the offset of a known anchor box; that model was a complex part of the previous YOLO models [21]. Anchorless detection reduces the number of box predictions, which speeds up Non-Maximum Suppression (NMS) and, consequently, the identification. The initial convolution was changed from 6×6 to 3×3 and the kernel size from 1×1 to 3×3, optimizing the classification. YOLOv8 augments the images during online training. In each epoch, the model identifies variations of the images that were provided to it. Overall, these changes have proven to be positive and timely changes to act in cases like this work that needs an efficient and effective classifier, and YOLOv8 stood out for these reasons. It was our choice as a state-of-the-art object detection model which we adapted to detect root rot in beans.

(a) Image example of Crown with disease and Healthy root.

(b) Image example of Healthy root and crown.

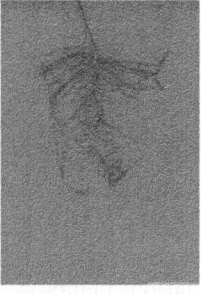

(c) Image example of Healthy crown and Root with disease.

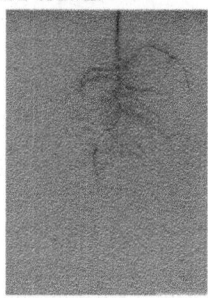

(d) Image example of Root and Crown with disease.

Fig. 2. Image samples of diagnostics for Crowns and Roots from bean plants.

2.3   Training and Validation

The images were randomly mixed, and four distinct classes were identified during the labeling process: Crown with disease (CrownDisease), Healthy Crown (HealthyCrown), Root with Disease (RootDisease), Healthy Root (HealthyRoot) and Background class (false-positive predictions). Thus, 125 images were manually labeled for training and 70 for validation. No preprocessing was performed on the images.

The root crown, also known as the root collar, is the region of the root system of a plant where the stem emerges, which may or may not be affected by fungi at the time of collection. Therefore, this separation between crown and root can demonstrate the stage of the disease since the spread of some fungi can occur from the crown to the root [12].

3   Results

We organized and prioritized the most relevant data from the test execution with 100 epochs, 8 images per batch and 8 workers. Initially, we made markings with bounding boxes spread across the roots. The results were poor, so we delimited the root box to the maximum possible root coverage area in a single box, and adjust hyperparameters again with new values, and a total of 40 epochs were used, with 16 images per batch, and 8 workers, the images used have a size of 640 × 480 pixels and were used augmentations degrees 0.3. All data and code for this research is available at github.com/renatocristianotorres/Yolov8_Root/

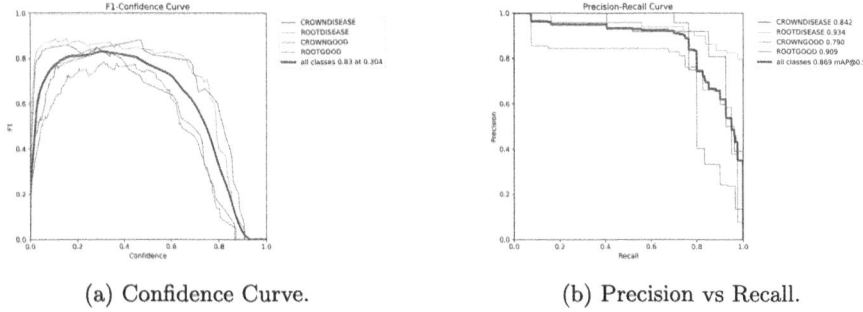

(a) Confidence Curve.

(b) Precision vs Recall.

Fig. 3. Confidence and Precision vs Recall curves of experiments.

The confidence curves in Fig. 3a) show the relationship between the confidence of the model's prediction and the metric that balances precision and recall. This analysis is important to determine the optimal decision threshold for the model. For confidence values above 0.6, the F1-score decreases for all classes, indicating that the model is becoming more selective and discarding many correct predictions. The F1 curve reaches a maximum value of 0.83 at a confidence threshold of 0.304. Initially, the F1-score rises rapidly for low confidence values, as the model classifies almost all samples as accurate. As confidence increases, the F1-score peaks and then decreases, indicating that fewer samples are classified but with greater precision. The best F1-score (0.83) occurs when confidence is close to 0.304, indicating that a very high threshold can lead to a loss of recall. In contrast, a very low threshold can compromise precision, so we must keep the decision threshold close to 0.3, which is appropriate.

The Precision-Recall curve (Fig. 3b)) evaluates the model's performance considering the relationship between precision and recall (positive examples correctly identified), which is important to assess a class imbalance. The mAP@0.5 for all classes is 0.869. The curves show that for higher recall values, precision tends to decrease as the model begins to classify more samples as positive, including some that are incorrect.

Analyzing the Confusion Matrix (Fig. 4), the *CrownDisease* class obtained 73% of the samples correctly classified. However, 17% were confused with the *background* class, which indicates a problem differentiating between these categories. The issue is due to similarities between the background and the classes or insufficient training data for the model to learn the differences, we chose to work with the same dataset of [14] to compare the efficiency of the models. The *Root-Disease* class obtained 88% of the samples correctly classified, with low rates of

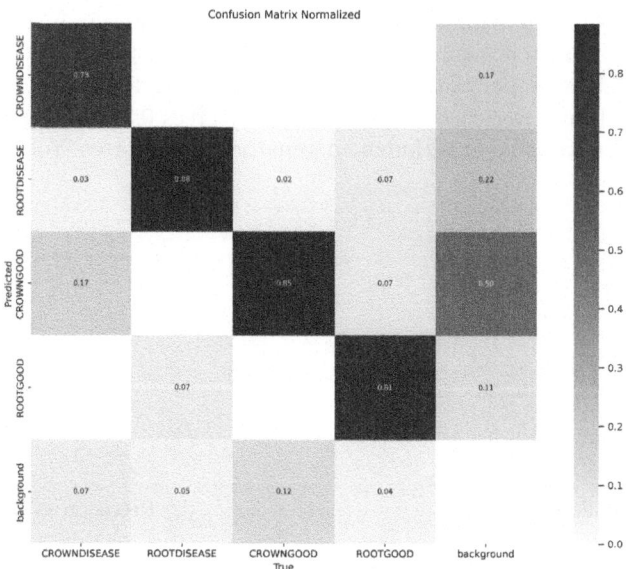

Fig. 4. Confusion Matrix.

confusion with other classes, but an error of 22% is attributed to false-positive predictions in the background class. The *HealthyCrown* class obtained 85% of correct classifications, but 17% were confused with *CrownDisease* and 7% with *HealthyRoot*, indicating an overlap between these classes. The *HealthyRoot* class obtained 81% accuracy, but there was confusion with Background (11%) and *HealthyCrown* (7%).

As shown in Fig. 5 (First line training data and the second line validation data), we can see that the bounding box detection loss is gradually decreasing, indicating that the model is learning to adjust the bounding boxes better. The classification loss is also decreasing, showing that the model is learning to make the appropriate distinction between classes. The smooth downward trend in losses indicates that the model was trained stably.

During validation the classification loss is decreasing, Fig. 5, but it also presents some oscillations, which may suggest a slight instability in the validation data, which may indicate that the model is not yet fully optimized for all classes, despite all the optimization efforts. The accuracy increases over the epochs, with some variations, but stabilizes at values above 0.8, which is satisfactory. The average accuracy at IoU $\geq$ 0.5 rises rapidly and stabilizes above 0.8, indicating that the model can make good predictions.

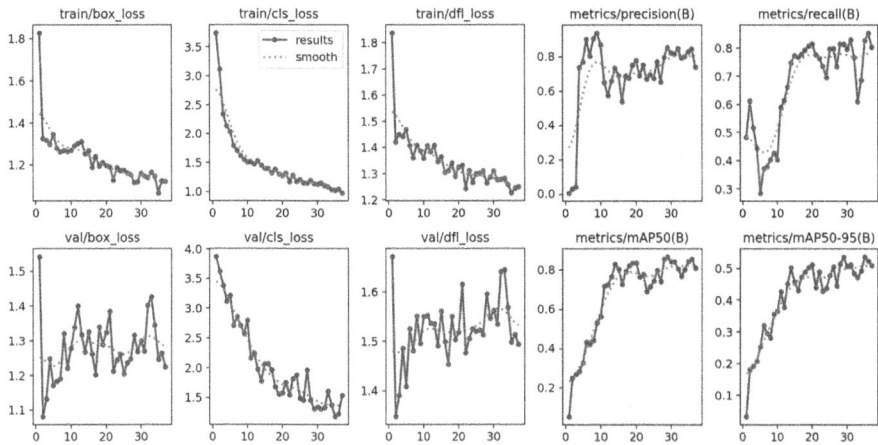

Fig. 5. Training and Validation Results.

4   Discussion

High-performance phenotyping is a challenge in many aspects of scientific research, especially in plant breeding [14], the works of Pierz et al. (2023) [14] and Sabety et al. (2024) [18] use a process that uses a set of software and a tool called PlantCV, which is composed of algorithmic functions to be used on a variety of plant types and imaging systems using segmentation and classification methods.

Pierz et al. (2023) [14] reached a correlation of $R^2 = 0.89$. In our case, we reached mAP = 0.869, comparing the data, we can say that both are reliable data or the model's ability to explain the observed data has a very close value. However, the use of YOLOv8 provided the pipeline with a much less complex one.

Processing through convolutions of the YOLOv8 network, associated with the activation function generates a satisfactory result for problems related to segmentation and identification, especially in this case where the fungus infestation causes a difference in spatialized or gradient coloration. When training the YOLO network, each input image finds many grid cells that do not contain objects in the prediction. It causes the confidence level of these cells to be almost zero, exceeding the levels of the cells that contain objects in their bounding boxes, and this causes instability in the network training result. This adversity is dealt with by increasing the loss of the bounding boxes that predicted objects and decreasing the loss for the boxes that do not contain objects [15], determined by the loss function.

The YOLO loss function is divided into three parts: the first is responsible for finding the coordinates of the bounding box, the second is responsible for predicting the marking of a box, and the last is responsible for predicting the marking of some classes. To perform the classification, it is necessary to compare the identified object with the determined classes. Solutions and algorithms, such as SIFTing, among others, perform this task; the vast majority are based on finding key features to obtain the degree of similarity.

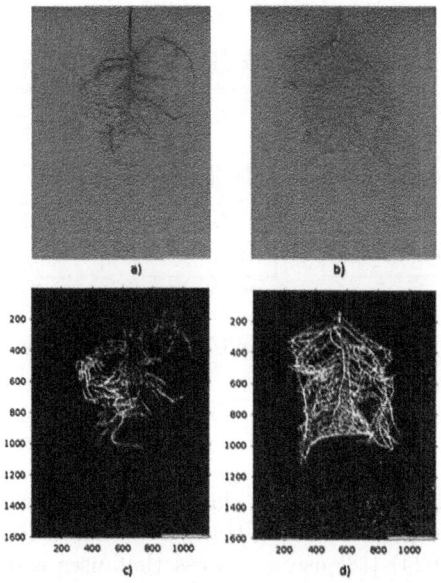

Fig. 6. Binaries Images - a) and c) Disease Root; b) and d) Healthy Root.

The YOLOv8 employs state-of-the-art backbone and neck architectures, resulting in improved feature extraction and object detection performance, and adopts an anchor-free split head, contributing to better accuracy and a more efficient detection process than anchor-based approaches. The paper de [13] examines the performance of YOLOv5 and YOLOv8 on trichiasis object detection tasks, focusing on key metrics such as Average Precision (mAP), specificity, recall, F1-score, and frames per second (FPS). Using a PC equipped with a Core i5-10500H CPU and an input size of 640*640, the results reveal the remarkable superiority of YOLOv8 over YOLOv5. YOLOv8 exhibits substantial improvements in mAP (31.8%), specificity (37.2%), recall (36.3%), and F1-score (34.4%), signifying its improved accuracy in identifying and classifying relevant instances. Since identifying spots is the goal of this work, edge-based algorithms can quickly determine similarity and, in this case, compare a diseased root with a healthy one. The degree of infestation can be determined by applying high-pass filters to the images in Fig. 6, which will be translated by the degree of similarity using contour and colors, which are important components in this process. The severity score of the disease, in this case, could be object output prediction in percentage.

5 Conclusion

The BACKGROUND class presents a high level of false positives, as it absorbs samples from other classes. It may indicate difficulty for the model in differentiating noise from real samples or possible imbalance in the dataset and characteristics of the BACKGROUND class similar to other categories. The CROWNDISEASE class is confused with CROWNGOOD (17%), suggesting that the model does not discriminate the signals between disease and normality well. Evaluating the results of the performance of the classes, we conclude that the ROOTDISEASE class has the best performance with AP = 0.934, presenting high precision and consistent recall. The ROOTGOOD also presents an AP = 0.909, showing that the class is well recognized, although there is a slight drop in precision at high levels of recall. CROWNDISEASE has a lower performance (AP = 0.842), suggesting that the class may be confused with others. CROWNGOOD has the lowest performance (AP = 0.790), indicating difficulties in differentiating this class from the others.

The mAP@0.5 (mean average precision) metric for all classes is 0.869, as overall performance. The curves show that precision tends to decrease for higher recall values âĂŞ which is expected since the model begins to classify more samples as positive. We intend to compare with other YOLO versions and change image dataset backgrounds, to improve false positives and use data augmentation techniques.

We conclude that using YOLOv8 is promising for identifying diseases caused by fungi, specifically those that cause crown and root rot. Although we used beans as a sample, there is potential to train the network with other crops. As an advantage, we highlight that this method uses a single tool and can be embedded in a portable device for use in the field [10].

References

1. Agnew, E., et al.: Whole-plant manual and image-based phenotyping in controlled environments. Curr. Protoc. Plant Biol. **2**, 1–21 (2017)
2. Diwan, T., Anirudh, G., Tembhurne, J.V.: Object detection using YOLO: challenges, architectural successors, datasets and applications. Multimed Tools Appl **82**, 9243–9275 (2023)
3. Furbank, R.T., Tester, M.: Phenomics – technologies to relieve the phenotyping bottleneck. Trends Plant Sci. **16**(12), 635–644 (2011)
4. Gehan, M.A., et al.: PlantCV v2.0: Image analysis software for high-throughput plant phenotyping. PEERJ 2088 (2017)
5. Isakeit, T., Gao, X., Kolomiets, M.: Exserohilum pedicellatum Root Rot of Corn in Texas. Plant Dis. **91**(5), 634–634 (2007)
6. Jiang, C.-J., Xie, X.: Soybean red crown rot: current knowledge and future challenges. Plant. Pathol. **72**, 1557–1569 (2023)
7. Jiang, P., Ergu, D., Liu, F., Cai, Y., Ma, B.: A review of yolo algorithm developments. Procedia Comput. Sci. **199**, 1066–1073 (2022)
8. He, K., Zhang, X., Ren, S., Sun, J.: Spatial pyramid pooling in deep convolutional networks for visual recognition. IEEE Trans. PAMI **37**(9), 1904–1916 (2015)

9. He, K., Gkioxari, G., Dollar, P., Girshick, R.: Mask R-CNN. In: Proceedings IEEE International Conference on Computer Vision (ICCV) (2017)

10. Kryvenchuk, Y., Stupnytskyi, R., Shevchuk, A., Kushka, B., Shevchuk, I.: UAV detection with YOLO on a standalone Raspberry Pi 5 system. In: Proceedings 2nd International Conference on Smart Automation and Robotics for Future Industr, Lviv, Ukraine, pp 1–10 (2025). session 1

11. Li, L., Zhang, S., Wang, B.: Plant disease detection and classification by deep learning–a review. IEEE Access **9**, 56683–56698 (2021)

12. Lopes, U., Michereff, S.: Challenges of management of root diseases caused by fungi (2018). ISBN: 978-85-7946-321-1

13. Wang, M.H., et al.: Optimizing real-time trichiasis object detection: a comparative analysis of YOLOv5 and YOLOv8 performance metrics. In: 2023 9th International Conference on Systems and Informatics (ICSAI), China, pp. 1–5 (2023)

14. Pierz, L.D., Heslinga, D.R., Buell, C.R., Haus, M.J.: An image-based technique for automated root disease severity assessment using PlantCV. Appl. Plant Sci. **11**(1), e11507 (2023)

15. Redmon, J., Divvala, S., Girshick, R., Farhadi, A.: You only look once: unified, real-time object detection. In: Proceedings of IEEE Conference CVPR, pp. 779–788 (2016)

16. Girshick, R.: "Fast R-CNN" In: 2015 IEEE International Conference on Computer Vision (ICCV), pp. 1440–1448. Santiago, Chile (2015)

17. Girshick, R., Donahue, J., Darrell, T., Malik, J.: Rich feature hierarchies for accurate object detection and semantic segmentation. In: IEEE Conference CVPR, Columbus, USA, vol. 2014, pp. 580–587 (2014)

18. Sabety, J., Serrano, A., Khan, A.: A root system architecture measurement pipeline for apple rootstocks. Plant Phenome J. **7**, e70011 (2024)

19. Seethepalli, A., Dhakal, K., Griffiths, M., Guo, H., Freschet, G., York, L.: Rhizo-Vision Explorer: open-source software for root image analysis and measurement standardization, AoB PLANTS, **13**(6), plab056 (2021)

20. Ren, S., He, K., Girshick, R., Sun, J.: Faster R-CNN: towards real-time object detection with region proposal networks. IEEE Trans. on PAMI **39**(6), 1137–1149 (2017)

21. Terven, J., Córdova-Esparza, D., Romero-González, J.: A comprehensive review of yolo architectures in computer vision: from yolov1 to yolov8 and yolo-nas. Mach. Learn. Knowl. Extr. **5**(4), 1680–1716 (2023)

22. Tovar, J.C., et al.: Raspberry Pi–powered imaging for plant phenotyping. Appl. Plant Sci. **6**(3), e1031 (2018)

23. Lee, W., Kim, T., Park, G.: Enhancements to YOLO for resource-constrained drone detection. Sensors **21**(2), 489 (2021)

24. Wang, X., Lu, L., Chen, D.: Refined YOLO architectures for high-speed UAV recognition. Comp. Vis. Image Underst. **207**, 120–130 (2021)

25. Cai, Z., Vasconcelos, N.: Cascade R-CNN: high quality object detection and instance segmentation. IEEE Trans. on PAMI **43**(5), 1483–1498 (2021)

Vision on the Move: Automated Hazardous Material Plate Detection in Freight Transport

Melissa Tijink[✉], Stanislav Levendeev, Ewaldo Nieuwenhuis,
Luuk Spreeuwers, Nicola Strisciuglio, and Estefanía Talavera

University of Twente, Enschede, The Netherlands
m.l.tijink@utwente.nl

Abstract. Enhancing the logistic efficiency and safety of freight transport requires fast, reliable identification of hazardous materials (hazmat). In this work, we explore how computer vision can automate the detection and reading of hazmat number plates on freight trains and trucks. We benchmark two object detection models for hazmat localization, YOLOv11x and Faster R-CNN, across a private freight train dataset and HazTruck, our newly introduced public dataset. For reading the detected plates, we evaluated three Optical Character Recognition (OCR) methods: the widely used Tesseract, EasyOCR, and the recent vision-language model Idefics2. Integrating YOLOv11x and Idefics2 into a unified pipeline achieved the state-of-the-art performance, with over 90% accuracy on freight train data, showcasing a powerful and scalable solution for automated hazmat identification in transport logistics. The code and datasets are available via https://github.com/Robust-Rail.

Keywords: Hazmat recognition · Transport Digitalization · OCR

1 Introduction

Monitoring dangerous goods in freight transport is an important aspect of the logistic process. This can be done by registering the RID (*Règlement Concernant le Transport International Ferroviaire Marchandises Dangereuses*, or ADR (*Accord relatif au transport international de marchandises Dangereuses par Route*) numbers indicating hazardous content [14,20]. The numbers are registered on orange plates with two lines, the first line (2 or 3 digits) indicating the RID/ADR number that indicates the type/group of hazardous material (hazmat); the second line is the UN number (4 digits) indicating a specific hazardous material [21]. We will refer to this set of numbers as RID/ADR numbers. Examples are shown in Fig. 1. In this paper, we consider data from cameras placed next to the train tracks and record passing freight trains, after which the relevant information is extracted.

An important step in reading the numbers from the plates is Optical Character Recognition (OCR). To this purpose, we deploy a recent architecture, namely

M. Castrillón-Santana et al. (Eds.): CAIP 2025, LNCS 15621, pp. 259–270, 2026.
https://doi.org/10.1007/978-3-032-04968-1_22

a)   b)   c)   d)

Fig. 1. Examples of an RID/ADR plates that are included in the HazTruck dataset (a, b) and the private dataset (c, d), corresponding sources can be found in the dataset.

a Vision-Language model (VLM). Such models take an image and a text prompt as inputs and output a text. The authors of [7] created an open VLM that can handle various resolutions and was designed for OCR-related tasks. We compare the performance of this new model to benchmark OCR models.

In this work, we present a novel pipeline for automatically detecting and reading hazmat codes from real-world videos of moving freight trains. This setting presents challenges, such as motion blur, variable lighting, and weather conditions, which remain largely unaddressed in current literature. To our knowledge, there is limited work focused on reading hazmat plates, particularly from (moving) freight trains. We propose a pipeline that demonstrates the feasibility of using a VLM for this task. To support further research, we introduce HazTruck, a novel dataset that can be used as a public benchmark and will be used to demonstrate the performance of our proposed pipeline. Both the dataset and the baseline method are available via https://github.com/Robust-Rail.

2   Related Works

Computer Vision and OCR in Traffic Applications. OCR in traffic is often applied for the detection and reading of license plates, which consists of two or three stages: vehicle detection (optional), plate detection and OCR [9, 11,17,22]. Most methods implement a YOLO network for detection tasks, and even the OCR can be based on a YOLO network [11].

The plate detection does not only apply for license plates, but for hazmat placards as well. Hazmats are not only represented in RID/ADR numbers, but also as diamond-shaped placards with symbols. Several studies have investigated the automatic detection and recognition of these placards [10,16,23]. All methods implement either a YOLO or a Faster R-CNN model for the detection of these plates. In this research, we will also use and compare a YOLO and Faster R-CNN model for the detection of the number plates. Although the diamond-shaped hazmat placards carry similar information as the RID/ADR number plates, they are not registered as they are intended for easy recognition of hazmats in emergencies, while the RID/ADR numbers are intended for registration and monitoring.

In automatic license plate recognition, often EasyOCR or TesseractOCR are used as the OCR methods [4, 22]. The authors of [22] find that EasyOCR outperforms Tesseract. This is also explained by [4] as EasyOCR is especially strong in handling diverse environmental conditions, while Tesseract was originally designed for document scanning. In this research, we compare EasyOCR and Tesseract with Idefics2 [7], a novel vision-language model that presents promising results.

Reading RID/ADR Numbers. The works of [18, 19] investigate the automatic reading of RID/ADR numbers, although their focus is on trucks rather than railway context. In [19] the algorithms and performance metrics on the reading of the RID/ADR numbers are not specified, however, the authors mention that their setup consists of two laser scanners and that the output of the system is later analyzed with a logistic chain in real-time.

Passing vehicles carrying hazmats in tunnels are being monitored in [18]. Their proposed pipeline consists of three stages: detection of vehicles, detection of the RID/ADR contours (numbers), and the last stage is the OCR. For the first stage, both YOLO and MobileNET are used; predictions were retained only when both models agreed. The RID/ADR contours are detected using a Faster-RCNN with the TensorFlow framework, and for the OCR the EasyOCR algorithm is used. The authors verified their algorithm with existing license plate recognition tools and noted a mean accuracy of 92% on the full pipeline on 250 photographs with and without ADR plates. Our proposed framework for hazardous material plate reading follows a similar structure, without the need of detecting the vehicle first.

The authors of [15] use railway data and reach a performance of over 95% accuracy for the task of automatic detection and reading of dangerous goods plates on trucks and trains. However, their code and data are not publicly available, meaning the models could be not be equally tuned to serve as a benchmark. This research used classic computer vision techniques as Histogram of Oriented Gradients (HOG) descriptors and Hough transformations, and an older version of Tesseract for OCR. Here, we use a VLM for OCR and explore deep learning techniques to detect the plates. The VLM is benchmarked against a modern version of Tesseract.

3   Method

The method consists of two major components: placard detection and OCR. An overview of our proposed pipeline is shown in Fig. 2. The pipeline starts with an input image. The first step is placard detection, which uses an object detection model. This will output a bounding box of the predicted location of the placard. Using this box, the plate can be extracted and provided to an OCR model. This will result in a final prediction of the RID/ADR number. The placard detection and OCR will be evaluated separately, and the best-performing models will be combined as our proposed pipeline.

Fig. 2. Overview pipeline of RID/ADR number recognition. The input is a single frame/image. Following by 1.) placard detection with an object detection model, the found placard is extracted using the bounding boxes. And finally, 2.) OCR can be applied to produce the recognized number.

3.1   Placard Detection

We consider two different models for placard detection: YOLO and Faster R-CNN. We select the available YOLOV11 model with the largest number of parameters, totalling 56.9 million parameters, which was shown to achieve better detection performance [5]. The chosen architecture for Faster R-CNN [13] uses ResNET-101 with a Feature Pyramid Network (FPN) as its backbone to combine the depth of ResNet-101 with FPN's multiscale detection capabilities [8]. We fine-tune both models on a custom dataset. The models generate bounding boxes that serve as input for the OCR stage. To improve the performance of the fine-tuned YOLO model, standard augmentations available in the Ultralytics training framework [5] were used, such as shifting, resizing, flipping and adjusting color intensity. We developed a custom data augmentation pipeline to implement data augmentation for the Faster R-CNN model. Four different augmentation strategies were included: horizontal flipping, adjusting brightness and contrast, applying Gaussian blur, and random zooming. An occurrence possibility is set for all individual data augmentation options, meaning they are non-exclusive.

3.2   OCR in Placards

Idefics2 [7] is implemented OCR method in the pipeline. This is a foundational VLM and takes an image and a text prompt as input and outputs a text. The model consists of a vision encoder, after which the visual features are mapped to a Large Language Model (LLM), where they are combined with the text input. The LLM predicts output text tokens. The Idefics2 model was trained on interleaved data of text and image tokens, image-text pairs, and PDF documents. The latter makes the model especially suitable for OCR-related tasks [7]. The input text prompt needs to be carefully designed. We decided to use a form of few-shot prompting, where the model is instructed on the desired output in edge cases [2]. The used prompt is shown in Listing 1.1. Using the bounding boxes from the Placard Detection methods (either YOLO or Faster R-CNN), the placard is extracted from the images.

Listing 1.1. The prompt used for extracting the RID/ADR numbers using Idefics2 [7]. The provided examples are random.

```
Analyze the image and extract two key values:
    - The UN number visible on the upper part of the placard.
    - The code visible on the lower part of the placard, located below the horizontal line
      separating the two sections.

Both codes are printed in black. If either the upper or lower part cannot be detected,
    replace the missing value with "0." Output the extracted values as plain text,
    separated by a comma if multiple codes are present. No additional context or formatting
    is needed.

Input Examples:                                      Desired Output:
    {98 {line} 4567}                                 98, 4567
    (not found, {line}, 8901)                        0, 8901
    {101 {line} 3345}                                101, 3345
    (not found, {line}, {not found})                 0, 0
    {45 {line} 2789}                                 45, 2789
    {22 {line} 5678}                                 22, 5678

Expected Transformation:
    For each input example, extract the UN number and the code below the horizontal line. If
        either part is missing (i.e., "not found"), replace it with 0. Output the extracted
        values as plain text, separated by a comma, without any additional context or
        formatting.
```

4   Experimental Framework

4.1   Datasets

In this study, we used two datasets: a private dataset called *Private Data* and a public dataset called *HazTruck*. Both datasets are labeled using the Computer Vision Anotation Tool (CVAT)[1]. The plates are annotated using bounding boxes, and the ADR/RID and UN numbers are manually annotated.

Private Dataset. All videos originate from a camera that is located next to a rail track and captures passing freight trains. The videos have varying lengths and a variety of illumination conditions, enhancing the dataset's diversity. Only frames with a hazmat plate are considered. In total, the dataset contains a total of 9,560 frames extracted from 24 annotated videos. The dataset includes 13 unique UN hazmat codes. The dataset was divided into training, validation, and testing subsets following an 80%-10%-10% split, resulting in 7,648 frames for training, 956 frames for validation, and 956 frames for testing.

HazTruck Dataset. This dataset is created[2] to provide a public annotated benchmark of hazmat number plates. Images from public sources are selected when they have a readable (by human eyes) hazmat number plate. After selecting, a web scraping tool is used to import the images. The available dataset consists of links to the used images and the annotation of all hazmat number plates present. Examples are shown in Fig. 1. Most public photos with the plates

---

[1] Available via https://github.com/cvat-ai/cvat.

[2] Available via https://github.com/Robust-Rail.

are from trucks, the total dataset consists of 210 images with 238 number plates, with variation in distance to the camera, angle, background, lighting, and more aspects. This dataset will be used for evaluation only.

4.2 Experiments and Evaluation Metrics

We evaluate the proposed framework and its individual steps. First, we assess the performance of the models we use for placard detection and OCR seperately. Then we evaluate the performance of the proposed pipeline.

Placard Detection. Both models, YOLO11x and Faster R-CNN are trained using the train and validation sections of the private dataset. Different training strategies are applied. YOLO11x is trained with data augmentation [5], with and without early stopping, and two different learning rates (0.01 and 0.002). YOLO11 without early stopping is trained for 10 epochs. The early stopping is defined by setting the number of epochs with no improvement (patience), which is set at 5 and 10 for the high and low learning rates (7 and 30 epochs), respectively. Faster R-CNN is trained with and without data augmentation and early stopping with a learning rate of 0.005. The most basic implementation of Faster R-CNN is trained for 9 epochs, with data augmentation for 13 epochs. The early stopping is set with a patience of 7, resulting in 11 epochs.

Evaluation: The Intersection over Union (IoU) of the predicted bounding box and the labeled ground truth is calculated. The mean Average Precision (mAP) can be computed for different IoU thresholds. The mAP represents the average area under the precision-recall curve for all classes. The IoU threshold determines whether a prediction is considered correct or false.

OCR. The implemented OCR method, Idefics2 [7], will be compared with two baseline methods: EasyOCR and Tesseract. EasyOCR is build on Pytorch and Tesseract is an open source OCR engine owned by Google [12,22]. Tesseract and EasyOCR are available as Python libraries, called `pytesseract` and `easyocr`, respectively [3,6]. In the implementation of EasyOCR and Tesseract the orange block is split horizontally on the black line. The OCR methods are evaluated only, and no finetuning is applied, so the full Private dataset is used, as well as the HazTruck data.

Evaluation: We rely on the Character Error Rate (CER) and Word Error Rate (WER), shown in Eq. 1 and 2, respectively,

$$CER = \frac{\text{Number of incorrect characters}}{\text{Total number of characters}}, \tag{1}$$

$$WER = \frac{\text{Number of incorrect words}}{\text{Total number of words}}. \tag{2}$$

Both metrics are based on the minimal number of operations (deletions, insertions, and substitutions) needed to transform one text into another. CER measures errors at the character level, assessing individual characters, while WER evaluates accuracy at the word level. In the case of code or numeric sequences, CER examines single digits, whereas WER determines the correctness of entire sequences (e.g. 83 and 2789 in Fig. 1) are considered separate words.

Hazmat Recognition. The best-performing placard detection and Idefics2 [7] are selected and evaluated as the proposed pipeline. The evaluation framework for the pipeline assesses performance across two key stages: detection and OCR. Placard detection is evaluated using the Intersection over Union (IoU) metric, where a detection is considered valid if its IoU with at least one ground truth (GT) bounding box exceeds a predefined threshold; otherwise, it is classified as a failure. Only placard detections with a confidence of 25% or greater are considered. If no detection is made, it is also considered an error. OCR is applied on successful detection to extract the UN and Hazard Identification (HIN) numbers, which are then compared to GT values. A prediction is correct only if both numbers match the ground truth, while any mismatch results in an error. Overall, errors occur in three cases: failure to detect the placard, insufficient IoU, or incorrect OCR extraction. This approach ensures a full assessment of the pipeline by combining spatial accuracy (IoU) with semantic accuracy (OCR validation).

5   Results and Discussion

The results are presented and discussed for placard detection and OCR, and the best-performing models will be combined as our proposed pipeline.

Placard Detection. The results of the hazmat number plate are shown in Table 1. This table shows the mAP50 and the mAP50-95 for both the test portion of the Private dataset and the HazTruck data for the different training strategies. The results are similar for the Private data for both YOLO11x and

Table 1. Performance metrics for YOLO11x and Faster RCNN under different training strategies on the two test sets. Training strategies are Early Stopping (E.S), Lower Learning rate (L.L) and Data Augmentation (D.A.)

Model	Private-Test		HazTruck	
	mAP50-95(%)	mAP50(%)	mAP50-95(%)	mAP50(%)
YOLO11x [5]	57.69	99.37	64.16	82.43
YOLO11x E.S.	50.72	98.77	55.29	85.40
YOLO11x E.S.+L.L.	56.46	99.38	62.43	89.46
Faster-RCNN [13]	56.34	98.92	46.1	70.57
Faster-RCNN D.A.	61.67	98.98	53.89	79.40
Faster-RCNN E.S.+D.A.	61.25	98.97	50.83	82.38

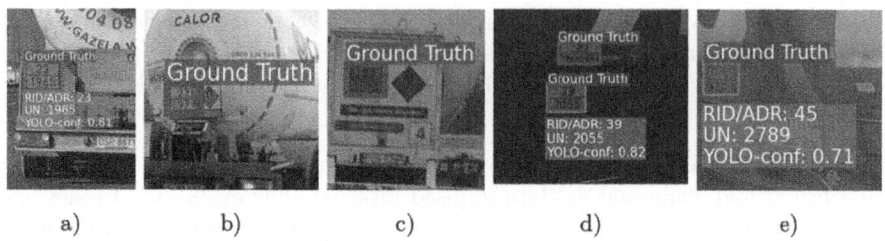

Fig. 3. Qualitative results of our proposed pipeline with YOLO11x [5] and Idefics2 [7]. Ground truth bounding box is shown in blue, the predicted bounding box and text are shown in purple, and below the plate. A) A correctly detected plate. B) A missed plate due to a small plate and a letter appearing. C) A missed plate due to low quality of the image. D) A correctly detected plate during nighttime, and a missed plate due to the skewed surface. E) A detected plate, but incorrect OCR due to rain. A, B, and C originate from the Haztruck Dataset, D and E are from the Private Dataset. All images are cropped.

Faster-RCNN. YOLO11x outperforms Faster-RCNN on the HazTruck data; this indicates a better generalizability of the YOLO11x model, as it shows higher mAP in all situations. However, HazTruck is a small dataset, and there is more variation in aspects as distance to the camera, angle, and resolution than in the Private data, which explains the performance drop. Ideally, the performance is evaluated on more data, and a cross-validation on all data types is performed.

Figure 3 shows qualitative results of the YOLO11x model with Early Stopping and the Lower Learning rate. Image (A), (D), and (E) show correctly detected plates, even during nighttime and rain. Figure 3 also shows missed plates, in one case, a letter appeared in the code, which is an unseen situation for the model. In (C), the plate is small in the original image, and in (D) the image appears on a skewed surface. In the training data, the freight trains have a fixed distance from the camera. To mitigate this issue, we implemented random zooming in the data augmentation steps, however, it can be concluded that this does not fully cover the real-world data. In future implementations, the model could be trained with a larger range of zooming or more varied data.

OCR. The CER and WER scores on the three OCR models are shown in Table 2. Idefics2 outperforms the other two models on both datasets, scoring higher on the Private dataset. This could be because the Private data has a variation in angle in one axis only, as the distance of the rail track to the camera is fixed. In the HazTruck data, there is more variation in angle and distance to the camera. Table 3 shows examples of the OCR results. The number in Plate 1 are read correctly by all methods, Tesseract fails on all other examples. Tesseract is sensitive to variations in data, as it was originally designed for document scanning. This data shows variation, which explains the low performance of Tesseract. EasyOCR makes mistakes in misreading numbers, for example misreading a vertical line as a 1 or 7 in Plate 2 or replacing a 1 with a 4 in Plate

Table 2. Performance comparison of Tesseract, EasyOCR, and Idefics2 models on Private and Haztruck datasets on the Character Error Rate (CER) and Word Error Rate (WER). Lower CER and WER mean better performance.

Model	Private		HazTruck	
	CER	WER	CER	WER
Tesseract [3]	56.3	182.19	45.88	129.70
EasyOCR [6]	29.99	98.6	44.46	130.12
Idefics2 [7]	4.85	15.20	13.28	39.74

Table 3. Examples of OCR results from Tesseract, EasyOCR, and Idefics2 on different cropped images. The crops are based on the ground truth labels from the HazTruck dataset. A not found code is indicated with the - sign.

Model	Plate 1	Plate 2	Plate 3	Plate 4	Plate 5	Plate 6
Tesseract [3]	90, 3082	-, 3082	-, 11965	-, 1965	-, -	-, -
EasyOCR [6]	90, 3082	-, 130827	23, 1965	23, 4965	-, -	-, 6207
Idefics2 [7]	90, 3082	90, 3082	23, 1965	23, 1965	33, -	-, 1203

4. EasyOCR might not be trained on these types of fonts, this could be fixed by fine-tuning EasyOCR and training it on reading Hazmat plates. Idefics2 shows good performance but fails when the image of the plate is of low resolution (Plate 6) or there is font variation (Plate 5) within the plate. The Idefics2 model was trained on reading PDF text for OCR purposes [7], however, this task has more variation in font, style, and background than plain PDF text documents. This is, for example, done by [1], showing an increase in performance of reading license plates when finetuning the VLM on the task. In future research, Idefics2 could be fine-tuned on the hazmat plate data to improve performance; challenges such as within-plate font variation should be included in the training data.

Hazmat Recognition. The results of the evaluation of our proposed pipeline are summarized in Table 4 and Fig. 3. The pipeline uses the two best-performing methods: YOLO11x and Idefics2. An accuracy of over 90% is reached on the Private-Test dataset. Contrastingly, an accuracy of 63.5% is reached on the HazTruck data. However, 1) the YOLO11x model was not trained on this data and 2) there is a larger variation in data, such as angle and distance to the camera, which can influence the plate detection and OCR. Figure 3 also shows that the OCR fails when the plate is heavily obscured by rain. However, the results show the potential of using YOLO11x and Idefics2 for Hazmat plate reading on freight trains. It is also shown that generalization can be improved, and in

Table 4. Accuracy of our proposed pipeline on the Private and Haztruck datasets at different IoU thresholds. The pipeline uses the YOLO11x [5] and Idefics2 [7] models.

Threshold IoU	Accuracy (%)	
	Private-Test	HazTruck
40 %	90.9	63.5
50 %	90.3	63.5
70 %	54.2	56.1

future research, using a combined dataset of freight trains and trucks can be investigated. Furthermore, as mentioned in Sect. 2 the hazardous materials are represented as diamond-shaped placards with symbols too. Recognition of these placards can be used as a verification of the numbered plates. This can help to reduce false positives or wrongly identified hazmats.

6   Conclusion

This study proposes a new framework for reading hazardous material number plates and a novel dataset composed of images of plates on trucks, which we make publicly available. We evaluate our model on images of freight trains and trucks. The results highlight the higher performance of the VLM-based model Idefics2 for OCR, 90.9% on freight train data and 63.5% on HazTruck data. The lower performance on images of trucks indicates a limitation in the representation of the training data. Further research should go into training the object detection stage with a more varied dataset. Furthermore, to reach a higher performance on both datasets, the VLM model could be fine-tuned for the task of reading hazmat plates. We release a publicly available pipeline based on using VLMs that serves as a strong baseline for detecting and reading Hazmat plates on real-world data from freight trains.

Acknowledgments. The research is part of project FP5-TRANS4M-R and is supported by the Europe's Rail Joint Undertaking and its members. Funded by the European Union. Views and opinions expressed are however those of the author(s) only and do not necessarily reflect those of the European Union or the Europe's Rail Joint Undertaking. Neither the European Union nor the granting authority can be held responsible for them.

References

1. AlDahoul, N., et al.: Advancing vehicle plate recognition: Multitasking visual language models with vehiclepaligemma (2024). https://arxiv.org/abs/2412.14197
2. Brown, T.B., et al.: Language models are few-shot learners. In: Advances in Neural Information Processing Systems, vol. 2020 (2020)

3. Hoffstaetter, S., Lee, M.: Pytesseract PyPI. https://pypi.org/project/pytesseract/. Accessed June 2025
4. Hussain, A.A.Z., Gowd, B.N.: Automatic Number Plate Recognition using Optical Character Recognition and EasyOCR. Technical reports, Visvesvaraya Technological University (2024). https://doi.org/10.13140/RG.2.2.27255.43684
5. Jocher Glenn: Ultralytics Yolo 11. https://github.com/ultralytics/ultralytics. Accessed June 2025
6. Kittinaradorn, R.: Easyocr PyPI. https://pypi.org/project/easyocr/. Accessed June 2025
7. Laurençon, H., Tronchon, L., Cord, M., Sanh, V.: What matters when building vision-language models? (2024). https://doi.org/10.48550/arXiv.2405.02246
8. Lin, T.Y., Dollár, P., Girshick, R., He, K., Hariharan, B., Belongie, S.: Feature pyramid networks for object detection (2017). https://arxiv.org/abs/1612.03144
9. Liu, Q., Liu, Y., Chen, S.L., Zhang, T.H., Chen, F., Yin, X.C.: Improving multi-type license plate recognition via learning globally and contrastively. IEEE Trans. Intell. Transp. Syst. **25**(9), 11092–11102 (2024)
10. Puumala, J., Rahtu, E.: Monitoring Dangerous Goods on Roads Using Computer Vision. Ph.D. thesis, Tampere University (2023)
11. Ramajo Ballester, A., Armingol Moreno, J.M., de la Escalera Hueso, A.: Dual license plate recognition and visual features encoding for vehicle identification. Robot. Auton. Syst. **172**, 104608 (2024). https://doi.org/10.1016/j.robot.2023.104608
12. Reddy, P.P., Shruthi, P.S., Himanshu, P., Singh, T.: License plate detection using yolo v8 and performance evaluation of easyocr, paddleocr and tesseract. In: 2024 15th International Conference on Computing Communication and Networking Technologies (ICCCNT), pp. 1–6 (2024). https://doi.org/10.1109/ICCCNT61001.2024.10725878
13. Ren, S., He, K., Girshick, R., Sun, J.: Faster R-CNN: towards real-time object detection with region proposal networks. IEEE Trans. Pattern Anal. Mach. Intell. **39**(6), 1137–1149 (2015)
14. Rijksinstituut voor Volksgezondheid en Milieu: Gevaarsnummers | Risico's van stoffen, https://rvs.rivm.nl/onderwerpen/gevaarsindeling/ADR/gevaarsnummers
15. Roth, P.M., Köstinger, M., Wohlhart, P., Bischof, H., Birchbauer, J.A.: Automatic detection and reading of dangerous goods plates. In: Proceedings - IEEE International Conference on Advanced Video and Signal Based Surveillance, AVSS, vol. 2010, pp. 580–585 (2010)
16. Sharifi, A., Zibaei, A., Rezaei, M.: A deep learning based hazardous materials (HAZMAT) sign detection robot with restricted computational resources. Mach. Learn. Appl. **6**, 100104 (2021)
17. Shi, H., Zhao, D.: License plate recognition system based on improved YOLOv5 and GRU. IEEE Access **11**, 10429–10439 (2023)
18. Sisias, G., Konstantinidou, M., Kontogiannis, S.: Deep learning process and application for the detection of dangerous goods passing through motorway tunnels. Algorithms **15**(10), 370 (2022). https://doi.org/10.3390/a15100370
19. Soussi, A., Tomasoni, A.M., Zero, E., Sacile, R.: An ICT-Based Decision Support System (DSS) for the Safety Transport of Dangerous Goods along the Liguria and Tuscany Mediterranean Coast. In: Lecture Notes in Networks and Systems, vol. 579, pp. 629–638 (2023)

20. UNECE: About the ADR - Agreement concerning the International Carriage of Dangerous Goods by Road (2025). https://unece.org/transport/standards/transport/dangerous-goods/adr-2023-agreement-concerning-international-carriage. Accessed June 2025

21. United-Nations: Recommendations on the TRANSPORT OF DANGEROUS GOODS - Model Regulations Volume I. Tech. rep., United Nations Publications, New York and Geneva (2023). https://unece.org/transport/dangerous-goods/un-model-regulations-rev-23

22. Vedhaviyassh, D.R., Sudhan, R., Saranya, G., Safa, M., Arun, D.: Comparative analysis of EasyOCR and TesseractOCR for automatic license plate recognition using deep learning algorithm. In: 6th International Conference on Electronics, Communication and Aerospace Technology, ICECA 2022 - Proceedings, pp. 966–971 (2022)

23. Zhang, R., Bahrami, Z., Feng, K., Liu, Z.: A visual and textual information fusion-based zero-shot framework for hazardous material placard detection and recognition. IEEE Trans. Artif. Intell. 5(4), 1755–1768 (2024)

VIGIA-E: Density-Aware Patch Selection for Edge-Based Small Object Detection with PTZ Cameras

Jonay Suárez-Ramírez[1] , Kiara Sánchez-Cordero[2] , and Nelson Monzón[2][(✉)]

[1] Qualitas Artificial Intelligence and Science, Las Palmas de Gran Canaria, Spain
jsuarez@qaisc.com
[2] CTIM, Instituto Universitario de Cibernética, Empresas y Sociedad,
University of Las Palmas de Gran Canaria, Las Palmas de Gran Canaria, Spain
{kiara.sanchez,nelson.monzon}@ulpgc.es
https://qaisc.com/, https://iuces.ulpgc.es/

Abstract. We present **VIGIA-E**(Vision-based Inference for Guided Interest Attention on Edge devices), a lightweight object detection strategy for online surveillance in wide-area scenes using resource-constrained edge hardware. Designed for typical surveillance settings, such as public spaces or coastal infrastructures monitored by drones or elevated cameras, it addresses the challenge of detecting small, distant objects caused by perspective and scale. The method adopts a two-stage pipeline: a global low-resolution pass followed by selective high-resolution inference over dense regions identified via a grid-based estimator. This strategy reduces redundant computation while preserving detection accuracy, enabling efficient deployment on embedded platforms. Unlike conventional multi-inference approaches, VIGIA-E brings online feasibility to density-driven small object detection on edge systems, a crucial requirement for modern smart surveillance.

We evaluate VIGIA-E on two complementary datasets: the domain-specific Anfi dataset and the VisDrone benchmark. In both cases, it achieves favorable trade-offs between accuracy and computational cost compared to representative multi-inference baselines. Additionally, VIGIA-E has been deployed in a real-world coastal surveillance system, demonstrating operational viability. While representing an initial iteration of our framework, it establishes a solid foundation for more advanced systems under development.

Keywords: Deep Learning · Small object detection · sliced inference · windowed inference · visdrone

1 Introduction

In recent years, surveillance systems have increasingly adopted object detection technologies to enable real-time monitoring in large and open environments. Such systems often use UAVs or high-mounted cameras to monitor public, coastal, or

M. Castrillón-Santana et al. (Eds.): CAIP 2025, LNCS 15621, pp. 271–282, 2026.
https://doi.org/10.1007/978-3-032-04968-1_23

urban areas, producing wide-angle views where objects appear small and distant. These perspectives pose challenges related to scale variation, clutter, and long-range visibility.

In parallel, the proliferation of intelligent video analytics has driven the adoption of Edge AI [18], where inference is performed locally at the data source. This is particularly relevant in coastal and touristic settings, where privacy, unstable connectivity, and latency constraints limit the feasibility of remote processing architectures. Edge platforms equipped with Pan-Tilt-Zoom (PTZ) cameras offer a flexible and autonomous alternative for monitoring dynamic outdoor environments.

However, the limited compute, memory, and power of embedded devices [17, 19] restrict the use of large deep models—especially for small object detection, which is crucial in scenes with people, vessels, umbrellas, and other dense elements. Although multi-shot inference can improve detection, its computational cost makes it impractical for low-latency online use on edge hardware. In this regard, we propose VIGIA-E (**V**ision-based **I**nference for **G**uided **I**nterest **A**ttention on **E**dge devices), a lightweight patch-based method that identifies high-density regions from a global low-resolution inference and focuses processing on selected areas. This targeted strategy reduces computation while maintaining high detection performance.

Developed as part of the first deliverable of the Smart Coast Surveillance System[1], VIGIA-E has been deployed in the coastal area of Anfi del Mar and Patalavaca (Gran Canaria, Spain), covering popular beach zones and waterfronts. The surveillance setup consists of two PTZ cameras and an embedded edge computing platform, as illustrated in Fig. 1, which shows both the physical installation and representative scene views from different angles.

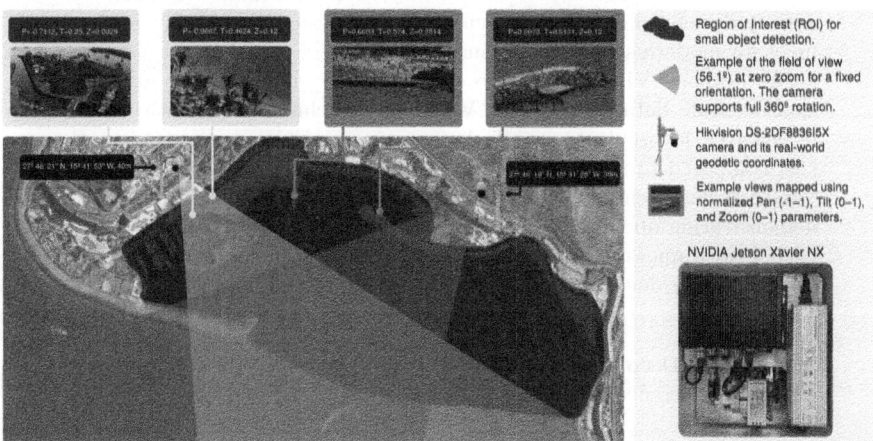

Fig. 1. Physical setup with two PTZ cameras and a Jetson Xavier NX device, including example views with normalized PTZ values. Visualization elements are explained in the right legend. Camera and system details are in Table 1.

---

[1] https://qaisc.com/projects-qaisc/smart-coast/.

The image acquisition unit is based on Hikvision DS-2DF8836I5X cameras, mounted at elevated positions to provide wide coverage. These devices offer pan, tilt, and zoom capabilities and capture high-resolution frames up to 4K. Edge-side processing is performed on an NVIDIA Jetson Xavier NX (8GB), a compact GPU-based platform optimized for real-time inference in embedded AI applications. Tables 1(a) and 1(b) summarize the key specifications of both components.

Table 1. Hardware specifications of the surveillance setup: (a) the Hikvision DS-2DF8836I5X PTZ camera used for data acquisition, and (b) the NVIDIA Jetson Xavier NX device used for edge-based inference.

(a) **Hikvision DS-2DF8836I5X Camera Specifications**

Feature	Description
Image sensor	2/3" CMOS
Shutter time	1/1 s – 1/30000 s
Focal length	7.5 mm – 270 mm
Optical zoom	x36
Pan range	360°
Tilt range	−20° to 90°
Maximum resolution	4K
Dimensions	Θ 266.6 mm × 410 mm

(b) **NVIDIA Jetson Xavier NX Specifications**

Feature	Description
AI Performance	21 TOPS
GPU	384-core Volta GPU + 48 Tensor Cores
CPU	6-core Carmel 64-bit
Memory	8 GB LPDDR4x (128-bit)
Power	10 / 15 / 20 W modes
Dimensions	69.6 mm × 45 mm

This operational configuration imposes practical constraints in terms of processing power, memory, and energy consumption, which directly influence the feasibility of deploying multi-inference pipelines. To assess the effectiveness of VIGIA-E under these conditions, we conduct a comparative evaluation using a custom dataset collected from this real-world setup, and extend the analysis to the VisDrone benchmark to validate generalization in dense urban aerial scenes. In the experimental section, VIGIA-E is benchmarked against state-of-the-art region selection strategies such as SAHI [1] and GLSAN [3], using multiple YOLO-based detectors to ensure consistent and meaningful comparisons in terms of speed and accuracy.

2   Related Works

Object detection is a core task in computer vision, significantly advanced by Deep Convolutional Neural Networks (DCNNs). Detection models are commonly categorized into two paradigms: two-stage approaches, such as R-CNN [6], which rely on a Region Proposal Network (RPN) followed by classification, and one-stage models like YOLO [16], which perform localization and classification in a single pass to enable faster inference.

A persistent challenge across both paradigms is the detection of small objects, particularly in aerial, satellite, and surveillance imagery. These instances are often underrepresented in large-scale benchmarks such as COCO [12], and their features tend to degrade due to spatial downsampling in deep networks [9]. This leads to weak activations, especially under conditions of occlusion, clutter, or extreme scale variation.

To mitigate these issues, five major strategies have been explored in the literature: (1) *targeted data augmentation*, including techniques such as copy-paste and GAN-based synthesis to enrich small or underrepresented classes [7,8,13]; (2) *multi-scale feature fusion*, improved by hierarchical and attention-based modules such as BiFPN, HTDet, and MS Transformer [15,23]; (3) *contextual reasoning* via auxiliary branches or semantic attention mechanisms [5,26]; (4) *training schemes* that incorporate focal loss variants and size-aware anchor assignment [13]; and (5) *multi-shot inference*, where detection is performed over multiple regions or resolutions using adaptive patch selection.

Within the last category, early works rely on uniform or random tiling [2,28], while more recent approaches guide patch selection using clustering techniques [25], difficult-region estimation [27], reinforcement learning [24], or density-based heuristics [3,11,14]. Focus-and-Detect [10] exemplifies this trend by clustering dense areas via Gaussian Mixture Models and assigning specialized detectors per region. In parallel, research on efficient model design has produced lightweight backbones, neural architecture search (NAS) strategies, and modular region-aware pipelines aimed at enabling small object detection on resource-constrained platforms, including drones and edge devices [7,15].

3   Our Method

Multi-inference approaches have shown strong performance in small object detection and are easy to integrate, but their high computational cost leads to increased execution time. This is especially critical for exhaustive methods [2,28], and remains a concern even for region-guided strategies [24,25,27], which still rely on auxiliary inference or costly clustering. These limitations are particularly restrictive on low-power edge devices. To address this, we propose the VIGIA-E module, designed to complement standard detection pipelines while maintaining a favorable balance between efficiency and detection quality.

The standard detection pipeline typically consists of three steps. First, the input image is passed to a detector—usually a neural network—which produces a set of candidate detections defined by their image coordinates and associated confidence scores. These predictions often include numerous low-confidence results that are unreliable; thus, detections below a predefined confidence threshold (C_{TH}) are discarded. Additionally, since most detectors rely on localization proposals, it is common for multiple detections to overlap the same object. To handle this redundancy, a Non-Maximum Suppression (NMS) algorithm is applied to retain only the most confident detections. This widely used pipeline is illustrated with a white background in Fig. 2.

Our proposed VIGIA-E module, depicted with red background in Fig. 2, is a plug and play module that can be integrated with any object detector without the need of entire pipeline training and leverages an efficient region search procedure to enhance the detection of small objects.

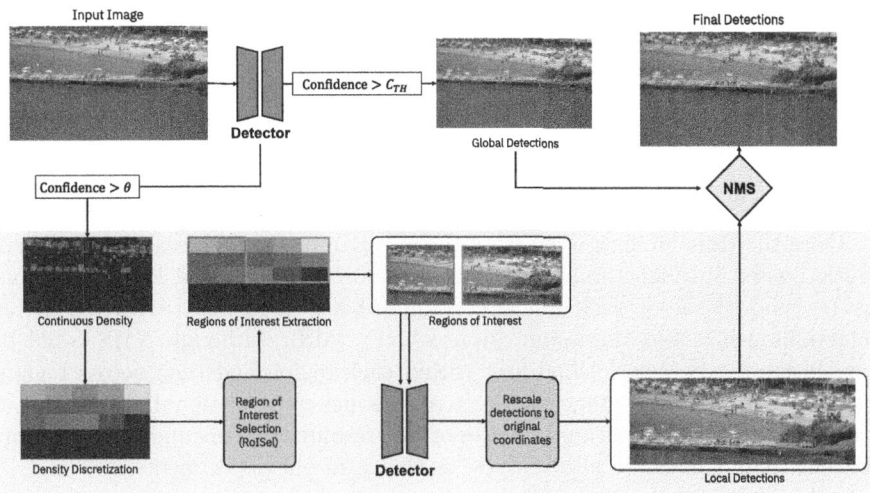

Fig. 2. Overview of the proposed method. The red region highlights the VIGIA-E module integrated into the standard detection pipeline. (Color figure online)

3.1 VIGIA-E Module

In multi-inference approaches, predictions are typically made over the entire image to detect large objects that may be present in the scene. We adopt this strategy and further exploit the resulting detection proposals to identify and select regions of interest (RoIs).

The approach begins by filtering detection proposals using a relaxed confidence threshold $\theta \leq C_{TH}$, resulting in a set of low-confidence yet spatially informative bounding boxes, referred to as *continuous density*. This density is discretized into a regular $N \times N$ grid to produce a Density Grid (DG), using a 2D histogram-like aggregation scheme. Each bounding box $b_p \in B_p \subset \mathbb{R}^{M \times 4}$ contributes by incrementing the cells corresponding to the positions of its corners. These coordinates are projected into grid cells of size $(\frac{H}{N}, \frac{W}{N})$, where H and W denote the image height and width, respectively. The result is a coarse but informative spatial map that highlights high-density regions, enabling efficient selection while minimizing computational overhead.

The Density Grid is passed to our Region of Interest Selection (RoISel) algorithm, described in Algorithm 1. In typical surveillance scenarios, the grid resolution N does not need to exceed 20. First, all candidate regions C are generated, consisting of rectangular, contiguous groups of grid cells with dimensions ranging from 1 to $N-1$. An interest mask I^{mask} is computed by marking all cells whose density is greater than or equal to the median value of DG. Next, the densest candidate regions are selected according to the equation in Step 6 of Algorithm 1.

This scoring function enhances the mean density by incorporating a scaling factor $\sqrt{KL}$, promoting the selection of larger regions to achieve broader coverage with fewer inferences. This adjustment is critical, as relying solely on the mean cell density would cause the algorithm to favor small, high-density cells, given that the average density of a region is always lower than its maximum. To avoid redundant processing, the cells within each selected candidate are cleared from I^{mask} and penalized in DG by subtracting the corresponding region's density estimate CD_s. This penalization reduces the likelihood of reselection, unless those cells remain significantly denser than their surroundings.

Once the RoIs are extracted from the original image, each region is independently passed through the same object detector. The resulting local detections are rescaled to the original image coordinates and then fused with the global detections using Non-Maximum Suppression (NMS). Although NMS could be applied jointly, it is performed first within each region and then across regions for efficiency. This two-stage inference adds some computational overhead, but it helps to improve detection in dense or low-resolution areas and remains more efficient than exhaustive alternatives, as shown in our experiments.

Algorithm 1: Region of Interest Selection (RoISel) Algorithm

Require: Density grid $DG \in \mathbb{R}^{N \times N}$, maximum number of RoIs R
Ensure: Selected patches S
1: $C \leftarrow$ all candidate regions of size $K \times L$ with $K, L \in \{1, 2, \ldots, N-1\}$
2: Interest threshold $I_{TH} \leftarrow \text{median}(DG)$
3: $I_{ij}^{\text{mask}} \leftarrow 1$ if $DG_{ij} \geq I_{TH}$, else 0
4: $S \leftarrow \emptyset$
5: **while** $C \neq \emptyset \wedge |S| < R \wedge \sum I^{\text{mask}} > 0$ **do**
6:     Compute candidate densities:

$$CD = \left[d_c \mid d_c = \left(\frac{1}{KL} \sum_{(i,j) \in c} DG_{ij} \right) \times \sqrt{KL} \right]$$

7:     Sort C based on CD
8:     Pick c_s with highest density CD_s
9:     Add c_s to S, remove from C
10:    Zero I_{ij}^{mask} and penalize the cells of c_s in DG: $DG_{ij} \leftarrow DG_{ij} - CD_s \ \forall (i,j) \in c_s$
11: **end while**
12: **return** S

---

4   Experiments

In this section, we evaluate VIGIA-E on two complementary datasets–VisDrone, a widely used benchmark, and Anfi, a domain-specific coastal dataset–to assess its effectiveness in detecting small objects under edge computing constraints. The experiments compare our method against multi-inference baselines such as SAHI and GLSAN, using multiple YOLO-based detectors. Results show that VIGIA-E consistently improves the trade-off between accuracy and inference speed, supporting its suitability for practical deployment in resource-limited scenarios.

4.1   Datasets

Our experiments were conducted on two datasets: the publicly available VisDrone-DET2019 [4] and a custom coastal dataset, referred here as Anfi.

VisDrone-DET2019 is a widely adopted benchmark for object detection in drone-based imagery. It comprises thousands of annotated images captured over urban and suburban areas in multiple Chinese cities. The dataset poses significant challenges, including occlusion, scale variation, and background clutter, and includes diverse object categories such as pedestrians, vehicles, and bicycles—making it a strong reference for evaluating detection performance in aerial scenarios.

The Anfi dataset, introduced in this work, consists of 375 images captured with a PTZ camera system deployed in the coastal touristic area of Anfi del Mar Bay, Gran Canaria (Spain). The dataset includes images at various resolutions: 144 in 4K, 59 in 2K, and 172 in Full HD, depicting typical beach scenes with dense and dynamic activity. The dataset was manually annotated with over 30,000 instances across four categories: person, boat, umbrella, and chair. Anfi provides a realistic, domain-specific benchmark for assessing object detectors in coastal surveillance and tourism-oriented monitoring applications.

4.2 Evaluation Results

Table 2. Performance comparison of baseline, GLSAN, SAHI, and our VIGIA-E approach across YOLO-based detectors on Anfi and VisDrone datasets. Baselines are the fastest approaches but least accurate, their results are in italic for reference. VIGIA-E achieves superior mAP and inference speed than exhaustive slice method like SAHI, demonstrating an efficient trade-off between accuracy and runtime.

Method	Model	Anfi @ conf=0.01			Visdrone @ conf=0.25		
		mAP_{50}	mAP	FPS	mAP_{50}	mAP	FPS
Baseline	YOLOR	53.4	26.9	*7.50*	42.4	27.3	*7.64*
	YOLOv7	40.1	19.5	*14.32*	40.0	23.9	*14.48*
	YOLOv9	37.3	18.2	*9.73*	32.4	21.4	*7.37*
GLSAN [3]	YOLOR	50.3	28.7	0.57	44.3	27.8	0.77
	YOLOv7	39.8	21.9	0.83	**46.8**	27.7	1.31
	YOLOv9	39.9	21.5	0.85	42.1	26.6	0.88
SAHI [1]	YOLOR	54.2	32.7	0.16	44.3	28.0	0.70
	YOLOv7	52.3	30.8	0.49	41.3	24.4	1.94
	YOLOv9	53.7	32.2	0.25	41.5	26.5	0.71
VIGIA-E	YOLOR	**56.7**	**33.1**	1.22	46.1	**29.0**	1.15
(Ours)	YOLOv7	53.1	29.4	**2.12**	46.2	27.1	**2.56**
	YOLOv9	52.7	29.5	1.83	42.4	26.7	1.65

Table 2 summarizes the performance of YOLOR [21], YOLOv7 [20], and YOLOv9 [22] on the Anfi and VisDrone-DET2019 validation sets using their best hyperparameter configurations. The default training hyperparameters for

BASELINE   **SAHI**   **VIGIA-E**

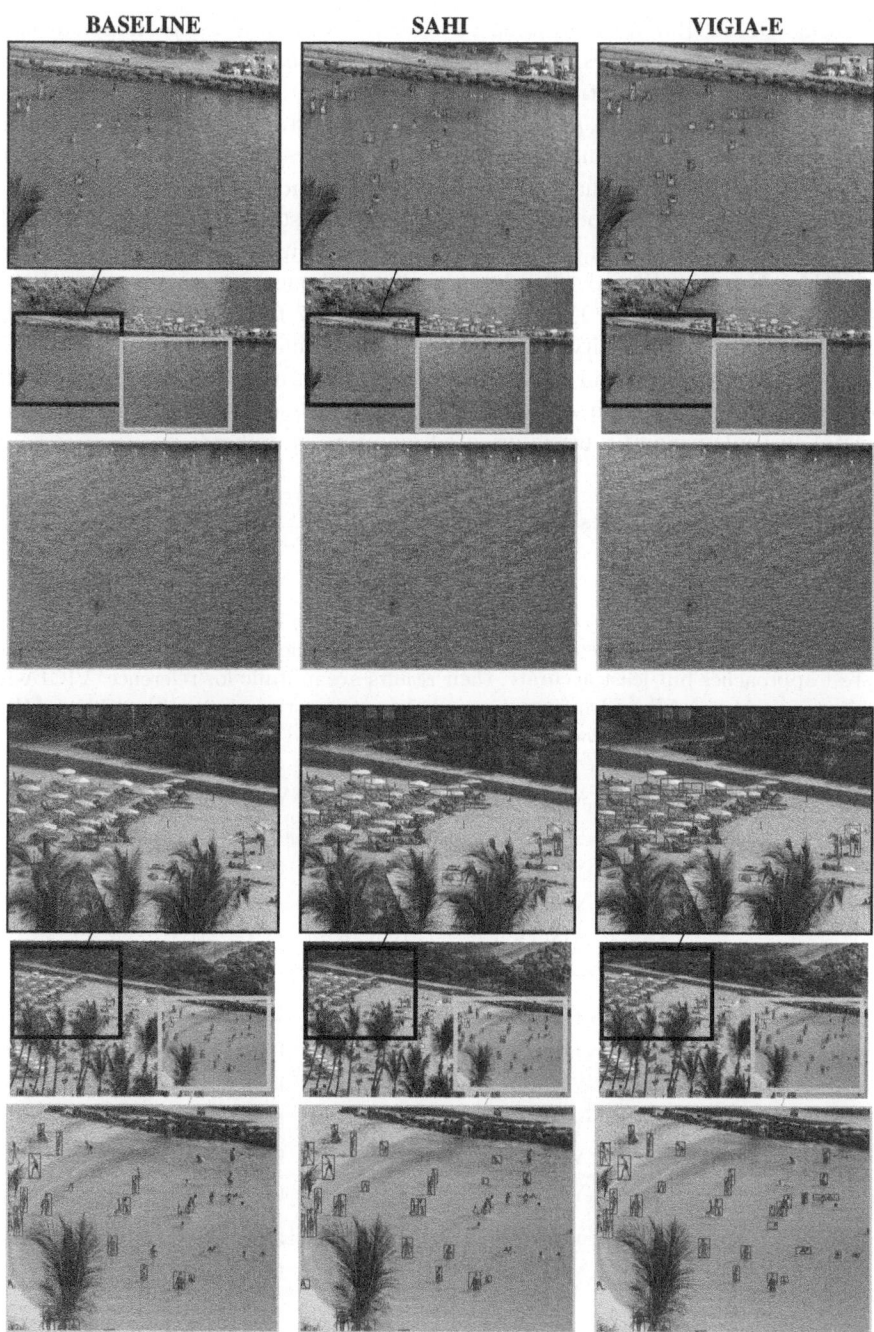

Fig. 3. Detections on two frames of the Anfi dataset (top and bottom). Zoomed regions show differences in crowded areas. The color code indicates the detection categories ▮ person ▮ boat ▮ chair ▮ umbrella.

each model were used for training. In the testing stage, for Anfi, a confidence threshold (C_{TH}) of 0.01 was used, while VisDrone performed best with a threshold of 0.25. SAHI was configured with fixed 640×640 patches and 25% overlap in both datasets, including full-image inference for VisDrone. In contrast, VIGIA-E employs a dynamic patching strategy with up to four adaptive regions on a 5×5 grid with a relaxed confidence threshold (θ) set to 0.01. In every case, the IoU threshold for NMS used was 0.45.

Across both datasets, VIGIA-E consistently outperforms the baseline and achieves accuracy comparable to or better than SAHI and GLSAN in most cases, while offering significantly higher inference speed. Notably, VIGIA-E maintains detection quality on Anfi while operating several times faster than SAHI and GLSAN. Although the frame rates remain below standard real-time thresholds, they are sufficient for typical edge-based surveillance scenarios, and represent a favorable trade-off between accuracy and efficiency. On VisDrone, it achieves both higher mAP and improved FPS across all evaluated detectors.

Figure 3 presents qualitative detection results on the Anfi dataset using YOLOR, comparing the baseline with the two best-performing methods from Table 2, namely SAHI and the proposed VIGIA-E. To aid interpretation, the figure includes zoomed-in views of regions with higher object density—typically distant parts of the scene—where small or clustered objects are more challenging to detect. The baseline often misses relevant objects or produces fragmented detections, particularly in dense or low-resolution areas. SAHI improves coverage through exhaustive slicing, but at the cost of redundant boxes, spatial misalignments, and increased inference time. VIGIA-E, by contrast, accurately localizes targets with fewer false positives, achieving comparable detection while operating significantly faster. These visual results support the quantitative findings, highlighting VIGIA-E's ability to focus computation on the most informative regions while preserving accuracy in edge-constrained scenarios.

5   Conclusions

This work presents VIGIA-E, a lightweight and modular patch-based detection strategy for edge-based surveillance of small and distant objects. Its two-stage design, combining global low-resolution inference with density-guided region selection, achieves high detection accuracy with low computational cost. Validated on both a domain-specific and a public benchmark, it outperforms representative multi-inference baselines in accuracy–speed trade-offs.

Deployed as part of the Smart Coast Surveillance System, VIGIA-E has proven effective in a real-world coastal monitoring setting. This operational deployment reinforces its practical relevance and highlights its suitability for embedded surveillance scenarios under resource constraints. Despite its two-stage pipeline and evaluation on a moderately sized dataset, it offers strong adaptability and efficiency. Future work includes refining region selection via continuous density cues, reducing latency, and extending deployment to multi-camera and mobile edge settings.

Acknowledgements. This work stems from the collaboration between the company Qualitas Artificial Intelligence and Science (QAISC) and the Imaging Technology Center (CTIM) at the University of Las Palmas de Gran Canaria, under the research contract C2024/54, signed between the company and the Canarian Science and Technology Park Foundation of the University of Las Palmas de Gran Canaria.

It has also been supported by Vicepresidencia Primera, Consejería de Vicepresidencia Primera y de Obras Públicas, Infraestructuras, Transporte y Movilidad from Cabildo de Gran Canaria (reference *Resolution No. 45/2021*).

The PORTS 4.0 equity fund also supports this work through a grant to the precommercial project *SMART COAST AI SOLUTIONS 4.0*, in which the Oceanic Platform of the Canary Islands (PLOCAN, https://plocan.eu/en), QAISC, and CTIM are partners, as part of the Spanish State Port Authorities initiative to promote the transition to Economy 4.0.

References

1. Akyon, F.C., Altinuc, S.O., Temizel, A.: Slicing aided hyper inference and fine-tuning for small object detection. In: 2022 IEEE International Conference on Image Processing (ICIP), pp. 966–970 (2022).https://doi.org/10.1109/ICIP46576.2022. 9897990

2. Akyon, F.C., Onur Altinuc, S., Temizel, A.: Slicing aided hyper inference and fine-tuning for small object detection. In: 2022 IEEE International Conference on Image Processing (ICIP), pp. 966–970 (2022).https://doi.org/10.1109/ICIP46576. 2022.9897990

3. Deng, S., et al.: A global-local self-adaptive network for drone-view object detection. IEEE Trans. Image Process. **30**, 1556–1569 (2021). https://doi.org/10.1109/ TIP.2020.3045636

4. Du, D., et al.: VisDrone-DET2019: The vision meets drone object detection in image challenge results. In: Proceedings of the IEEE/CVF International Conference on Computer Vision (ICCV) Workshops, pp. 213–226 (2019)

5. Gidaris, S., Komodakis, N.: Object detection via a multi-region and semantic segmentation-aware CNN model. In: 2015 IEEE International Conference on Computer Vision (ICCV), pp. 1134–1142 (2015).https://doi.org/10.1109/ICCV.2015. 135

6. Girshick, R., Donahue, J., Darrell, T., Malik, J.: Rich feature hierarchies for accurate object detection and semantic segmentation (2014). https://arxiv.org/abs/ 1311.2524

7. Hussain, M., Yang, L.: A comprehensive review of deep learning-based tiny object detection methods. Artif. Intell. Rev. (2025). https://doi.org/10.1007/s10462-024- 10659-2, early Access

8. Kisantal, M., Wojna, Z., Murawski, J., Naruniec, J., Cho, K.: Augmentation for small object detection (2019).https://doi.org/10.48550/ARXIV.1902.07296, https://arxiv.org/abs/1902.07296

9. Kong, Y., Liu, K., Liang, Z., Liu, T., Huang, Y., Qin, M.: Research on small object detection methods based on deep learning. In: 2022 IEEE 4th International Conference on Power, Intelligent Computing and Systems (ICPICS), pp. 680–686 (2022).https://doi.org/10.1109/ICPICS55264.2022.9873614

10. Koyun, O.C., Keser, R.K., İbrahim Batuhan Akkaya, Töreyin, B.U.: Focus-and-Detect: a small object detection framework for aerial images. Sig. Proc. Image Commun. **104**,pp. 1–9 (2022).https://doi.org/10.1016/j.image.2022.116675, https://www.sciencedirect.com/science/article/pii/S0923596522000273

11. Li, C., Yang, T., Zhu, S., Chen, C., Guan, S.: Density map guided object detection in aerial images. In: proceedings of the IEEE/CVF conference on computer vision and pattern recognition workshops, pp. 190–191 (2020)

12. Lin, T.-Y., et al.: Microsoft COCO: common objects in context. In: Fleet, D., Pajdla, T., Schiele, B., Tuytelaars, T. (eds.) ECCV 2014. LNCS, vol. 8693, pp. 740–755. Springer, Cham (2014). https://doi.org/10.1007/978-3-319-10602-1_48

13. Liu, W., Liu, Y., Zhang, Y., Ji, X.: A review of small object detection based on deep learning. Neural Comput. Appl. (2024). https://doi.org/10.1007/s00521-024-09422-6, online First

14. Meethal, A., Granger, E., Pedersoli, M.: Cascaded zoom-in detector for high resolution aerial images. In: 2023 IEEE/CVF Conference on Computer Vision and Pattern Recognition Workshops (CVPRW), pp. 2046–2055 (2023).https://doi.org/10.1109/CVPRW59228.2023.00198

15. Mehmood, I., Mehmood, I., Lee, S.: Small object detection in diverse application land: A survey. IEEE Access **12**, 88637–88660 (2024). https://doi.org/10.1109/ACCESS.2024.3390781

16. Redmon, J., Divvala, S., Girshick, R., Farhadi, A.: You only look once: unified, real-time object detection. In: 2016 IEEE Conference on Computer Vision and Pattern Recognition (CVPR), pp. 779–788 (2016).https://doi.org/10.1109/CVPR.2016.91

17. Satyanarayanan, M.: The emergence of edge computing. Computer **50**(1), 30–39 (2017). https://doi.org/10.1109/MC.2017.9

18. Shi, W., Cao, J., Zhang, Q., Li, Y., Xu, L.: Edge computing: vision and challenges. IEEE Internet Things J. **3**(5), 637–646 (2016). https://doi.org/10.1109/JIOT.2016.2579198

19. Suárez-Ramírez, J., Betancor-Del-Rosario, A., Santana-Cedrés, D., López, N.M.: Exploring deep learning capabilities for coastal image segmentation on edge devices. In: VISIGRAPP (2023). https://api.semanticscholar.org/CorpusID:257358815

20. Wang, C.Y., Bochkovskiy, A., Liao, H.Y.M.: Yolov7: trainable bag-of-freebies sets new state-of-the-art for real-time object detectors. In: 2023 IEEE/CVF Conference on Computer Vision and Pattern Recognition (CVPR), pp. 7464–7475 (2023). https://doi.org/10.1109/CVPR52729.2023.00721

21. Wang, C., Yeh, I., Liao, H.M.: You only learn one representation: unified network for multiple tasks. CoRR **abs/2105.04206** (2021). https://arxiv.org/abs/2105.04206

22. Wang, C.Y., Yeh, I.H., Mark Liao, H.Y.: Yolov9: learning what you want to learn using programmable gradient information. In: Leonardis, A., Ricci, E., Roth, S., Russakovsky, O., Sattler, T., Varol, G. (eds.) Computer Vision - ECCV 2024, pp. 1–21. Springer Nature Switzerland, Cham (2025)

23. Wang, G., Xiong, Z., Liu, D., Luo, C.: Cascade mask generation framework for fast small object detection. In: 2018 IEEE International Conference on Multimedia and Expo (ICME), pp. 1–6. IEEE (2018)

24. Xu, J., Li, Y., Wang, S.: AdaZoom: adaptive zoom network for multi-scale object detection in large scenes. arXiv preprint arXiv:2106.10409 (2021)

25. Yang, F., Fan, H., Chu, P., Blasch, E., Ling, H.: Clustered object detection in aerial images. In: Proceedings of the IEEE/CVF International Conference on Computer Vision, pp. 8311–8320 (2019)

26. Zagoruyko, S., et al.: A multipath network for object detection. arXiv preprint arXiv:1604.02135 (2016)
27. Zhang, J., Huang, J., Chen, X., Zhang, D.: How to fully exploit the abilities of aerial image detectors. In: 2019 IEEE/CVF International Conference on Computer Vision Workshop (ICCVW), pp. 1–8 (2019).https://doi.org/10.1109/ICCVW.2019.00007
28. Ünel, F.O., Özkalayci, B.O., Çiğla, C.: The power of tiling for small object detection. In: 2019 IEEE/CVF Conference on Computer Vision and Pattern Recognition Workshops (CVPRW), pp. 582–591 (2019).https://doi.org/10.1109/CVPRW.2019.00084

Exploring Open-Vocabulary Models
for Category-Free Detection

Pablo Garcia-Fernandez$^{(\boxtimes)}$ ⓘ, Daniel Cores ⓘ, and Manuel Mucientes ⓘ

University of Santiago de Compostela, Santiago, Spain
{pablogarcia.fernandez,daniel.cores,manuel.mucientes}@usc.es

Abstract. Object detection models typically rely on a predefined set of categories, limiting their applicability in real-world scenarios where object classes may be unknown. In this paper, we propose a novel, training-free framework that enables off-the-shelf open-vocabulary object detectors (OvOD) to perform category-free detection—localizing and classifying objects without any prior category knowledge. Our approach leverages image captioning to dynamically generate descriptive terms directly from the image content, followed by a WordNet-based filtering process to extract semantically meaningful category names. These discovered categories are then embedded and matched with visual region features using a frozen OvOD model to perform detection. We evaluate our method on the COCO dataset in a fully zero-shot setting and demonstrate that it significantly outperforms strong multimodal large language model baselines, achieving an improvement of over 30 AP points. This highlights our method as a promising direction for more adaptive solutions to real-world detection challenges.

Keywords: category-free · open-vocabulary object detection · captioning

1 Introduction

Object detection has become a cornerstone of modern computer vision, with advances in deep learning enabling highly accurate detectors across a wide range of applications. However, the success of most object detectors hinges on a fundamental assumption: the complete set of object categories is known and fixed during training. This assumption is deeply embedded in the construction of popular benchmarks such as COCO and LVIS, where models are trained to detect a limited number of predefined categories and evaluated accordingly. While effective in controlled environments, this closed-world assumption breaks down in real-world settings, where objects of interest may not be part of the training vocabulary.

In many practical applications—such as autonomous driving, surveillance, or general-purpose scene understanding—the relevant object categories may be unknown, ambiguous, or context-dependent. Relying on a fixed set of known

M. Castrillón-Santana et al. (Eds.): CAIP 2025, LNCS 15621, pp. 283–293, 2026.
https://doi.org/10.1007/978-3-032-04968-1_24

categories in those cases severely limits the flexibility and generalization ability of traditional detectors. Ideally, we would like to build object detectors that do not rely on any prior knowledge of the category space, and can instead adaptively discover and detect objects based on the content of the image itself.

Open-vocabulary object detection (OvOD) represents a significant step toward this goal. These methods leverage vision-language alignment techniques to enable object detection for arbitrary textual categories at inference time. Typically, a detector is trained to align region-level features with textual embeddings (e.g., from CLIP [25]), allowing it to respond to category names outside of the training set. However, despite their impressive flexibility, OvOD methods still implicitly assume access to a relevant set of object names at test time—whether provided manually, sampled from a predefined vocabulary, or selected from a prompt. In other words, while the vocabulary is technically open, its selection is still guided by prior knowledge. This raises an important question: **can we detect objects in an image without assuming any prior knowledge of the possible categories?**

In this work, we propose a training-free framework that enables *off-the-shelf* open-vocabulary object detectors to operate without relying on predefined category priors. Our central idea is to discover a relevant vocabulary directly from the image using an image captioning model. Captioning provides a natural way to surface human-interpretable terms associated with objects in an image. We extract candidate object names from generated captions and use them to guide an open-vocabulary detector. This allows our system to perform object detection in a zero-prior setting, where no fixed label set, prompt, or external vocabulary is assumed.

One consequence of working without a fixed category set is that traditional baselines are no longer applicable. Most supervised detectors are not designed to operate without a predefined vocabulary. As such, we compare our method against multimodal large language models (MLLMs) trained for grounding and spatial reasoning, such as KOSMOS-2 [24], as well as general-purpose MLLMs like GPT-4o [1] and LLaVA-Video [17], which are not explicitly designed for object localization. These approaches represent reasonable references for detection guided purely by image understanding and free-form language. We show that our method outperforms state-of-the-art methods by over ↑30 AP points in this challenging zero-prior setting. To summarize, our key contributions are:

- We demonstrate that, although multimodal large language models (MLLMs) are currently the only models capable of operating in a zero-category prior setting, they exhibit significant limitations in reliably detecting objects.
- We propose a training-free approach that adapts *off-the-shelf* OvOD detectors to operate without any predefined or prompted category priors, by automatically deriving a relevant vocabulary from image captions.
- We show that our approach achieves state-of-the-art performance, surpassing the strongest baseline by over ↑30 AP points.

2   Related Work

Open Vocabulary Object Detection. (OvOD) [26,38] has made significant progress driven by the emergence of vision-language models (VLMs), which enable detectors to move beyond fixed category sets and handle novel concepts at inference time. Unlike traditional zero-shot object detection, which relies solely on textual semantics to recognize unseen classes, OvOD methods incorporate various forms of weak supervision to improve both classification and localization performance. These methods can generally be grouped into four main strategies: (i) region-aware training, (ii) pseudo-labeling, (iii) knowledge distillation, and (iv) transfer learning.

Region-aware approaches aim to improve the alignment between image regions and textual descriptions during training, enhancing the detector's ability to generalize to unseen categories. Works like DetCLIP [33] , DetCLIPv2 [32], CORA [29], and VLDet [15] refine region-level feature-text correspondence to improve localization and semantic alignment. **Pseudo-labeling** methods expand the training data by using large-scale VLMs to generate object labels automatically. RegionCLIP [36], PromptDet [6], CoDET [21], GLIP [14], Detic [37], and Grounding DINO [18] follow this paradigm to build richer supervision from unlabeled images. **Knowledge distillation** techniques transfer the representational power of VLMs into detection models by using them as teachers. Approaches like BARON [27], DK-DETR [13], CLIPSelf [28], and SIC-CADS [5] distill knowledge from the VLM into the detector to improve its generalization ability. Finally, **transfer learning-based** methods directly incorporate pretrained vision-language encoders into the detection pipeline, either through fine-tuning (OWL-ViT [23]) or by freezing the encoder and learning lightweight heads (F-VLM [12]).

While classical OvOD approaches have significantly expanded the flexibility of object detectors, they still rely on externally supplied vocabularies—in the form of text prompts, or dataset-specific category lists. In contrast, our method removes this dependency entirely by discovering object categories dynamically through image captioning. This enables the detector to operate without any prior assumptions about which object classes might appear at inference time, representing a more flexible and realistic scenario.

Open-set and Open World Detection. The problem of detecting objects without prior knowledge of all possible categories has been studied under several paradigms. **Open-Set Object Detection** (OSOD) addresses scenarios in which a detector must correctly classify instances belonging to known categories while also identifying and localizing objects from unknown categories—without assigning them specific semantic labels. These unknown instances are typically grouped under a generic unknown class. Thus, the main objective of OSOD is to enable robustness against out-of-distribution (OOD) categories, focusing on their rejection rather than discovery. Dhamija et al. [4] were the first to formally define the OSOD setting, showing that the performance of conventional object detectors is often significantly overestimated under open-set conditions. Subsequent works have explored methods to improve OOD robustness through various mechanisms such as background expansion [9], adaptive classification thresholds [19], or uncertainty modeling using Bayesian dropout [22], yet they still treat unknown objects as undifferentiated outliers and do not aim to recover their category semantics.

Open World Object Detection (OWOD) extends the open-set setting by introducing a continual learning framework in which novel categories are progressively encountered and incorporated over time. The open-world paradigm was first introduced in image classification by Bendale et al. [2], who proposed a model capable of rejecting unknown classes at test time and incrementally integrating them once labeled. Joseph et al. [11] later adapted this paradigm to object detection, formalizing the OWOD task and proposing a method based on exemplar replay to enable the model to learn new object categories while mitigating catastrophic forgetting. However, despite these contributions, OWOD models [8,30,31,39] remain limited in their ability to autonomously explore or infer the semantics of unknown categories; they still require explicit human annotation to incorporate new classes. This reliance on supervision poses a major bottleneck for scalability in realistic open-world scenarios, where novel objects frequently appear and manual labeling is impractical.

To alleviate this issue, Zheng et al. [35] propose a method to automatically discover categories of unknown objects based on their visual appearance. Positioned between the open-set and open-world paradigms, their approach clusters unknown instances into a fixed number of generic categories—each potentially corresponding to a novel class, without requiring labeled data. While this method enables unsupervised discovery, a fundamental limitation remains: it does not capture the semantics of these categories. The discovered groups lack meaningful and interpretable labels. Our approach addresses this gap by introducing a language-driven mechanism for both the discovery and semantic grounding of unknown categories. Rather than relying on manual supervision (as in OWOD) or unsupervised clustering without semantic interpretation (as in [35]), we leverage image captioning models to extract rich, contextual object-level descriptions directly from images. As a result, our system can not only localize previously unseen objects but also assign them human-interpretable labels.

3   Method

We propose a *training-free* method that enables *off-the-shelf* open-vocabulary object detectors to operate without predefined category priors, by discovering a relevant vocabulary directly from image captions. As illustrated in Fig. 1, our approach comprises two main components: Vocabulary Discovery and Category-Free Object Detection. In the **Vocabulary Discovery** stage, a captioning model generates textual descriptions for the input images. Using an external corpus, we filter these terms to retain only those that are valid object category candidates. This step allows the model to construct its vocabulary directly from the data, without relying on external supervision. In the **Category-Free Object Detection** stage, the discovered vocabulary is embedded using a text encoder, and the resulting text embeddings are matched with region-level visual embeddings to detect objects corresponding to the discovered categories.

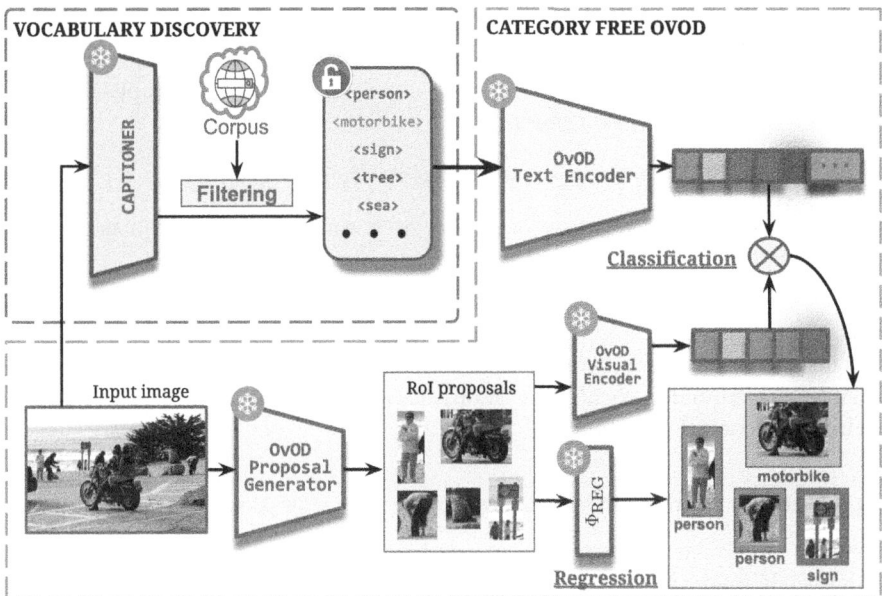

Fig. 1. Our framework consists of two main stages. In **Vocabulary Discovery** (3.1), we prompt a captioning model to produce object-centric descriptions, from which we extract candidate categories using WordNet-based noun filtering. In **Category-Free OVOD** (3.2), we use a frozen open-vocabulary detector to match visual region features with the discovered categories via cosine similarity in a shared embedding space.

3.1 Vocabulary Discovery

The vocabulary discovery stage aims to extract a set of candidate object categories $\mathcal{V}$ from a collection of unlabeled images $\{\mathbf{I}_i\}_{i=1}^{N}$, without relying on predefined labels or external supervision. This is achieved by generating object-centric textual descriptions using a vision-language model, followed by a filtering step to retain only valid nouns.

Caption Generation. For each image $\mathbf{I}_i$, we prompt a vision-language model to produce a structured caption consisting of visible object names, constrained to a semicolon-separated format. Specifically, the model is guided via prompting to output:

$$T_i = \mathcal{C}(\mathbf{I}_i), \tag{1}$$

where T_i is a string of candidate object terms (e.g., "car; tree; road; sign").
These strings are parsed into token lists $W_i = \text{Split}(T_i, ";")$, and aggregated
across the dataset to form a candidate term set:

$$W = \bigcup_{i=1}^{N} W_i \tag{2}$$

WordNet-Based Noun Filtering. To ensure semantic validity, we filter the candidate terms using WordNet. A word $w \in \mathcal{W}$ is retained if it appears in the
WordNet lexical corpus as a noun:

$$\text{IsNoun}(w) = (\exists s \in \text{WordNet}(w) \text{ s.t. } \text{POS}(s) = \text{'n'}), \tag{3}$$

where $\text{POS}(w)$ denotes the part of speech of w, and 'n' indicates a noun. The
final vocabulary is defined as:

$$\mathcal{V} = \{w \in \mathcal{W} \mid \text{IsNoun}(w)\}. \tag{4}$$

This simple yet effective strategy allows us to construct a domain-relevant,
semantically grounded vocabulary of object categories directly from the dataset,
without external priors.

3.2   Category-Free Object Detection

Once the vocabulary $\mathcal{V} = \{v_1, v_2, \ldots, v_K\}$ has been discovered, we leverage an
off-the-shelf open-vocabulary object detector to localize and classify instances
of these categories. Our approach does not require **any retraining or fine-
tuning**, relying entirely on the alignment between visual and textual embeddings
in a shared feature space.

Text Embedding. Each category name $v_k \in \mathcal{V}$ is embedded into a shared semantic
space using the OvOD pretrained text encoder ϕ_{text}:

$$\mathbf{z_k}^{\text{text}} = \phi_{\text{text}}(v_k), \quad \forall k \in \{1, \ldots, K\}. \tag{5}$$

This yields a matrix of category embeddings $\mathbf{Z}_{\text{text}} \in \mathbb{R}^{K \times d}$, where d is the
dimensionality of the joint embedding space.

Visual Embedding and Region Proposal. Given an input image $\mathbf{I}$, the detector
extracts a set of region proposals $\{z_j\}_{j=1}^{M}$, each associated with a visual embedding computed via a visual encoder ϕ_{vis}:

$$\mathbf{z_j}^{\text{vis}} = \phi_{\text{vis}}(z_j), \quad \forall j \in \{1, \ldots, M\}. \tag{6}$$

These embeddings $\mathbf{Z}_{\text{vis}} \in \mathbb{R}^{M \times d}$ are aligned with the text embeddings.

Similarity-Based Classification. To assign a category label to each region proposal, we compute the similarity between visual and textual embeddings using
cosine similarity, $\langle \cdot, \cdot \rangle$:

$$\hat{y}_j = \arg\max_k \langle \mathbf{z_j}^{\text{vis}}, \mathbf{z_k}^{\text{text}} \rangle. \tag{7}$$

Each region is thus classified as the most similar discovered category, and detections are scored according to the similarity values. This approach allows the detector to operate without category priors, using only the vocabulary discovered from captions and a frozen *off-the-shelf* open-vocabulary detector.

4   Experiments

4.1   Experimental Setup

We evaluate our method on the MS COCO dataset [16], a widely used benchmark for object detection that contains more than 120,000 images, and over 860,000 annotated object instances spanning 80 object categories. The images in COCO are collected from complex everyday scenes. Since our method is training-free, we do not make use of the training split. All evaluations are performed on the COCO-2017 validation set, which contains 5,000 images and approximately 36,000 annotated object instances.

We assess the performance of our method in a fully zero-shot setting, without access to predefined category priors. This poses a challenge for evaluation, as our discovered vocabulary—obtained through the image captioning process—may not align directly with the canonical COCO category names. To address this, we employ a large language model, specifically LLaMA 3.3 [7], to map each discovered term to its most semantically similar COCO class. This mapping is performed automatically and is used solely for evaluation, without influencing the detection process. For performance assessment, we use the standard COCO evaluation protocol and report the average precision (AP), as well as AP at Intersection-over-Union (IoU) thresholds of 0.5 (AP_{50}) and 0.75 (AP_{75}).

4.2   Implementation Details

For the vocabulary discovery stage, we use LLaVA-Video-7B-Qwen2 [17,34] as our captioning model $\mathcal{C}$. To guide the model towards producing object-centric outputs, we employ the following prompt:

```
Analyze the image and provide a list of all object categories
present. Then, based on your understanding of the scene,
extend this list by including other categories of objects
that might normally appear in similar contexts. Return all
the category names, separated by semicolons (;).
```

This prompt not only encourages precise object name extraction but also implicitly introduces a form of vocabulary augmentation by leveraging the model's contextual understanding to suggest additional plausible categories beyond those explicitly visible.

In the detection stage, we use Grounding DINO-SwinT [18] as our open-vocabulary object detector. It employs a dual-encoder architecture, where the visual encoder, ϕ_{vis}, is Swin-Tiny [20], and the textual encoder, ϕ_{text}, is BERT-Base [3].

Table 1. SOTA comparison. Our method significantly outperforms general-purpose and localization-grounded Multimodal Large Language Models (MLLMs) under vocabulary-free conditions. G-DINO with access to the category vocabulary is reported as an upper bound.

	Method	AP	AP_{50}	AP_{75}
Free vocab.	LLaVA-Video-7B-Qwen2 [34] + CLIP [25] scoring	0.4	1.8	0.1
	Idefics3-8B-Llama3 [10] + CLIP [25] scoring	0.2	0.3	0.1
	KOSMOS-2 [24]	9.6	15.0	10.2
	OURS	**40.5**	**51.9**	**44.6**
Known vocab.	G-DINO (upper limit)	55.7	72.8	61.3

4.3   Results

The results in Table 1 highlight a clear performance gap between our method and existing MLLM-based baselines. Both LLaVA-Video and Idefics3, when paired with CLIP-based scoring, achieve very low AP scores (0.4 and 0.2, respectively). Since these models do not produce native confidence scores—which are essential for ranking detections in AP computation— we calculated them by computing the cosine similarity between the predicted region and its label embedding using CLIP, normalized between 0 and 1. While this allows for rough comparability, it does not overcome the fundamental limitations of these models in precise object localization.

KOSMOS-2, which is explicitly trained for localization-grounding, performs notably better with an AP of 9.6. However, it still falls far short of our method, which combines caption-based vocabulary discovery with G-DINO and achieves an AP of 40.5. This constitutes over a 4× improvement over KOSMOS-2, demonstrating the strength of dynamically adapting the detection vocabulary to the image content via captioning. The improvement is consistent across AP_{50} and AP_{75}. We also report the performance of G-DINO with access to the full ground-truth vocabulary, serving as an upper bound. This oracle setting achieves the highest AP (55.7), but our method recovers a substantial portion of this performance—despite having no prior knowledge of the object categories.

Finally, Fig. 2 provides a qualitative analysis showcasing the vocabulary discovery process and the final OvOD detections. In overall, these results validate our method as an effective way to equip open-vocabulary object detectors with the ability to perform detection without predefined categories in a training-free manner.

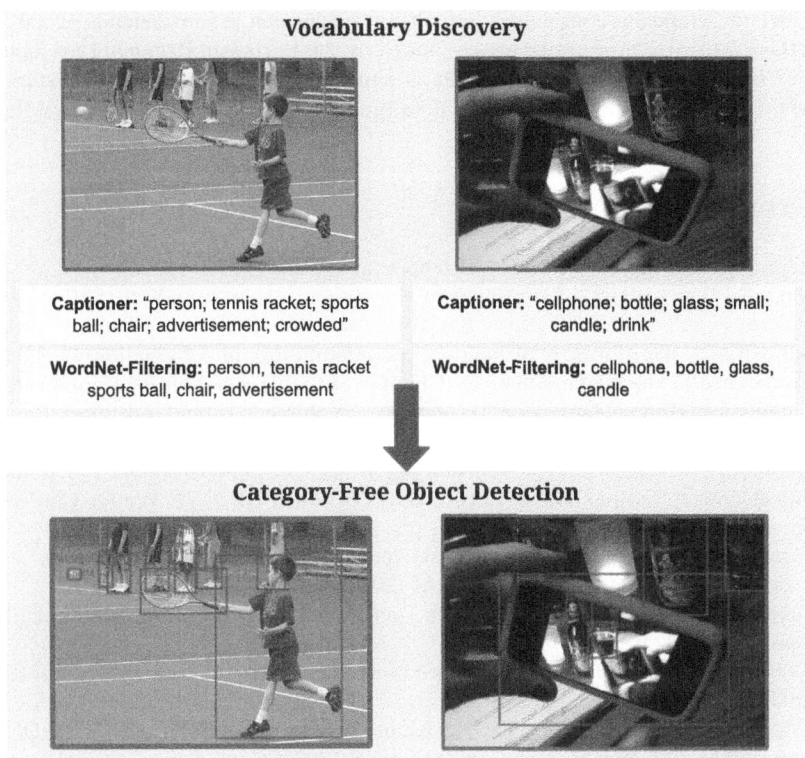

Fig. 2. Qualitative analysis showing both the vocabulary discovery process and the final OvOD detections with a confidence greater than 0.5.

5   Conclusions

In this work, we have presented a novel, training-free framework that adapts open-vocabulary object detectors to operate without relying on predefined category priors. By leveraging image captioning to dynamically derive a relevant vocabulary for each image, our approach enables object detection in a fully category-free, zero-prior setting. We demonstrate that this strategy not only removes the need for external supervision but also significantly outperforms strong multimodal baselines in zero-shot detection tasks, recovering a substantial portion of the upper-bound performance achieved by models with access to ground-truth vocabularies. Our results highlight the potential of using language-driven methods for both discovering and semantically grounding unknown object categories, pointing toward more scalable and adaptive solutions for real-world detection challenges.

Acknowledgements. This work was partially supported by the Spanish Ministerio de Ciencia e Innovación (grant numbers PID2020-112623GB-I00, PID2023-149549NB-

I00), and the Galician Consellería de Cultura, Educación e Universidade (2024-2027 ED431G-2023/04). These grants are co-funded by the European Regional Development Fund (ERDF). Pablo Garcia-Fernandez is supported by the Spanish Ministerio de Universidades under the FPU national plan (grant number FPU21/05581).

References

1. Achiam, J., et al.: Gpt-4 technical report. arXiv:2303.08774 (2023)
2. Bendale, A., Boult, T.: Towards open world recognition. In: CVPR (2015)
3. Devlin, J., Chang, M.W., Lee, K., Toutanova, K.: BERT: pre-training of deep bidirectional transformers for language understanding. In: Proceedings of the 2019 Conference of the North American Chapter of the Association for Computational Linguistics: Human Language Technologies, Volume 1 (Long and Short Papers), pp. 4171–4186 (2019)
4. Dhamija, A., Gunther, M., Ventura, J., Boult, T.: The overlooked elephant of object detection: open set. In: Proceedings of the IEEE/CVF Winter Conference on Applications of Computer Vision, pp. 1021–1030 (2020)
5. Fang, R., Pang, G., Bai, X.: Simple image-level classification improves open-vocabulary object detection. In: AAAI (2024)
6. Feng, C., et al.: Promptdet: Towards open-vocabulary detection using uncurated images. In: ECCV (2022)
7. Grattafiori, A., et al.: The llama 3 herd of models. arXiv preprint arXiv:2407.21783 (2024)
8. Gupta, A., Narayan, S., Joseph, K., Khan, S., Khan, F.S., Shah, M.: OW-DETR: open-world detection transformer. In: CVPR (2022)
9. Han, J., Ren, Y., Ding, J., Pan, X., Yan, K., Xia, G.S.: Expanding low-density latent regions for open-set object detection. In: CVPR (2022)
10. Hugging Face: Introducing idefics: An open reproduction of state-of-the-art visual language model (2023). https://huggingface.co/blog/idefics. Accessed 30 Apr 2025
11. Joseph, K., Khan, S., Khan, F.S., Balasubramanian, V.N.: Towards open world object detection. In: CVPR (2021)
12. Kuo, W., Cui, Y., Gu, X., Piergiovanni, A., Angelova, A.: F-VLM: open-vocabulary object detection upon frozen vision and language models. arXiv:2209.15639 (2022)
13. Li, L., et al.: Distilling DETR with visual-linguistic knowledge for open-vocabulary object detection. In: ICCV (2023)
14. Li, L.H., et al.: Grounded language-image pre-training. In: CVPR (2022)
15. Lin, C., et al.: Learning object-language alignments for open-vocabulary object detection. In: ICLR (2023)
16. Lin, T.Y., et al.: Microsoft coco: common objects in context. In: ECCV (2014)
17. Liu, H., Li, C., Wu, Q., Lee, Y.J.: Visual instruction tuning. In: Advances in Neural Information Processing Systems, vol. 36, pp. 34892–34916 (2023)
18. Liu, S., et al.: Grounding DINO: marrying DINO with grounded pre-training for open-set object detection. In: ECCV (2024)
19. Liu, Y.C., et al.: Open-set semi-supervised object detection. In: ECCV. Springer (2022)
20. Liu, Z., et al.: Swin transformer: hierarchical vision transformer using shifted windows. In: ICCV (2021)
21. Ma, C., Jiang, Y., Wen, X., Yuan, Z., Qi, X.: CoDet: co-occurrence guided region-word alignment for open-vocabulary object detection. In: NeurIPS (2023)

22. Miller, D., Nicholson, L., Dayoub, F., Sünderhauf, N.: Dropout sampling for robust object detection in open-set conditions. In: 2018 IEEE International Conference on Robotics and Automation (ICRA), pp. 3243–3249. IEEE (2018)
23. Minderer, M., et al.: Simple open-vocabulary object detection. In: ECCV (2022)
24. Peng, Z., et al.: Kosmos-2: grounding multimodal large language models to the world. arXiv preprint arXiv:2306.14824 (2023)
25. Radford, A., et al.: Learning transferable visual models from natural language supervision. In: ICML (2021)
26. Wu, J., et al.: Towards open vocabulary learning: a survey. In: PAMI (2024)
27. Wu, S., Zhang, W., Jin, S., Liu, W., Loy, C.C.: Aligning bag of regions for open-vocabulary object detection. In: CVPR (2023)
28. Wu, S., et al.: Clipself: vision transformer distills itself for open-vocabulary dense prediction. In: ICLR (2024)
29. Wu, X., Zhu, F., Zhao, R., Li, H.: Cora: adapting clip for open-vocabulary detection with region prompting and anchor pre-matching. In: CVPR (2023)
30. Wu, Y., Zhao, X., Ma, Y., Wang, D., Liu, X.: Two-branch objectness-centric open world detection. In: Proceedings of the 3rd International Workshop on Human-Centric Multimedia Analysis, pp. 35–40 (2022)
31. Wu, Z., Lu, Y., Chen, X., Wu, Z., Kang, L., Yu, J.: UC-OWOD: unknown-classified open world object detection. In: ECCV. Springer (2022)
32. Yao, L., et al.: Detclipv2: Scalable open-vocabulary object detection pre-training via word-region alignment. In: CVPR (2023)
33. Yao, L., et al.: Detclip: dictionary-enriched visual-concept paralleled pre-training for open-world detection. In: NeurIPS (2022)
34. Zhang, Y., et al.: Video instruction tuning with synthetic data, 2024g. https://arxiv.org/abs/2410.02713
35. Zheng, J., Li, W., Hong, J., Petersson, L., Barnes, N.: Towards open-set object detection and discovery. In: CVPR (2022)
36. Zhong, Y., et al.: Regionclip: region-based language-image pretraining. In: CVPR (2022)
37. Zhou, X., Girdhar, R., Joulin, A., Krähenbühl, P., Misra, I.: Detecting twenty-thousand classes using image-level supervision. In: ECCV (2022)
38. Zhu, C., Chen, L.: A survey on open-vocabulary detection and segmentation: past, present, and future. IEEE TPAMI (2024)
39. Zohar, O., Wang, K.C., Yeung, S.: PROB: probabilistic objectness for open world object detection. In: CVPR (2023)

Synthetic Data for Robust Runway Detection

Estelle Chigot[1,2]([✉]) [iD], Dennis G. Wilson[1], Meriem Ghrib[2], Fabrice Jimenez[2], and Thomas Oberlin[1]

[1] Fédération ENAC ISAE-SUPAERO ONERA, Université de Toulouse,
Toulouse, France
{estelle.chigot2,dennis.wilson,thomas.oberlin}@isae.fr
[2] Airbus, Toulouse, France
{meriem.ghrib,fabrice.jimenez}@airbus.com

Abstract. Deep vision models are now mature enough to be integrated in industrial and possibly critical applications such as autonomous navigation. Yet, data collection and labeling to train such models requires too much efforts and costs for a single company or product. This drawback is more significant in critical applications, where training data must include all possible conditions including rare scenarios. In this perspective, generating synthetic images is an appealing solution, since it allows a cheap yet reliable covering of all the conditions and environments, if the impact of the synthetic-to-real distribution shift is mitigated. In this article, we consider the case of runway detection that is a critical part in autonomous landing systems developed by aircraft manufacturers. We propose an image generation approach based on a commercial flight simulator that complements a few annotated real images. By controlling the image generation and the integration of real and synthetic data, we show that standard object detection models can achieve accurate prediction. We also evaluate their robustness with respect to adverse conditions, in our case nighttime images, that were not represented in the real data, and show the interest of using a customized domain adaptation strategy.

Keywords: Object detection · Synthetic data · Domain adaptation · Runway detection · Vision-based landing

1 Introduction

With the tremendous progress of artificial intelligence, deep vision models are increasingly integrated into critical systems. This paper considers the context of commercial aircraft, where there are major opportunities for both airlines and airports to increase safety and optimize routing or ground operations. In this work, we consider the task of runway detection from onboard cameras.

Deep learning models require large amounts of data to achieve high performance, but in industrial applications, data are often difficult or expensive to obtain. It may be difficult to reliably collect data, impossible to do so safely,

M. Castrillón-Santana et al. (Eds.): CAIP 2025, LNCS 15621, pp. 294–304, 2026.
https://doi.org/10.1007/978-3-032-04968-1_25

or expensive to gather and label data. Aircraft operations make data gathering even more challenging: data collection may be heavily regulated and dangerous to perform outside of nominal conditions. On the other hand, flight simulators seem to provide the ideal solutions. They are cheap, customizable to include adverse conditions such as night, snow or fog, and, most of all, they can generate corresponding precise labels, thanks to the underlying scene modeling process. However, when using simulated data as the only source of data to train a detection model, a performance drop usually occurs when operating the model in real conditions, since the model cannot generalize to real images. This is known as the domain gap, synthetic-to-real gap, or sim-to-real problem. In this context, we study the reliability and performance gains from using synthetic data in a critical runway detection system.

The problem of runway recognition for aircraft has already been studied in the literature, either as a problem of object detection [11,14,27] or semantic segmentation [26]. There are now multiple publicly available datasets for runway detection [1,2,5]. However, these datasets usually include few to no real images, limiting the range of applicable methods and the evaluation on real use cases. In this study, we have access to a small private dataset of real images, and a large dataset of synthetic images; in both real and synthetic datasets, we have images representing different environmental conditions. We use these data to explore domain adaptation methods and to test models' robustness to unseen situations. In this work, we focus on runway detection at night as an underrepresented case, as we have access to nighttime images in both datasets, enabling us to evaluate the robustness of runway detection models on real images.

We make the following contributions, within the case of runway detection: (1) we demonstrate the benefits of using synthetic data to train a detection model in an industrial use case; (2) we customize a state-of-the-art domain adaptation method to the runway detection problem; (3) we show the advantages of using a domain adaptation approach to enhance model's robustness under adverse conditions, in our case nighttime.

The paper is structured as follows: Sect. 2 reviews the literature in the field of synthetic data for object detection. Section 3 details our data collection and the methods compared, while results are presented in Sect. 4. Section 5 concludes the study and draws some perspectives.

2 Related Work

2.1 Object Detection

Object detection is the computer vision task of determining the location and type of objects in a given image. Recent advances in deep learning have significantly improved object detection, especially in safety-critical industrial applications. One-stage detectors, such as YOLO [18] and SSD [15], prioritize speed over accuracy by directly predicting object location and class probabilities from the image. Two-stage models like Faster R-CNN [19] perform object detection in two steps. First, an image classification model extracts features from the image;

this first model is often referred to as the "backbone." Then, a region proposal network (RPN) generates potential object detection proposals from the features. Generally, the backbone is pretrained on an image classification task, such as ImageNet [3], to improve the model's performance on the object detection task. In this work, we focus on Faster R-CNN due to its prevalence in the domain adaptation literature and its good performance.

Runway detection is critical for vision-based aircraft landing systems, in order to properly position the aircraft during the procedure. It has been treated as an object detection task in various studies. Faster R-CNN, in particular, has been used for the detection of runways [14]. A fusion of visual and infrared images has been proposed to improve runway detection at night, also using Faster R-CNN [27]. Federated learning, a distributed means of training models, was shown to improve the robustness of runway detection models [11]. One of the key challenges in the application of deep-learning based models for runway detection is the lack of sufficient real-world data, which synthetic data can help address.

2.2 Synthetic Data

In order to train deep learning models and especially computer vision models, it is necessary to gather a large quantity of data. Large public datasets like COCO [13] or ImageNet [3] have accelerated object detection research, but their utility in industrial contexts, such as aviation, is limited. However, the cost, safety or legal concerns and time-consuming labeling process of real-world data collection make datasets difficult to build. Synthetic data are then a valuable alternative to create more data.

Two main paradigms exist for data generation. The first one, domain randomization [7,24,25], implies that photorealism doesn't matter as it tries to make data as diverse as possible. Simulators or 3D game engines are used to generate objects with large variations of environmental conditions, lighting or textures, improving model robustness across a wide range of scenarios. The aim is to train the model to perceive real images as just another variation of these conditions, leading to strong detection performance.

The second approach is to generate images as close to the real world as possible. This can be achieved by using computer graphics and simulators to build world replicas, showing realistic physics and real world behaviors. This method is very popular in the automotive industry, where video games like GTA-V [8,20] or world models developed in 3D engines [21] have been used to generate widely utilized datasets. Additionally, whole driving simulators [4,12] have been released for this purpose.

Generative models [6,9,23] are also a powerful way to create photorealistic pictures. Generative adversarial networks (GANs) and variational autoencoders learn to approximate the data distribution of their training dataset, while diffusion models are trained to denoise step-by-step random noises into coherent data samples. GANs especially have been leveraged to generate data for industrial applications [10,16,29]. However, generative models need lots of real-world input data to produce satisfactory results. Also, they can't generate labels automatically. In this work we will focus on simulators for those two reasons.

In the aviation industry, due to the lack of real open data, datasets have been released to facilitate research in this field. LARD [5] provides a generator for runway detection, based on Google Earth, as well as 1800 real images taken from YouTube videos and manually labeled. On other tasks than runway detection, FS2020 [1] provides a synthetic dataset generated with Microsoft Flight Simulator, meant for runway segmentation and lines detection. Finally, Rareplanes [22] integrates 3D models of aircraft into real images for remote sensing applications. Logically, the works published for runway detection use mainly synthetic data. They either create their own with flight simulators [27], or use available open datasets [11] such as LARD. In a try to assess the benefits of synthetic images, Linden et al. [14] compare the impact on detection performance of several environmental conditions in a synthetic dataset and examines a style transfer approach on automotive data.

2.3   Synthetic to Real Domain Adaptation

Domain adaptation (DA) methods attempt to minimize the synthetic to real gap by bringing closer either the images directly or the feature distributions between domains. Following the first option, SC-UDA [28] uses a neural style transfer method to apply a realistic style on synthetic images, and then uses Faster R-CNN to generate bounding boxes to augment the unlabeled real dataset. On the other hand, CARE [17] tries to align the instances features between synthetic and real images by modifying the traditional Faster R-CNN loss function. Using the available labels, the loss term introduced by CARE computes the difference between the features of a synthetic object and a real one, and tries to minimize it along the other object detection loss terms. CARE also introduces two reweighting terms to adapt to the difference of object sizes and frequency between the two domains.

In the context of runway detection, DA remains underexplored, likely due to the scarcity of real-world data. In this work we focus on studying DA methods to enhance model generalization from synthetic to real images.

3   Methodology

3.1   Data Collection

In our use case, real images are limited in quantity due to their high cost of acquisition. The images are captured using an aircraft equipped with cameras and sensors, navigating across various airports in the U.S. Data is acquired by photographing runways at selected airports during landing sequences. In this study, we have access to a database of real images from 27 airports with their associated metadata. However, the images of the same airports tend to be quite similar, showing the same runways from close points during the flight phases, resulting in limited diversity within the real data. Labeling is performed automatically based on the GPS positions of the aircraft and the runways, leading to potential errors in labels due to factors like miscalibration or cloud occlusions.

In contrast, we generate our own synthetic dataset using the commercial flight simulator XPlane12, which allows us to specify the aircraft's position, orientation, and various environmental conditions such as weather and time. This customization capacity enables us to collect diverse images from 207 airports worldwide. One of the key benefits of synthetic images is the ability to achieve automatic and accurate labeling of bounding boxes and other parameters that typically require human input in real-world scenarios. We use a custom pipeline rather than relying on existing datasets because it enables the generation of a large volume (>5,000) of high-quality images, across a wide variety of airports and under specific environmental conditions.

Taking advantage of this generation capacity, we want to trial synthetic data and assess if using simulated images enables a model to be robust under unseen conditions. To do so, we use the nighttime scenario. At night, runways are very different than during the day. The light conditions differ a lot as they are outlined by lights, in order to enable the pilot to see the runway. The shape is conserved but the texture and colors of the image are not, which is a good way to assess the generalization capabilities of a model. With this idea in mind, we generate nighttime synthetic data from the same airports as the daytime ones.

In Fig. 1 we have examples of real and synthetic images for the runway detection task. For copyright reasons, we show synthetic images from [1] which are not the one used to train the model in this study, but closely resemble.

(a) Real runway (Source: Airbus) (b) Synthetic runway (Source: [1])

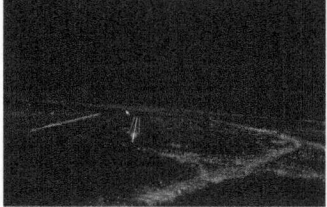

(c) Synthetic runway at night (Source: [1])

Fig. 1. Examples of real data and synthetic data for runway detection.

3.2   Mixing Strategies

In this work, our objective is to assess the use of synthetic data for runway detection. We also want to study the impact of synthetic data in unseen situations. To this end we design several training strategies, applicable for both the nominal and the nighttime studies.

We explore the utility of synthetic data by employing various configurations of real and synthetic datasets. Our study begins with a baseline experiment using only real images of runways (REAL), allowing us to establish a performance benchmark based solely on real data.

Next, we create a synthetic dataset using a flight simulator software (SYNTH). For all datasets, we ensure that each airport is represented by the same number of images, preventing imbalances in both training and validation sets.

To integrate both real and synthetic data during training, we assess different mixing strategies. The standard approach involves randomly sampling from the combined dataset (MIX), resulting in a variable ratio of synthetic to real data at each training step, dependent on the dataset composition.

To mitigate the domain shift between synthetic and real data, we also examine the CARE [17] domain adaptation method (CARE) tailored to our use case. This method introduces an additional loss term during training, encouraging the detection model to represent real and synthetic objects similarly by minimizing the Euclidean distance between their feature maps.

Furthermore, the CARE framework employs a specific sampling strategy that ensures an equal number of synthetic and real images are used at each training step in the minibatch. While this sampling method is necessary to their loss terms, we also investigate its effectiveness independently, without the accompanying loss terms (SAMPLER).

3.3   Experimental Setup

Model Overview. We consider a standard object detector, Faster R-CNN [19], to follow the recent literature in domain adaptation for object detection [17,28] and previous works in airport runway detection [11,14,27].

The objective function used in this study, which the model tries to minimize, is the standard Faster R-CNN loss function, ℓ_{F-RCNN}.

In the experiments using CARE, we add to the standard loss the alignment term proposed. CARE also incorporates reweighting factors that are unnecessary in our monoclass setting where object size and aspect ratio are consistent. We do not include them in our loss function. The overall training loss then becomes:

$$\ell_{det} = \ell_{F-RCNN} + \lambda\ell_{align}. \tag{1}$$

λ is a balancing factor, which controls the contribution of the alignment loss in the overall loss. ℓ_{align} is the alignment loss proposed by CARE, a cycle loss computing the difference between features of synthetic and real objects.

Implementation Details. For all the experiments using the real dataset, our data being redundant, we use a small training dataset of 1,000 images from 19 airports. Then, for the simulated dataset we generate 10,000 images to keep a reasonable ratio of synthetic to real images. Those images are from 199 airports, including the ones in the real dataset.

In order to study the robustness to domain shift of each model, we introduce the nighttime condition. For this purpose, we generate another synthetic dataset composed of 5,000 daytime images and 5,000 nighttime images from the same 199 airports.

Our detection model is a Faster R-CNN with FPN and a ResNet50 backbone, pretrained on COCO2017 [13]. The training configuration includes a batch size of 8, a learning rate of 0.002, and is trained during 10,000 iterations. We use $\lambda = 0.1$ for the CARE loss (Eq. (1)). We perform the validation at the end of the training.

Evaluation. Following previous work [14,28] we report COCO Average Precision (AP) for each strategy, AP being the average precision over several Intersection over Union values, from 0.5 to 0.95 by 0.05 steps.

To assess the training methods and the impact of training datasets, we build an evaluation dataset of 200 real daytime images from 8 airports. Those airports are not represented in any training set. In order to evaluate robustness to the nighttime condition, we build an additional validation dataset of 200 real night-time images from the same 8 airports. Real nighttime images are never seen during training.

4   Results

4.1   Comparison of Training Strategies

Table 1. Values of AP for each mixing strategy over daytime validation datasets. Models were trained *without* synthetic nighttime image.

	REAL	SYNTH	MIX	SAMPLER	CARE
DAY AP	58.60	59.04	**65.25**	64.01	63.66

In this section, we evaluate models trained on daytime images only. Table 1 compares the results of the different training strategies and datasets on the daytime validation set. We can observe that all methods outperform the model trained on real data, including the synthetic only experiment (+0,44%). We could explain this behavior with the model's pretraining; having already seen real images in COCO2017, it acquired features relevant to real images before the synthetic training. In this setting, adding synthetic data into the training dataset seems to bring better performance on the validation dataset, and a simple method such as MIX seems to bring the best results (+6.65% compared to REAL only). This conclusion was expected, as bringing more data with reliable labeling should lead to a better model.

4.2   Results on Night Domain Shift

We analyze the results of models integrating nighttime images in their synthetic training set, compared to the same models trained on daytime images only.

Table 2. Values of AP for each mixing strategy over daytime and nighttime validation datasets. Models were trained *with* synthetic nighttime image.

	REAL	SYNTH	MIX	SAMPLER	CARE
DAY AP	58.60	57.55	66.77	64.73	**67.50**
NIGHT AP	15.21	43.22	43.38	42.00	**43.75**

In Table 2, we first compare the models performance on the daytime validation dataset. The REAL model keeps the same training set throughout all experiments and therefore gets the same results. The synthetic only model SYNTH suffers from the addition of nighttime condition. On the contrary, while we expected the same results from the other training strategies, but MIX (+8,17%) and especially CARE (+8,90%) seem to benefit from the diversification of conditions. On average SAMPLER attains the same results, showing little use of the nighttime images. Overall all mixing methods still outperform the REAL experiment. The CARE method exhibits the best score showing the benefits of this domain adaptation strategy when diversifying the training dataset.

We also evaluate our models over the nighttime validation dataset. The REAL model experiences a significant drop in performance (−43,39%), and is unable to generalize properly on this unseen condition. All the other methods also experience a drop in performance. However, they are all able to maintain an AP score similar to the SYNTH model (−14,33% compared to daytime validation), way higher than the REAL model on both validation set. In this scenario, adding synthetic data seems to guarantee a correct level of performance, compared to a completely unseen condition. Our customized CARE method still exhibits the best score in this difficult scenario by a small margin.

In Fig. 2 we show inference results for all models on one validation image. This picture example is representative of the type of nighttime images we can find in the dataset. The REAL model is unable to see the runway whereas all the models trained on synthetic data could detect it.

Overall, mixing synthetic and real data seems to lead to more robustness to unseen environmental conditions, as we obtain good AP scores in both nominal and adverse conditions. Adding the target data into the synthetic datasets helps getting reasonable performance in difficult conditions. In our case, domain adaptation seems to improve performance with the customized loss of CARE allowing

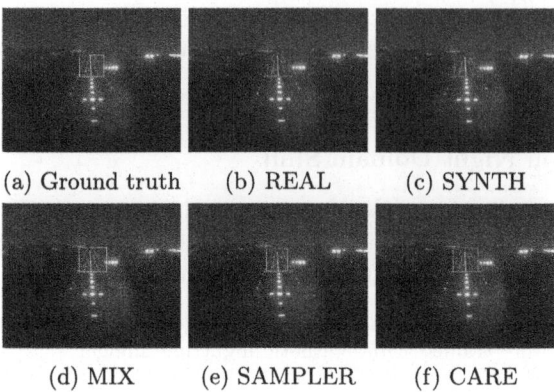

(a) Ground truth   (b) REAL   (c) SYNTH

(d) MIX   (e) SAMPLER   (f) CARE

Fig. 2. A real image of an airport runway at night from the validation dataset, and the bounding box determined by inference through each model. The REAL model does not predict any bounding box, while all other models correctly predict the bounding box covering the runway.

the model to generalize better. While the sampling method introduced in CARE seems to lower the detection performance, its alignment loss compensates this drop and improves the final results when getting enough diversity in the training dataset.

5 Conclusion and Perspectives

In this work, we studied the use of synthetic data for runway detection, specifically how data sampling schemes and domain adaptation can improve model performance. We find that, in general, synthetic data improves model performance and robustness in adapting to rare conditions.

In our first study on the nominal case of regular daytime conditions we find that synthetic images can improve model accuracy, but the presence of both real and synthetic images is necessary.

Synthetic data can be especially valuable to represent rare cases that are underrepresented in real data, in order to ensure model robustness. We found that synthetic images of nighttime conditions can greatly improve model performance on runway detection at night. Even training a model on only synthetic data vastly outperforms models trained only on real data. We consider that this is due to the lack of real nighttime images and the difficulty of adapting runway detection from daytime to nighttime in the absence of nighttime examples.

Domain adaptation seems to be a promising approach to enhance runway detection models. While the benefits of CARE compared to a simple mixing approach seem modest, the customized loss helps overcoming the decline introduced by the sampler, and CARE achieves the best performance overall.

In a future direction we aim to tackle other adverse conditions, such as snowy or foggy weathers that are more complicated to capture in reality or render in simulation. It would also be of interest to explore the impact of the CARE sampler, to further improve the benefits of this domain adaptation approach.

In summary, we show that synthetic data can aid in the task of airport runway recognition. Images from simulation software are especially useful in the representation of rare cases, such as nighttime data. By including synthetic data from diverse cases, we can ensure and improve computer vision model robustness, especially with the help of our customized CARE domain adaptation strategy.

Disclosure of Interests. Estelle Chigot, Meriem Ghrib and Fabrice Jimenez are employees of the Airbus company. Dennis G. Wilson and Thomas Oberlin have no competing interests to declare that are relevant to the content of this article.

References

1. Chen, M., Hu, Y.: An image-based runway detection method for fixed-wing aircraft based on deep neural network. IET Image Process. (2024). https://www.kaggle.com/datasets/relufrank/fs2020-runway-dataset/data
2. Chen, W., Zhang, Z., Yu, L., Tai, Y.: BARS: a benchmark for airport runway segmentation. Appl. Intell. **53**(17), 20485–20498 (2023)
3. Deng, J., Dong, W., Socher, R., Li, L.J., Li, K., Fei-Fei, L.: Imagenet: a large-scale hierarchical image database. In: 2009 IEEE Conference on Computer Vision and Pattern Recognition, pp. 248–255 (2009)
4. Dosovitskiy, A., Ros, G., Codevilla, F., Lopez, A., Koltun, V.: CARLA: an open urban driving simulator. In: Proceedings of the 1st Annual Conference on Robot Learning, pp. 1–16. PMLR (2017)
5. Ducoffe, M., et al.: LARD–Landing Approach Runway Detection–Dataset for Vision Based Landing. arXiv preprint arXiv:2304.09938 (2023)
6. Goodfellow, I., et al.: Generative adversarial nets. In: Proceedings of the 27th International Conference on Neural Information Processing Systems - Volume 2, vol. 27 (2014)
7. Hinterstoisser, S., Pauly, O., Heibel, H., Martina, M., Bokeloh, M.: An annotation saved is an annotation earned: using fully synthetic training for object detection. In: 2019 IEEE/CVF International Conference on Computer Vision Workshop (ICCVW) (2019)
8. Johnson-Roberson, M., Barto, C., Mehta, R., Sridhar, S.N., Rosaen, K., Vasudevan, R.: Driving in the matrix: can virtual worlds replace human-generated annotations for real world tasks? In: 2017 IEEE International Conference on Robotics and Automation (ICRA) (2017)
9. Kingma, D.P., Welling, M.: Auto-encoding variational bayes. In: International Conference on Learning Representations (2014)
10. Li, B., Zou, Y., Zhu, R., Yao, W., Wang, J., Wan, S.: Fabric defect segmentation system based on a lightweight GAN for industrial internet of things. Wirel. Commun. Mob. Comput. **2022**(1), 9680519 (2022)
11. Li, Y., Angelov, P., Yu, Z., Lopez Pellicer, A., Suri, N.: Federated adversarial learning for robust autonomous landing runway detection. In: International Conference on Artificial Neural Networks, pp. 159–173 (2024)
12. Lin, L., Liu, Y., Hu, Y., Yan, X., Xie, K., Huang, H.: Capturing, reconstructing, and simulating: the UrbanScene3D dataset. In: Computer Vision – ECCV 2022, pp. 93–109 (2022)

13. Lin, T.-Y., et al.: Microsoft COCO: common objects in context. In: Fleet, D., Pajdla, T., Schiele, B., Tuytelaars, T. (eds.) ECCV 2014. LNCS, vol. 8693, pp. 740–755. Springer, Cham (2014). https://doi.org/10.1007/978-3-319-10602-1_48

14. Lindén, J., et al.: Curating datasets for visual runway detection. In: IEEE/AIAA 40th Digital Avionics Systems Conference (DASC), pp. 1–9 (2021)

15. Liu, W., et al.: SSD: single shot MultiBox detector. In: Leibe, B., Matas, J., Sebe, N., Welling, M. (eds.) ECCV 2016. LNCS, vol. 9905, pp. 21–37. Springer, Cham (2016). https://doi.org/10.1007/978-3-319-46448-0_2

16. Niu, S., Li, B., Wang, X., Lin, H.: Defect image sample generation with GAN for improving defect recognition. IEEE Trans. Autom. Sci. Eng. **17**(3), 1611–1622 (2020)

17. Prabhu, V.U., et al.: Bridging the Sim2Real gap with CARE: supervised detection adaptation with conditional alignment and reweighting. Trans. Mach. Learn. Res. (2023)

18. Redmon, J., Divvala, S., Girshick, R., Farhadi, A.: You only look once: unified, real-time object detection. In: 2016 IEEE Conference on Computer Vision and Pattern Recognition (CVPR), pp. 779–788 (2016)

19. Ren, S., He, K., Girshick, R., Sun, J.: Faster R-CNN: towards real-time object detection with region proposal networks. IEEE Trans. Pattern Anal. Mach. Intell. **39**(6), 1137–1149 (2017)

20. Richter, S.R., Vineet, V., Roth, S., Koltun, V.: Playing for data: ground truth from computer games. In: Leibe, B., Matas, J., Sebe, N., Welling, M. (eds.) ECCV 2016. LNCS, vol. 9906, pp. 102–118. Springer, Cham (2016). https://doi.org/10.1007/978-3-319-46475-6_7

21. Ros, G., Sellart, L., Materzynska, J., Vazquez, D., Lopez, A.M.: The SYNTHIA dataset: a large collection of synthetic images for semantic segmentation of urban scenes. In: 2016 IEEE Conference on Computer Vision and Pattern Recognition (CVPR), pp. 3234–3243 (2016)

22. Shermeyer, J., Hossler, T., Etten, A.V., Hogan, D., Lewis, R., Kim, D.: RarePlanes: synthetic data takes flight. In: 2021 IEEE Winter Conference on Applications of Computer Vision (WACV), pp. 207–217 (2021)

23. Sohl-Dickstein, J., Weiss, E., Maheswaranathan, N., Ganguli, S.: Deep unsupervised learning using nonequilibrium thermodynamics. In: Proceedings of the 32nd International Conference on Machine Learning - Volume 37, pp. 2256–2265 (2015)

24. Tobin, J., Fong, R., Ray, A., Schneider, J., Zaremba, W., Abbeel, P.: Domain randomization for transferring deep neural networks from simulation to the real world. In: 2017 IEEE/RSJ International Conference on Intelligent Robots and Systems (IROS), pp. 23–30 (2017)

25. Tremblay, J., et al.: Training deep networks with synthetic data: bridging the reality gap by domain randomization. In: 2018 IEEE/CVF Conference on Computer Vision and Pattern Recognition Workshops (CVPRW), pp. 969–977 (2018)

26. Wang, Q., Feng, W., Zhao, H., Liu, B., Lyu, S.: VALNet: vision-based autonomous landing with airport runway instance segmentation. Remote Sens. **16**(12), 2161 (2024)

27. Wang, Z., Zhao, D., Cao, Y.: Visual navigation algorithm for night landing of fixed-wing unmanned aerial vehicle. Aerospace **9**(10), 615 (2022)

28. Yu, F., et al.: SC-UDA: style and content gaps aware unsupervised domain adaptation for object detection. In: 2022 IEEE/CVF Winter Conference on Applications of Computer Vision (WACV), pp. 382–391 (2022)

29. Zhang, L., Dai, Y., Fan, F., He, C.: Anomaly detection of GAN industrial image based on attention feature fusion. Sensors **23**(1), 355 (2022)

Identification of Buriti (*Mauritia flexuosa*) and Palmito Juçara (*Euterpe edulis*) Species Using RT-DETR Through High-Resolution Images Captured by UAV

Isaac Ambrosio da Silva[1] , Sanderson César Macêdo Barbalho[1] ,
Leonardo Lima Bergamini[3] , Frederico Scherr Caldeira Takahashi[3] ,
and Díbio Leandro Borges[2(✉)]

[1] Faculty of Technology, University of Brasília, Brasília, Brazil
{isaac.ambrosio,sandersoncesar}@unb.br

[2] Department of Computer Science, University of Brasília, Brasília, Brazil
dibio@unb.br

[3] Brazilian Institute of Geography and Statistics, Brasília, Brazil
{leonardo.bergamini,frederico.takahashi}@ibge.gov.br

Abstract. The use of Deep Learning tools for tree identification has
been steadily increasing in the last 10 years, driven by advances in computational power and image analysis techniques. Deep neural network
models, particularly architectures like CNNs (*Convolutional Neural Networks*), are effective for analyzing aerial, drone, and satellite images to
recognize distinct canopy patterns and specific tree species. This project
focuses on the identification of two native species of Brazilian flora found
in the Cerrado biome: Buriti (*Mauritia flexuosa*) and Palmito Juçara
(*Euterpe edulis*). To achieve this, were applied state-of-the-art Deep
Learning techniques, specifically the RT-DETR and YOLOv8 methods,
which excel in object detection tasks. A comprehensive data set was
created consisting of high-resolution images captured by aerial drone
surveys. Using this data set, the project's goal was to demonstrate the
potential of deep learning to automate the identification of plant species,
ultimately contributing to ecological monitoring and conservation efforts
in Brazil's rapidly changing ecosystems.

Keywords: Deep Learning · RT-DETR · Palm Trees identification ·
Mauritia flexuosa · *Euterpe edulis*

1 Introduction

The application of remote sensing (RS) technologies in conjunction with Deep
Learning tools has gained prominence in various areas of environmental research.
The use of satellite images, drones, and other remote sensors allows large-scale
data collection, facilitating real-time ecosystem monitoring with high precision.

© The Author(s), under exclusive license to Springer Nature Switzerland AG 2026
M. Castrillón-Santana et al. (Eds.): CAIP 2025, LNCS 15621, pp. 305–315, 2026.
https://doi.org/10.1007/978-3-032-04968-1_26

Furthermore, these technologies have been essential for the mapping and monitoring of deforestation, contributing to the identification of high-risk areas and the analysis of changes in land use. The use of drones equipped with high resolution cameras has also allowed the capture of detailed images of specific areas, facilitating to detect wildfire hotspots and the assessment of the resulting damage, as presented in the work of [4–6].

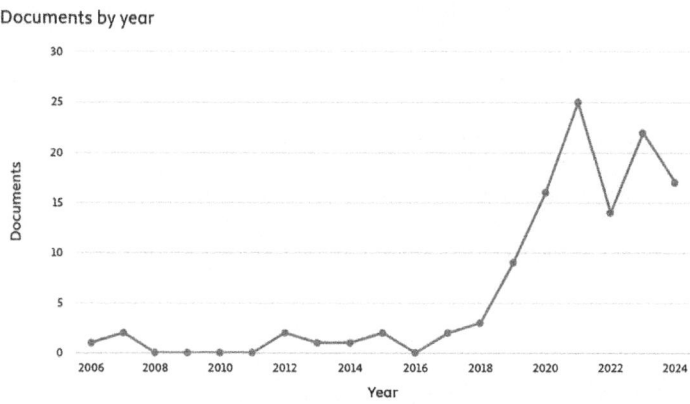

Fig. 1. Publications with the keywords Cerrado and Deep Learning/Machine Learning in Scopus.

A search in the Scopus journal database using the keywords "Cerrado", "Deep Learning", and "Machine Learning" retrieved 111 articles published in the last 10 years (2014 - 2024), showing a growing trend since 2018 (Fig. 1). Currently, forests are exposed to ever-growing pressure from human use as well as to a rapidly changing environment due to climate change. To understand changes in forest composition and structure as a response to these stressors, and successful forest management, scientists and forest managers require precise and detailed information on the current state of Brazilian forests.

The *Mauritia flexuosa* (Fig. 2a) is a native palm tree of tropical regions in South America, reaching heights of 30 to 40 m. It is mainly found in Brazil, Venezuela, Colombia, Peru and Bolivia. In Brazil, it stands out in ecosystems such as the Cerrado, Amazon, and Pantanal, often associated with wetland areas known as "veredas" [16].

This species plays important roles in biodiversity, economy, and nutrition. Its roots help retain soil moisture, preserving wetlands and water springs. It provides food and shelter for various bird, mammal, insect, and fish species. Moreover, it serves as an environmental indicator, as its presence is linked to fertile soils and water availability [13].

Additionally, *Mauritia flexuosa* oil, extracted from its fruit, is rich in carotenoids and is used in cosmetic industry (moisturizers, sunscreens) and food industry. It can also be sold in its natural form for pulp, juice, and sweets production. Its fibers are used for making mats, hats, and baskets, supporting traditional communities [15].

(a) *Mauritia flexuosa* specimens. Photo: Marcelo Kuhlmann.

(b) *Euterpe edulis* specimens. Photo: John DeMott.

Fig. 2. Representative specimens of the two target species analyzed in this study.

The (*Euterpe edulis*) (Fig. 2b) can reach heights of 5 to 12 m, with a single, straight, cylindrical stem of medium to tall stature. Unlike the Açaí palm, it does not produce offshoots and does not regenerate when cutted. Its leaves are alternate, pinnate, numbering 8 to 15, and can reach up to 3 m in length. It naturally occurs from southern Bahia and Minas Gerais to Rio Grande do Sul in the Atlantic Forest, as well as in Goiás, Mato Grosso do Sul, São Paulo, and Paraná in the riparian forests of the Paraná River basin [3].

Considered vulnerable to extinction due to deforestation and intense illegal harvesting, as recorded in the Red Book of Brazilian Flora [10], its preservation depends directly on the adequate conservation of native forests. Its main threat is the predatory extraction of its heart of palm, which involves the complete removal of the tree, compromising the natural regeneration of the Palmiteiro population.

Different approaches and methods can be used to identify plant species within dense vegetation. In [11], the objective was to identify *Mauritia flexuosa* specimens in the southern region of Peru using three different UAV (Unmanned Aerial Vehicle) models and a convolutional neural network (CNN) architecture based on Google's DeepLab v3+. The model achieved an accuracy of 0.9804 in the semantic segmentation of palm crowns in high-resolution aerial imagery.

In [9], the focus was also on identifying *Mauritia flexuosa*, but in the Amazon region of Colombia. In this case, a single UAV model was used, flying at an altitude of 60 m, along with an R-CNN object detection algorithm applied to 132 captured images. The model reached an average precision of 0.96.

Finally, in [8], the study targeted the segmentation of two Amazonian palm species – *Euterpe precatoria* and *Oenocarpus bataua*. A model based on the ResNet-18 architecture combined with DeepLabv3+ was used for segmentation, achieving an average accuracy of 0.958. The study did not differentiate between the two palm species.

Located 35 km south of downtown Brasília-DF, the Ecological Reserve of the Brazilian Institute of Geography and Statistics (IBGE), or Roncador Ecological Reserve (RECOR), covers an area of 1,350 hectares, encompassing all Cerrado physiognomic types (Fig. 3). Together with the Brasília Botanical Garden (JBB) and the Água Limpa Farm (FAL-UnB), it totals approximately 10,000 hectares of core preservation area within the Environmental Protection Area (APA) of *Ribeirões do Gama* and *Cabeça-de-Veado* and the Cerrado Biosphere Reserve [1].

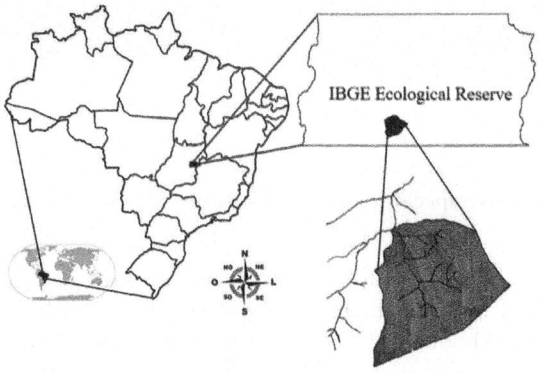

IBGE Ecological Reserve

Fig. 3. Location of the IBGE Ecological Reserve, (15°56'31" S, 47°52'47" W) [12].

This conservation area is dedicated to research and preservation of the Cerrado, significantly contributing to the study of the flora, fauna, and ecological dynamics of this biome. Additionally, it serves as a crucial space for environmental awareness and the development of sustainable practices that benefit both the local population and long-term conservation [2].

Yian Zhao *et al.* [17] proposed a real-time object detector called RT-DETR (*Real-Time Detection Transformer*), based on the *Transformers* architecture and tested on the COCO (Common Objects in Context) image dataset. According to the authors, this architecture eliminates the need for NMS (Non-Maximum Suppression) in post-processing, a technique used to select the best bounding box from a set of overlapping boxes. Another state-of-the-art object detection is the YOLOv8 model [14], which has been shown an efficient deep learning architecture for object detection in UAV images [7].

The purpose of this project was to identify two tree species found in the Cerrado and evaluating the RT-DETR, and the YOLOv8 methods, based on aerial images obtained with the aid of a drone in specific regions of the IBGE Ecological Reserve, located in Brasília-DF, with support from a team of biologists and botany specialists from the reserve.

2   Methodology

To obtain the necessary *dataset* for this project, images were captured using a DJI Mavic 3 drone during several field visits to the IBGE Ecological Reserve, with the support of biologists and botany specialists.

Figure 4a shows the regions mapped during the field visits to the Reserve, with yellow representing flights conducted at an altitude of 100 m and orange indicating lower flights at 40 m. The total mapped area is approximately 200 hectares, that equals to 15% of the Reserve's total area.

Flights were conducted in two regions that feature a type of vegetation typical of the presence of the two target species, known as Gallery Forest, which is characterized by following small rivers and streams. A total of 826 RGB images were captured at a resolution of 5280 × 3965 pixels, covering a mapped area of 13.3 hectares. These images, which comprise the dataset used in this study, were collected specifically in the region highlighted in (Fig. 4b).

(a) Overview of the mapped areas within the Reserve.

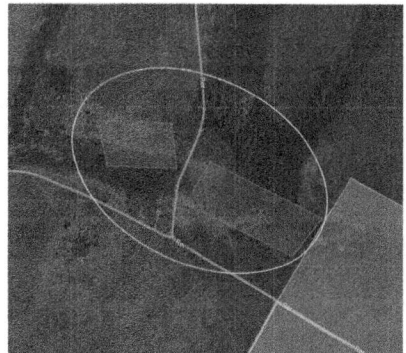

(b) Zoomed-in view showing more detail and flight coverage.

Fig. 4. Mapped regions considered in this study.

To label the images, the Python-based graphical tool *labelImg* was used, allowing users to create and edit bounding boxes around objects of interest. Considering only the images labeled with at least one of the two species, a total of 159 images were identified with 390 annotations, including 151 annotations for the *Mauritia flexuosa* species and 239 for the *Euterpe edulis*.

Taking into account the inequality in the number of annotations between classes, a normalization was performed to ensure that both classes had a similar number of instances – 148 for *Mauritia flexuosa* and 144 for *Euterpe edulis*. These images were randomly split into 60% for training, 20% for validation, and 20% for testing.

Due to the hardware demands of running deep learning algorithms, a cloud-based solution was selected to avoid the need for a local setup. For this study, Google's cloud platform, *Colab Pro*, was used, offering access to high-performance GPU resources, specifically a single NVIDIA T4 GPU, directly through the browser.

The first step was to create a script to convert the image resolution to 640×640 pixels to optimize computational resource usage, standardize the dataset, and ensure compatibility with popular Deep Learning architectures.

Additionally, a conversion script was implemented to convert annotation files from the YOLO format (.txt) into .json annotations following the COCO format, which is required by the RT-DETR method.

For this study, the RTDETR model from *RT-DETR PyTorch* and YOLOv8 were executed, using 100 epochs. The training was first performed with only one class, and subsequently with both classes together.

To ensure a robust and unbiased assessment of model performance, K-Fold Cross-Validation was applied. This technique partitions the dataset into five (5) equally sized folds, training the model five times with a different fold used for validation in each iteration.

This approach provides several advantages: it maximizes the use of limited annotated data, reduces performance variance caused by random data splits, and yields more reliable metrics such as precision, recall, and AP50. Furthermore, cross-validation helps to detect overfitting and allows a more accurate comparison between YOLO and RT-DETR architectures under consistent evaluation protocols.

3   Results

After completing the necessary configurations for code execution, the training of the model RTDETR is initiated. In this project, AP50 (*Average Precision* at IoU = 0.50) was used as the evaluation metric, where a detection is considered a True Positive if the predicted bounding box overlaps the *ground truth* by at least 50%. Training was conducted for a total of 100 epochs.

After running the 100 epochs, in one of the folds, the highest AP50 value was achieved at epoch 48. To generate the AP50 vs. number of epochs graph, a regex search was applied to the log file to extract key terms and record the highest AP values (Fig. 5). A rapid stabilization trend of the AP50 value could be observed as early as epoch 20, remaining above 0.80 until the end of the training.

The same training was conducted for the *Euterpe edulis* class, also for 100 epochs. In this case, the highest AP50 value was 0.715, reached at epoch 89. A different stabilization pattern was observed, with greater variation between training epochs (Fig. 6). This behavior may be related to the lower number of labeled images available for this class.

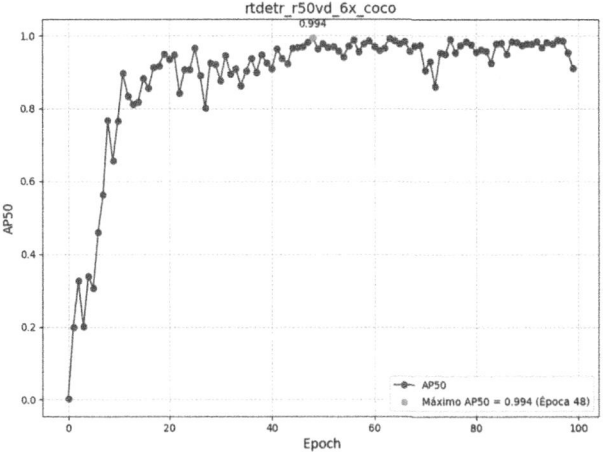

Fig. 5. AP50 over 100 epochs for the *Mauritia flexuosa* class

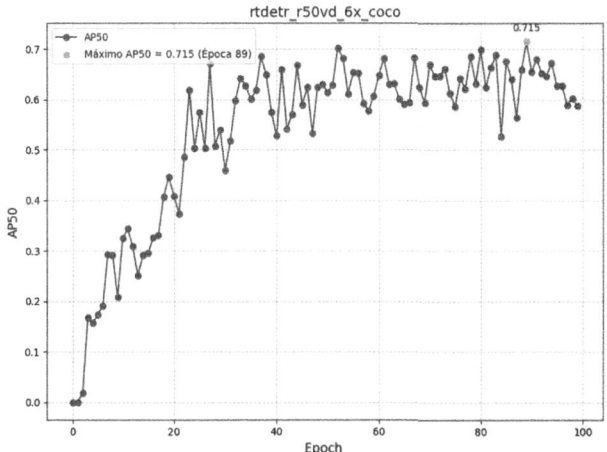

Fig. 6. AP50 over 100 epochs for the *Euterpe edulis* class

For comparison purposes, the same data set containing both classes was applied to the YOLOv8 model developed by Ultralytics [14] in the same Google Colab environment. For the *Mauritia flexuosa* class, the YOLOv8 model achieved highest AP50 value at epoch 99, reaching 0.955. (Fig. 7).

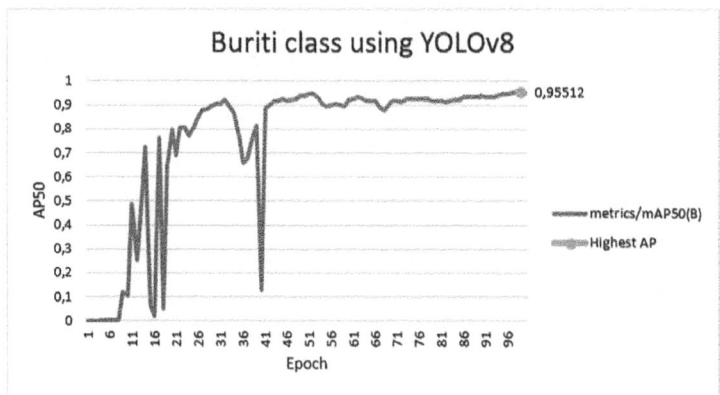

Fig. 7. AP50 over 100 epochs for the *Mauritia flexuosa* class using YOLOv8.

For the *Euterpe edulis* class, the highest AP50 value was 0.7571 at epoch 54, as shown in Fig. 8.

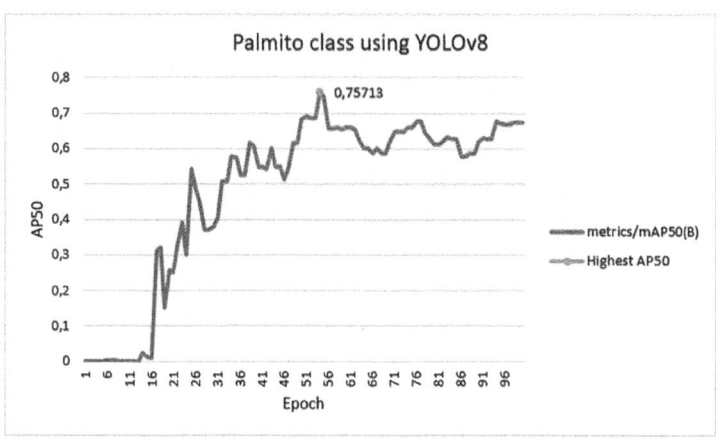

Fig. 8. AP50 over 100 epochs for the *Euterpe edulis* class using YOLOv8.

The same training was conducted using the RT-DETR and YOLOv8 models with both classes combined. Considering cross-validation, an average AP value of 0.939 was obtained using the RT-DETR model, whereas YOLOv8 achieved an average of 0.872.

Finally, to test the accuracy of the trained model, were used two images from the dataset. Figure 9 shows the identification of *Mauritia flexuosa*, and Fig. 10 presents the identification result for *Euterpe edulis*.

As observed, the model was able to identify accurately the tree individuals present in the image, including those of smaller size.

Fig. 9. Buriti (*Mauritia flexuosa*) identification result.

Fig. 10. Palmito (*Euterpe edulis*) identification result.

Following the execution of the RT-DETR and YOLOv8 models on the data set, a comparison table was constructed showing the average AP50 values across the five folds. (Table 1).

Table 1. K-Fold Cross Validation Results.

Class	average AP50 RTDETR 5 folds	average AP50 YOLOv8 5 folds
Mauritia flexuosa	0.991	0.9768
Euterpe edulis	0.897	0.718
Mauritia flexuosa and *Euterpe edulis*	0.939	0.872

4   Conclusion

In this project, the aim was to evaluate Deep Learning methods, specifically the RT-DETR and YOLOv8, for the identification of tree species present in the Cerrado bioma.

Using a data set obtained aided by drone flights, it was possible to compare the results from the *Mauritia flexuosa* and *Euterpe edulis* species. Positive outcomes was found in training, indicating the model's ability to detect these individuals even in dense vegetation.

When training with more than one class, a decrease in the AP50 confidence of the detector was observed, which was expected. This demonstrates that as more classes are added, the model must learn to differentiate between several categories, increasing complexity and the potential for errors.

In contrast to previous studies such as [8,9,11], the present work aimed to identify *Mauritia flexuosa* and *Euterpe edulis* species using the RT-DETR architecture. The model achieved a precision of 0.991 for *M. flexuosa* and 0.897 for *E. edulis*, demonstrating high effectiveness. For benchmarking purposes, the same dataset was evaluated using the YOLOv8 architecture, which resulted in a lower precision of 0.977 and 0.718 respectively.

Disclosure of Interests. The authors have no competing interests to declare that are relevant to the content of this article.

References

1. Brasil, U.: Vegetação no Distrito Federal: tempo e espaço. UNESCO Brasil (2002)
2. Câmara, P.E.A.S.: Musgos acrocárpicos das matas de galeria da reserva ecológica do ibge, recor, distrito federal, brasil. Acta Botanica Brasilica **22**, 1027–1035 (2008)
3. Carvalho, P., Carvalho, P.E.R., et al.: Espécies Arbóreas Brasileiras. Euterpe edulis. Embrapa/Ministério do Meio Ambiente, Palmiteiro (2003)
4. Gevaert, C.M., et al.: Explainable few-shot learning workflow for detecting invasive and exotic tree species. arXiv preprint arXiv:2411.00684 (2024)
5. Hiraguri, T., Kimura, T., Endo, K., Ohya, T., Takanashi, T., Shimizu, H.: Shape classification technology of pollinated tomato flowers for robotic implementation. Sci. Rep. **13**(1), 2159 (2023)
6. Huang, Y., Ou, B., Meng, K., Yang, B., Carpenter, J., Jung, J., Fei, S.: Tree species classification from uav canopy images with deep learning models. Remote Sensing **16**(20), 3836 (2024)

7. Jrondi, Z., Moussaid, A., Hadi, M.Y.: Exploring end-to-end object detection with transformers versus yolov8 for enhanced citrus fruit detection within trees. Systems and Soft Computing **6**, 200103 (2024)

8. Júnior, A.G., Ribas, R.P.: Identificação de palmeiras (arecaceae) nativas em áreas de floresta tropical baseado em rede neural convolucional com imagens de vant. Revista Brasileira de Geografia Física **16**(05), 2360–2374 (2023)

9. Marin, W., Mondragon, I.F., Colorado, J.D.: Aerial identification of amazonian palms in high-density forest using deep learning. Forests **13**(5), 655 (2022)

10. Martinelli, G., Moraes, M.A.: Livro vermelho da flora do Brasil. CNCFlora, Centro Nacional de Conservação da Flora Rio de Janeiro (2013)

11. Morales, G., Kemper, G., Sevillano, G., Arteaga, D., Ortega, I., Telles, J.: Automatic segmentation of mauritia flexuosa in unmanned aerial vehicle (uav) imagery using deep learning. Forests **9**(12), 736 (2018)

12. Roque, F.: Drosofilídeos (Insecta, Diptera) da mata do Pitoco : diversidade e distribuição vertical. Master thesis, Universidade de Brasilia, Brasilia, Brasil (2009). https://repositorio.unb.br/handle/10482/8279

13. da Silva Almeida, T., et al.: Structural and physiological responses to water availability provide insights into the maintenance of mauritia flexuosa (arecaceae) seedling banks. Forest Ecology and Management **561**, 121881 (2024)

14. Ultralytics: Yolov8: Ultralytics yolov8 documentation. https://docs.ultralytics.com (2023). Accessed 04 Aug 2025

15. Vieira, R.F., Camillo, J., Coradin, L.: Espécies nativas da flora brasileira de valor econômico atual ou potencial: Plantas para o futuro: Região Centro Oeste. Embrapa/Ministério do Meio Ambiente (2022)

16. Virapongse, A., Endress, B.A., Gilmore, M.P., Horn, C., Romulo, C.: Ecology, livelihoods, and management of the mauritia flexuosa palm in south america. Global Ecology and Conservation **10**, 70–92 (2017)

17. Zhao, Y., Lv, W., Xu, S., Wei, J., Wang, G., Dang, Q., Liu, Y., Chen, J.: Detrs beat yolos on real-time object detection. In: IEEE/CVF Conference on Computer Vision and Pattern Recognition (CVPR), pp. 16965–16974 (2024)

3D Vision and Reconstruction

Framework for Generation of Moment Invariants by Tensor Method

Tomáš Suk[1]([✉])[iD] and Roxana Bujack[2][iD]

[1] Czech Academy of Sciences, Institute of Information Theory and Automation,
Pod vodárenskou věží 4, 182 08 Praha 8, Czech Republic
suk@utia.cas.cz
[2] Data Science at Scale Team, Los Alamos National Laboratory, P.O. Box 1663,
Los Alamos, NM 87545, USA
bujack@informatik.uni-leipzig.de
https://www.utia.cas.cz

Abstract. The integrated application "Afinvtensors" can generate moment invariants for various types of data and transformations from grayscale images up to tensor fields both to rotation and to affine transformation and both in 2D and in 3D. It is based on the graph method and tensor contraction. Its abilities are demonstrated in the experiment with 2D affine transformation of coordinates and independent 3D affine transformation of values of color images.

Keywords: tensor fields · graph method · tensor contraction · moment invariants

1 Introduction

In our research, we needed to process various types of data over time. One of such type is grayscale images. Over time, other data types appeared: color images, surface data from a 3D scanner, full volume 3D data, vector fields describing e.g. wind blowing both in 2D and 3D, or tensor fields describing e.g. inner tensions in solid materials. The typically by translation, scaling, rotation, and affine transformation. The tensor method provides a way to compute the so-called *invariants*, that is, features that do not change in these transformations. They are computed from *moments* as type of data descriptor.

Formerly, we separately derived the invariants for each combination of data type and transformation, e.g. affine invariants of images [13], vector fields [10], tensor fields [3] or 3D data [14]. Finally, we found that some parts of the algorithms are similar or the same for all these combinations, therefore, we decided to make an integrated application yielding a unified framework for generation of various types of invariants. Some data-transformation combinations are new and have not been published yet.

M. Castrillón-Santana et al. (Eds.): CAIP 2025, LNCS 15621, pp. 319–329, 2026.
https://doi.org/10.1007/978-3-032-04968-1_27

2 Tensor Method

A tensor is a multidimensional array $\mathbf{T}^{i_1...i_n}_{j_1...j_m}$ that behaves under the affine transformation by the invertible matrix $\mathbf{A}^i_j \in \mathbb{R}^{d \times d}$ like

$$\mathbf{T}'^{i_1...i_n}_{j_1...j_m} = |\det(\mathbf{A}^{-1})|^w \mathbf{A}^{l_1}_{j_1} \cdots \mathbf{A}^{l_m}_{j_m} (\mathbf{A}^{i_1}_{k_1} \cdots (\mathbf{A}^{-1})^{i_n}_{k_n} \mathbf{T}^{k_1...k_n}_{l_1...l_m}. \tag{1}$$

It has covariant rank m, contravariant rank n, and weight w. The total rank is $m + n$. So, the tensor is an array of numbers and the rank of a tensor determines the dimensionality of this array. Special cases include scalars, which are tensors of rank zero, vectors, which are tensors of rank one, and matrices, which are tensors of rank two. An introduction to tensors can be found in [1] or in [7]. A good explanation can also be found in [8].

The basic tensor operations are addition, multiplication, and contraction. In the tensor addition, we add the corresponding components

$$(\mathbf{T_1} + \mathbf{T_2})^{i_1...i_n}_{j_1...j_m} := \mathbf{T_1}^{i_1...i_n}_{j_1...j_m} + \mathbf{T_2}^{i_1...i_n}_{j_1...j_m}. \tag{2}$$

In the multiplication, we multiply each component of one tensor by each component of the other tensor

$$(\mathbf{T_1} \otimes \mathbf{T_2})^{i_1...i_n i'_1...i'_{n'}}_{j_1...j_m j'_1...j'_{m'}} := \mathbf{T_1}^{i_1...i_n}_{j_1...j_m} \mathbf{T_2}^{i'_1...i'_{n'}}_{j'_1...j'_{m'}}. \tag{3}$$

The contraction is the sum over one index used twice, once as contravariant and once as covariant. When the covariant index is i_k and the contravariant index is j_l, the contraction is

$$\left(\sum_{(i_k = j_l)} \mathbf{T}' \right)^{i_1...i_{k-1} i_k i_{k+1}...i_n}_{j_1...j_{l-1} j_l j_{l+1}...j_m} := \mathbf{T}^{i_1...i_{k-1} i_{k+1}...i_n}_{j_1...j_{l-1} j_{l+1}...j_m}. \tag{4}$$

The result has both covariant and contravariant rank decreased by one. The total contraction is performed over all indices. The contraction over one index causes that the direct and inverse matrices of the affine transformation in the tensor definition (1) are multiplied and canceled. The total contraction cancels all the matrices and we obtain a relative affine invariant. If we leave one index without contraction, we obtain a vector that behaves linearly under the affine transformation of coordinates. It can be used for normalization.

Besides tensors, we can have also tensor fields. It means we have defined the tensor in each point of a space

$$\mathbf{T}'^{i_1...i_n}_{j_1...j_m} ((x')^1 \cdots (x')^d) = |\det(\mathbf{A}^{-1})|^{w_A} |\det(\mathbf{B}^{-1})|^{w_B} \mathbf{A}^{i_1}_{k_1} \cdots \mathbf{A}^{i_n}_{k_n} \times$$
$$\times (\mathbf{A}^{-1})^{l_1}_{j_1} \cdots (\mathbf{A}^{-1})^{l_m}_{j_m} \mathbf{T}^{k_1...k_n}_{l_1...l_m} (x^1 \cdots x^d), \tag{5}$$
$$\text{where } ((x')^1 \cdots (x')^d)^T = \mathbf{B}^{-1}(x^1 \cdots x^d)^T.$$

The affine transformation of the tensor values by the matrix $\mathbf{A}$ is called outer and the affine transformation of the coordinates by the matrix $\mathbf{B}$ is called inner.

2.1   Geometric Transformations

The linear transformation of coordinates can be described as

$$((x')^1 \cdots (x')^d)^T = \mathbf{A}(x^1 \cdots x^d)^T. \tag{6}$$

If there are no constraints on the matrix $\mathbf{A}$ of the size $d \times d$, it is called affine transformation. If the matrix $\mathbf{A} = \{a_{ij}\}$ is orthogonal, i.e. $\sum_{i=1}^d a_{ij}a_{ik} = 0$ and $\sum_{i=1}^d a_{ji}a_{ki} = 0$ for $j \neq k$, then the transformation is called orthogonal. If in addition $\det(\mathbf{A}) = 1$, then it is rotation, at least in 2D and 3D.

The inverse of the orthogonal matrix equals its transposition, i.e. $\mathbf{A}^{-1} = \mathbf{A}^T$. It implies that then we need not distinguish contravariant and covariant indices and perform contraction over arbitrary two indices.

2.2   Moment Tensors

Geometric moments have been introduced to pattern recognition in [9]. The geometric moment of order $o = \sum_{i=1}^d p_i$ of an image $f(\mathbf{x})$ is

$$m_{p_1 \ldots p_d} = \int_{\mathbb{R}^d} (x^1)^{p_1} \cdots (x^d)^{p_d} f(\mathbf{x}) \, \mathrm{d}^d \mathbf{x}, \tag{7}$$

where $(x^i)^{p_i}$ refers to the p_i-th power of ith coordinate.

The moment tensor $^o\mathbf{M}$ of order $o \in \mathbb{N}$ takes the form

$$^o\mathbf{M}^{k_1 \ldots k_o} = \int_{\mathbb{R}^d} x^{k_1} \cdots x^{k_o} f(\mathbf{x}) \, \mathrm{d}^d \mathbf{x}, \tag{8}$$

with $k_l \in \{1, \ldots, d\}$, $l \in \{1, \ldots, o\}$ and x^{k_l} representing the k_l-th component of $x \in \mathbb{R}^d$. The component $^o\mathbf{M}^{k_1 \ldots k_o}$ of a moment tensor equals the geometric moment $m_{p_1 \ldots p_d}$ iff the number of indices $k_1, \ldots, k_o$ equaling i is p_i for $i \in \{1, \ldots, d\}$. The moment tensors can be used for the construction of the moment invariants [4,5].

The tensor valued function [2,11] is generalization of the moment tensor

$$^o\mathbf{M}^{k_1 \ldots k_o i_1 \ldots i_n}_{j_1 \ldots j_m} = \int_{\mathbb{R}^d} x^{k_1} \cdots x^{k_o} \mathbf{T}^{i_1 \ldots i_n}_{j_1 \ldots j_m} (x^1 \cdots x^d) \, \mathrm{d}^d \mathbf{x}. \tag{9}$$

The moment tensor $^o\mathbf{M}^{k_1 \ldots k_o i_1 \ldots i_n}_{j_1 \ldots j_m}$ has o coordinate indices, n contravariant indices, and m covariant indices, i.e. $o+n$ upper and m lower indices. The upper left index o is written to distinguish the coordinate indices and the contravariant indices of the original tensor field.

The tensor fields usually have different number of covariant and contravariant indices, then the direct total contraction is not possible. We can use the permutation tensor ε [5]. In 2D, the permutation tensor ε_{ij} takes the form

$$\varepsilon = \begin{pmatrix} 0 & 1 \\ -1 & 0 \end{pmatrix}. \tag{10}$$

In 3D, $\varepsilon_{123} = \varepsilon_{231} = \varepsilon_{312} = 1$, $\varepsilon_{132} = \varepsilon_{213} = \varepsilon_{321} = -1$, other 21 components equal 0. The permutation tensor can be used both as covariant $\varepsilon_{1\dots d}$ for contraction of the contravariant indices and as the contravariant $\varepsilon^{i_1 \dots i_d}$ for contraction of the covariant indices, e.g.

$$I = {}^2\mathbf{M}_\ell^{ijk} \, {}^1\mathbf{M}_o^{mn} \, {}^0\mathbf{M}_q^p \varepsilon_{ikn}\varepsilon_{jmp}\varepsilon^{\ell oq}. \tag{11}$$

According to Einstein's notation, the sum symbol over all indices from 1 to d is omitted.

2.3 Graph Method

We use graph theory to generate all possible contractions of the moment tensor products. Every product contraction can be expressed by a graph, where each moment tensor corresponds to a node and each index corresponds to a connection edge–node. The edge then means either a contraction over an index used twice at two moment tensors or d contractions by means of the permutation tensor ($d = 2$ for 2D data and $d = 3$ for 3D data). In 3D, it means we have triple hyperedges connecting three nodes.

The type of the edge is distinguished by color of the edge, we use black color for coordinate indices, magenta for contravariant indices and green for covariant indices. So, if we perform contraction over one coordinate and one covariant index, the edge color is changed in the middle of the edge. If there are some non-contracted indices, they can be expressed by special "half-edges" leading to nowhere. An example of a graph corresponding to the contraction (11) is in Fig. 1(a). Another example can be contraction with graph in Fig. 1(b).

$$P^r = {}^3\mathbf{M}_i^{ijk\ell} \, {}^1\mathbf{M}_o^{mn} \, {}^0\mathbf{M}_p^q \, {}^1\mathbf{M}_q^{rs} \varepsilon_{jn}\varepsilon_{km}\varepsilon^{op}\varepsilon_{\ell s} \tag{12}$$

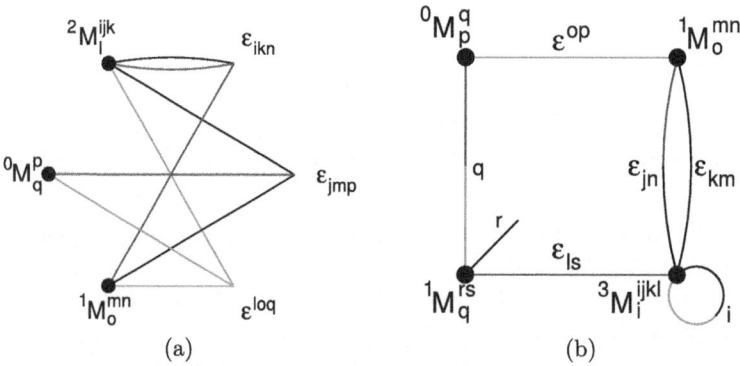

(a)   (b)

Fig. 1. Graph examples generating an invariant (a) with the zero-rank contraction ${}^2\mathbf{M}_\ell^{ijk} \, {}^1\mathbf{M}_o^{mn} \, {}^0\mathbf{M}_q^p \varepsilon_{ikn}\varepsilon_{jmp}\varepsilon^{\ell oq}$ of a vector field, (b) with the first-rank contraction ${}^3\mathbf{M}_i^{ijk\ell} \, {}^1\mathbf{M}_o^{mn} \, {}^0\mathbf{M}_p^q \, {}^1\mathbf{M}_q^{rs} \varepsilon_{jn}\varepsilon_{km}\varepsilon^{op}\varepsilon_{\ell s}$ of a tensor field with one covariant index and one contravariant index.

2.4   Graph Generation

We need to generate all graphs for the given case. The algorithm starts with the first graph and then generates the next graph from the current one until the last graph is reached. The graphs are described by a list of edges. The main parameter is the number of edges n_e, it is given by a user, the number of nodes n_d is computed from the graph. In the 2D case and also for 3D rotations, the first graph is $\begin{matrix} 1\,1\cdots 1\,1 \\ 1\,1\cdots 1\,1 \end{matrix}$ and the last graph is $\begin{matrix} 1\,3\cdots 2n_e-3\,2n_e-1 \\ 2\,4\cdots 2n_e-2 \quad 2n_e \end{matrix}$. If we generate 3D affine invariants, triple hyperedges are generated. The first graph is $\begin{matrix} 1\,1\cdots 1\,1 \\ 1\,1\cdots 1\,1 \\ 1\,1\cdots 1\,1 \end{matrix}$ and the last graph is in this case $\begin{matrix} 1\,2\cdots n_e-1\,n_e \\ 2\,3\cdots n_e \quad n_e+1 \\ 3\,4\cdots n_e+1\,n_e+2 \end{matrix}$. Then, the edges are colored, it is based on the similar principle. If there is no other possibility of the coloring, it is stopped, otherwise next color form of the edges is generated. If partial invariants with r_i contravariant indices and o_o covariant indices are generated, then we need $r_i + o_o$ half-edges. They are created so the last $r_i + o_o$ edges is filled by 1 in the first row and zeros in other rows in the first graph.

The generation of the next graph is divided to two parts. First, the next half-edge is tried to generate. Only if it is not possible, next standard edges and triple hyperedges are generated by the Algorithm 1. The algorithm is illustrated in Fig. 2. On the position v_1, there was value $v_1 - 1$. It was the first value from the end that can be increased to v_1. The new value v_n is maximum from the values v_1 and a_i. The value v_1 fills also the right bottom corner of the graph.

Algorithm 1. Generation of the next graph from the current one.

1: Row ← last row
2: **while** Row ≥ first row & no new graph found **do**
3:      Search Row of the edge list from behind.
4:      **if** any node label can be increased **then**
5:          Increase it to v_1.
6:          **if** Row ≠ first row **then**
7:              Fill the rest of Row with $\max(v_1, a_i)$, a_i is the node label above it.
8:          **else**
9:              Fill the rest of Row with v_1.
10:         **end if**
11:         Fill the node labels below and rest of these rows with v_1.
12:     **end if**
13:     Decrease Row by one to the beginning.
14: **end while**
15: **if** any node label was increased **then**
16:     Return the new graph.
17: **else**
18:     Stop.
19: **end if**

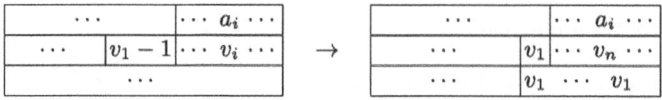

Fig. 2. Filling the rest of the graph after increasing a node label. The new value $v_n = \max\{v_1, a_i\}$ does not depend on the previous value v_i.

The algorithm for the half-edges is a loop that inserts the half-edges successively to all nodes. Similarly we move the colors of the edges going from the individual nodes, so each node would have r_i magenta edges and r_o green edges. The remaining o black edges define the moment order.

2.5   Alternative Approach

We can compare the tensor method e.g. with [12], where the authors propose an approach based on two fundamental primitives

$$\Gamma_{12} = F(X_1)^T \cdot F(X_2) \tag{13}$$

and

$$\Lambda_{12\ldots N} = |F(X_1), F(X_2), \ldots, F(X_N)|, \tag{14}$$

where $F(X)$ is multi-channel function, $X_1, X_2, \ldots, X_N$ are N independent points and $|\cdot|$ is the determinant.

At the first glance, the two primitives looks like advantage, but in our opinion, it is a complication. We must manage more combinations of the primitives and the result is less general.

2.6   Removing of Dependent Invariants

The graph method produces many dependent invariants that should be removed. We can distinguish a few types of dependencies. Some invariants are zero, there are pairs of the same invariants, some invariants are products of the others, some are linear combinations and there are also polynomial dependencies among the invariants.

The zero and identical invariants are removed immediately after their generation. The removing of the products is based on multiplication of each invariant by each and search on the product in the list of the previously generated invariants. If e.g. product of invariants I_{i_1} and I_{i_2} is an invariant I_{i_3} and we find it in the list, we know $I_{i_3} = I_{i_1} \cdot I_{i_2}$. If it is not in the list, we add it there. Finally, the products are used for the test of linear dependencies and then they are removed.

There is a standard procedure for elimination of linearly dependent invariants, see e.g. [6]. When some group of invariants have the same moment orders in each term (invariants with the same structure), they can be potentially linearly dependent. We compose a matrix of the coefficients; one coefficient from each term. The rank of this matrix equals the number of linearly independent (irreducible) invariants from this group. The other (reducible) invariants are finally removed together with the products. We use reduced row echelon form (RREF) to find a basis of invariants.

Recently, Langbein and Hagen [11] published a numerical test for elimination of the polynomial dependencies. If some invariants I_1, I_2, ..., I_n are dependent, it means there is a function f that $f(I_1, I_2, ..., I_n) = 0$. Since the invariants are polynomials of the moments, their dependence is also polynomial, i.e. smooth and the function f can be differentiated

$$\frac{\partial f(I_1, I_2, ..., I_n)}{\partial m_j} = \sum_{i=1}^{n} \frac{\partial f(I_1, I_2, ..., I_n)}{\partial I_i} \frac{\partial I_i}{\partial m_j} = 0. \tag{15}$$

The m_j is some moment occurring in the invariants. Equation (15) can be understood as a system of linear equations with known matrix $\mathbf{S} = \{s_{ij}\} = \partial I_i / \partial m_j$ and the unknown vector $\mathbf{b} = \{b_i\} = \partial f(I_1, I_2, ..., I_n)/\partial I_i$. If $\mathbf{S}$ has the full rank, the system has only one solution full of zeros. The invariants are then independent, otherwise the rank of it equals the number of the independent invariants.

Analytic solution in the whole space would be too demanding, therefore we compute the system only at one point. We choose some random numbers with uniform distribution from 0 to 1 as values of the moments, evaluate the derivatives for these moment values and compute a basis of independent invariants by the reduced row echelon form. The indices of the used pivots equal the indices of the independent invariants. The invariants should be sorted in $\mathbf{S}$ according to their complexity; the elimination then selects the set of independent invariants as simple as possible.

3   Application

The application "Afinvtensors" is intended for generation of moment invariants. The software is written in language C++ in the Microsoft Visual Studio Community 2017. The graphical user interface (GUI) is in Fig. 3.

In the top left part of GUI, we must set the parameters. We can choose, if we want to generate invariants to rotation or to affine transformation. We can also set special cases as inner rotation and outer affine transformation or inner affine transformation and outer rotation. Other option is identical inner and outer transformation. We can also choose either two-dimensional (2D) or three-dimensional (3D) data. If the inner and outer transformations are not identical, they can have different dimensionality, either 2D coordinates and 3D values (typical example is a color image) or 3D coordinates and 2D values.

Other parameters are ranks of the tensors. The contravariant rank r_i is the most important. For $r_i = 0$, we generate the invariants of the images, $r_i = 1$ means invariants of the vector fields, and $r_i > 1$ are tensor fields. The contravariant rank of result r_o is zero for the invariants. If $r_o > 0$, we obtain just the partial invariants, for $r_o = 1$, the result is a vector suitable for normalization. The covariant ranks should stay usually zeros, a non-zero value would lead to the invariants

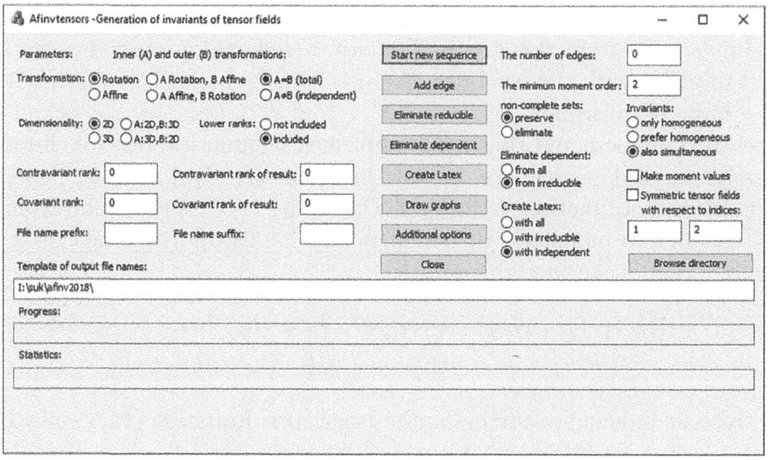

Fig. 3. Main dialog window (GUI).

to the inverse inner and outer transformations. The interesting options are also in the right part of GUI. When we use the zeroth-order moments for the scaling normalization and the first-order moments for the translation normalization, we need the minimum moment order 2. In the opposite case, it is better to compute the invariants from the zeroth order. We can also set symmetric tensors, what is often case in physics. The standard order of operations is following:

1. Fill the parameters (transformation, dimensionality, ranks and symmetry).
2. Press the "Start new sequence" button.
3. Press the "Add edge" button repeatedly to increase the moment order.
4. When the moment order is sufficient, press the "Eliminate reducible" button.
5. Press the "Eliminate dependent" button.
6. Press the "Create Latex" button.
7. Press the "Draw graphs" button.
8. Translate the made files to PDF.

The "Additional options" button shows auxiliary GUI, see Fig. 4, where additional parameters can be set. The most important parameters from this GUI is the tolerance RREF. It must be increased, if different sets of the random moment values yield different results (different "independent" sets of the invariants).

4   Experiment

The aim of this experiment is to demonstrate the generated invariants in practise. We chose images of two horologes, one from Prague and one from Venice, see Figs. 5(a) and (b). We generated invariants to 2D affine transformations of coordinates and independent 3D affine transformations of vector values. We used 21 graphs with from two to ten edges, the moment orders are from zero to nine.

Printing Options   ×

Latex
Characters in line: 83

Lines in page: 38

Independence
Number of tests: 5

Tolerance SVD: 0

Tolerance RREF: 0

Derivatives on disk:
◉ none
○ write
○ read

OK

Storno

Graphs
Horizontal size: 256

Vertical size: 256

Radius of a node: 5

Distance of edges: 8

Edge thickness: 1

☑ Include border   ☐ Describe nodes
Color of edges:
Coordinate: Contravariant Covariant:

Red: 0   255   0

Green: 0   0   255

Blue: 0   255   0

Fig. 4. Dialog window with additional parameters.

As an example, we show the invariant V_2, its graph is described by the list of edges, where V means value edge (magenta), the number without a letter is coordinate edge (black) and A0 means there is no third node connected by this edge.

$$
\begin{aligned}
V_2 = & m_{10}^{(0)} m_{20}^{(1)} m_{03}^{(2)} - m_{10}^{(0)} m_{20}^{(2)} m_{03}^{(1)} - 2m_{10}^{(0)} m_{11}^{(1)} m_{12}^{(2)} + 2m_{10}^{(0)} m_{11}^{(2)} m_{12}^{(1)} \\
& + m_{10}^{(0)} m_{02}^{(1)} m_{21}^{(2)} - m_{10}^{(0)} m_{02}^{(2)} m_{21}^{(1)} - m_{10}^{(1)} m_{20}^{(0)} m_{03}^{(2)} + m_{10}^{(1)} m_{20}^{(2)} m_{03}^{(0)} \\
& + 2m_{10}^{(1)} m_{11}^{(0)} m_{12}^{(2)} - 2m_{10}^{(1)} m_{11}^{(2)} m_{12}^{(0)} - m_{10}^{(1)} m_{02}^{(0)} m_{21}^{(2)} + m_{10}^{(1)} m_{02}^{(2)} m_{21}^{(0)} \\
& + m_{10}^{(2)} m_{20}^{(0)} m_{03}^{(1)} - m_{10}^{(2)} m_{20}^{(1)} m_{03}^{(0)} - 2m_{10}^{(2)} m_{11}^{(0)} m_{12}^{(1)} + 2m_{10}^{(2)} m_{11}^{(1)} m_{12}^{(0)} \\
& + m_{10}^{(2)} m_{02}^{(0)} m_{21}^{(1)} - m_{10}^{(2)} m_{02}^{(1)} m_{21}^{(0)} - m_{01}^{(0)} m_{20}^{(1)} m_{12}^{(2)} + m_{01}^{(0)} m_{20}^{(2)} m_{12}^{(1)} \\
& + 2m_{01}^{(0)} m_{11}^{(1)} m_{21}^{(2)} - 2m_{01}^{(0)} m_{11}^{(2)} m_{21}^{(1)} - m_{01}^{(0)} m_{02}^{(1)} m_{30}^{(2)} + m_{01}^{(0)} m_{02}^{(2)} m_{30}^{(1)} \\
& + m_{01}^{(1)} m_{20}^{(0)} m_{12}^{(2)} - m_{01}^{(1)} m_{20}^{(2)} m_{12}^{(0)} - 2m_{01}^{(1)} m_{11}^{(0)} m_{21}^{(2)} + 2m_{01}^{(1)} m_{11}^{(2)} m_{21}^{(0)} \\
& + m_{01}^{(1)} m_{02}^{(0)} m_{30}^{(2)} - m_{01}^{(1)} m_{02}^{(2)} m_{30}^{(0)} - m_{01}^{(2)} m_{20}^{(0)} m_{12}^{(1)} + m_{01}^{(2)} m_{20}^{(1)} m_{12}^{(0)} \\
& + 2m_{01}^{(2)} m_{11}^{(0)} m_{21}^{(1)} - 2m_{01}^{(2)} m_{11}^{(1)} m_{21}^{(0)} - m_{01}^{(2)} m_{02}^{(0)} m_{30}^{(1)} + m_{01}^{(2)} m_{02}^{(1)} m_{30}^{(0)}.
\end{aligned}
$$

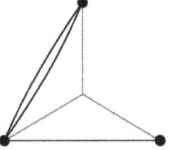

Generating graph:
V1 1 1 1
V2 2 2 3
V3 A0 A0 A0

Then we computed ten 2D affine transformations of coordinates and 3D affine transformations of the RGB color channels and the invariants of all the transformed images. An example of the transformed images is in Fig. 5(c). The moments were normalized to translation and scaling both in coordinates and in image values by the invariants V_1 and V_3. Their values are then 1, the relative errors of the other 19 invariants are in Table 1 with average 1.82%. The errors were caused by the resampling in the 2D affine transformation. The feature space of the two invariants V_2 and V_7 is in Fig. 5(d).

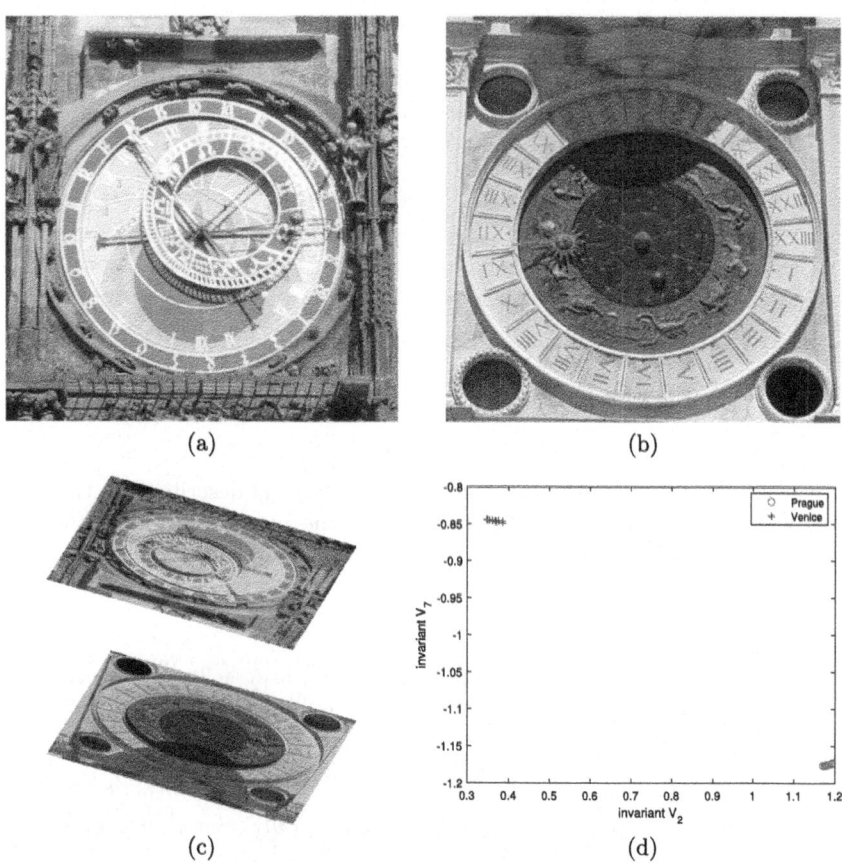

(a)   (b)

(c)   (d)

Fig. 5. The experimental images - horologe (a) in Prague, (b) in Venice; (c) examples of the distorted images, (d) feature space of invariants V_2 and V_7.

Table 1. Relative errors of the invariants in %.

V_2	V_4	V_5	V_6	V_7	V_8	V_9	V_{10}	V_{11}	V_{12}	V_{13}	V_{14}	V_{15}	V_{16}	V_{17}	V_{18}	V_{19}	V_{20}	V_{21}
2.2	5.7	0.5	1.3	0.1	0.8	1.2	0.7	0.6	1		0.4	0.9	6.8	0.9	0.7	0.7	0.6	1	8.5

5   Conclusion

The advantage of the application "Afinvtensors" is possibility of generation of moment invariants for various combinations of the data types and transformations from grayscale images up to tensor fields both to rotation and to affine transformation both in 2D and 3D.

Acknowledgement. This study was funded by Czech Science Foundation (grant number GA24-10069S). We also gratefully acknowledge the support of the U.S. Department of Energy through the LANL Laboratory Directed Research Development Program under project number 20250145ER for this work.

Disclosure of Interests. The authors have no competing interests to declare that are relevant to the content of this article.

References

1. Bowen, R., Wang, C.: Introduction to Vectors and Tensors. Dover Books on Mathematics, Dover Publications, New York (2008)
2. Bujack, R., Hagen, H.: Moment invariants for multi-dimensional data. In: Ozerslan, E., Schultz, T., Hotz, I. (eds.) Modelling, Analysis, and Visualization of Anisotropy. Mathematica and Visualization, Springer, Basel (2017)
3. Bujack, R., Zhang, X., Suk, T., Rogers, D.: Systematic generation of moment invariant bases for 2D and 3D tensor fields. Pattern Recogn. **123**, 108313 (2022)
4. Cyganski, D., Orr, J.A.: Object recognition and orientation determination by tensor methods. In: Huang, T.S. (ed.) Advances in Computer Vision and Image Processing, pp. 101–144. JAI Press, Greenwich (1988)
5. Dirilten, H., Newman, T.G.: Pattern matching under affine transformations. IEEE Trans. Comput. **26**(3), 314–317 (1977)
6. Flusser, J., Suk, T., Zitová, B.: 2D and 3D Image Analysis by Moments. Wiley, Chichester (2016)
7. Grinfeld, P.: Introduction to Tensor Analysis and the Calculus of Moving Surfaces. Springer, New York (2013)
8. Gurevich, G.B.: Foundations of the Theory of Algebraic Invariants. Nordhoff, Groningen (1964)
9. Hu, M.K.: Visual pattern recognition by moment invariants. IRE Trans. Inf. Theory **8**(2), 179–187 (1962)
10. Kostková, J., Suk, T., Flusser, J.: Affine invariants of vector fields. IEEE Trans. Pattern Anal. Mach. Intell. **43**(4), 1140–1155 (2021)
11. Langbein, M., Hagen, H.: A generalization of moment invariants on 2D vector fields to tensor fields of arbitrary order and dimension. In: Proceedings of 5th International Symposium Advances in Visual Computing, ISVC 2009, Part II. Lecture Notes in Computer Science, vol. 5876, pp. 1151–1160. Springer, Cham (2009)
12. Mo, H., Li, H., Zhao, G.: Gaussian-hermite moment invariants of general multichannel functions (2023). https://arxiv.org/abs/2201.00877
13. Suk, T., Flusser, J.: Graph method for generating affine moment invariants. In: Proceedings of the 17th International Conference on Pattern Recognition ICPR 2004, pp. 192–195. IEEE Computer Society (2004)
14. Suk, T., Flusser, J.: Tensor method for constructing 3D moment invariants. In: Computer Analysis of Images and Patterns CAIP 2011. LNCS, vol. 6854–6855, pp. 212–219. Springer, Cham (2011)

HAME-NeRF: High Accuracy Mesh Extraction Leveraging Neural Radiance Fields

Panagiotis Frasiolas(✉)📵, Grigorios-Aris Cheimariotis📵,
Panos K. Papadopoulos📵, and Dimitrios Zarpalas📵

Information Technologies Institute (ITI), Centre for Research and Technology Hellas
(CERTH), 69121 Thessaloniki, Greece
`frasiolas@iti.gr`

Abstract. Neural Radiance Fields (NeRFs) have transformed image-based 3D reconstruction through differentiable volumetric rendering, enabling high-quality novel view synthesis. However, their implicit volumetric nature is incompatible with the polygonal meshes needed for real-time graphics and simulation applications. The proposed model defines the volume density function as the Secant Hyperbolic Function applied to a signed distance function (SDF) representation. To enable accurate surface representation, the sharpness of the density transition is modulated by a spatially-varying parameter $\beta(x)$, which is learned through a multi-layer perceptron (MLP). Experimental results on the NeRF-Synthetic and Mip-NeRF 360 datasets demonstrate improved surface reconstruction accuracy and visual quality compared to NeRF2Mesh, highlighting the effectiveness of the proposed enhancements for efficient and high-fidelity real-time scene reconstruction.

Keywords: Neural Radiance Field · Signed Distance Function · Mesh

1 Introduction

Reconstruction of high-fidelity 3D surfaces from 2D images is a fundamental problem in computer vision, with applications spanning robotics, photogrammetry, and AR/VR. Many of these applications depend on precise 3D geometry to enable physics-based simulations, real-time visualization, and seamless integration into graphics pipelines. Recent advances in Neural Radiance Fields (NeRF)

The original version of the chapter has been revised. The author name is corrected. A correction to this chapter can be found at
https://doi.org/10.1007/978-3-032-04968-1_34

Supplementary Information The online version contains supplementary material available at https://doi.org/10.1007/978-3-032-04968-1_28.

[6] have significantly improved novel view synthesis (NVS), allowing for photo-realistic rendering of complex scenes. However, while NeRF excels in generating high-quality renderings, it is not inherently optimized for accurate 3D geometry extraction.

The volumetric representation in NeRF enables the synthesis of novel views through differentiable volumetric rendering, but the implicit nature of its scene encoding poses challenges for downstream applications. Unlike conventional polygonal meshes, which are widely used in 3D graphics due to their hardware efficiency and ease of manipulation, NeRF relies on ray marching and implicit functions that lack direct compatibility with standard rendering engines. Moreover, the underlying geometry in NeRF is not explicitly defined as a level-set surface, often resulting in approximations through dense volumetric regions rather than well-defined object boundaries. This makes extracting accurate and editable 3D meshes from NeRF particularly difficult [2,3].

This paper presents a framework for extracting high-fidelity, textured surface meshes from RGB images. First, a grid-based NeRF is trained to decompose appearance into diffuse and specular components and extract a coarse geometry via a density field. Then, the initial mesh is refined through differentiable rendering and an iterative mesh refinement algorithm, adjusting vertex positions and face density based on reprojected 2D rendering errors. The proposed implementation incorporates SDF-based density modeling. To the best of our knowledge, this method is the first to optimize the SDF parameter that controls the sharpness of the surface boundary, commonly referred to as β (beta) parameter. Thus, this parameter becomes learned and spatially-varying, enabling sharper surface representation and better geometric control.

The proposed method was evaluated on two benchmark datasets: NeRF-Synthetic [6] and Mip-NeRF 360 [1]. For the NeRF-Synthetic dataset, which provides access to ground truth meshes, geometric accuracy is evaluated using Chamfer Distance and Normal Consistency metrics. These measures specifically assess the quality of the reconstructed shape geometry, capturing both spatial alignment and surface normal fidelity between the predicted and reference meshes. Peak Signal-to-Noise Ratio (PSNR) was used to assess rendering quality on both datasets. Additionally, a qualitative analysis was conducted across all scenes. Experimental results show that the proposed method outperforms NeRF2Mesh [9] in both Chamfer Distance, Normal Consistency, PSNR and visual quality on the NeRF-Synthetic dataset, and achieves comparable, though slightly lower, performance on the more challenging Mip-NeRF 360 dataset.

2   Related Work

2.1   NeRF and Efficient Scene Reconstruction

Neural Radiance Fields (NeRF) have significantly advanced 3D scene reconstruction by representing scenes as continuous volumetric fields parameterized by neural networks. While achieving impressive photorealistic rendering, the original NeRF formulation suffers from slow optimization and difficulties handling unbounded scenes. Subsequent methods, such as Instant-NGP [7] and SNeRG

[4], introduce explicit 3D structures like multi-resolution hash grids and sparse voxel grids to accelerate training and enable real-time rendering. However, these approaches often compromise geometric fidelity, limiting their utility for high-precision reconstruction tasks.

2.2   Surface Reconstruction from Implicit Fields

An alternative to volumetric rendering is surface-based representation using Signed Distance Functions (SDFs), which enable high-quality mesh extraction through algorithms like Marching Cubes [5]. Techniques such as the NeuS [10] model SDFs with differentiable rendering frameworks, improving surface accuracy while maintaining learnability from 2D supervision. To further address challenges with thin structures and complex topologies, hybrid methods such as BakedSDF [11] and NVdiffrec [8] combine volumetric and surface-based representations, optimizing both geometry and appearance for real-time or high-fidelity mesh extraction.

2.3   Bridging Implicit and Explicit Representations

Transforming implicit volumetric models into explicit, textured surface meshes remains an active area of research. Approaches like NeRF2Mesh aim to bridge this gap by extracting surfaces directly from trained NeRFs while preserving photorealistic details. Recent methods increasingly leverage differentiable rendering pipelines to refine surface geometry with 2D supervision, improving mesh quality without sacrificing appearance. Despite these advances, achieving real-time, high-fidelity reconstruction that balances photorealism, geometric precision, and computational efficiency continues to be a core challenge, motivating further exploration of hybrid representations and optimization strategies.

3   Method

The proposed framework extracts a textured surface mesh from a set of RGB images, extending the NeRF2Mesh pipeline. The process comprises two main stages (Fig. 1): an initialization stage using a grid-based NeRF to obtain coarse geometry and appearance, followed by a refinement stage improving both accuracy and visual quality.

Unlike the baseline, which applies Marching Cubes directly to a density field, the proposed method introduces a Signed Distance Function (SDF) for mesh extraction. This provides a more accurate geometric proxy and better surface localization.

The SDF incorporates two key innovations: (a) volumetric density modeling through a hyperbolic secant function for sharper boundaries, and (b) spatially varying sharpness control, with the parameter β learned via an auxiliary MLP across (x, y, z) coordinates.

3.1   Initialization of Geometry and Appearance

A volumetric NeRF structure with an SDF component serves as the foundation for initializing geometry and appearance.

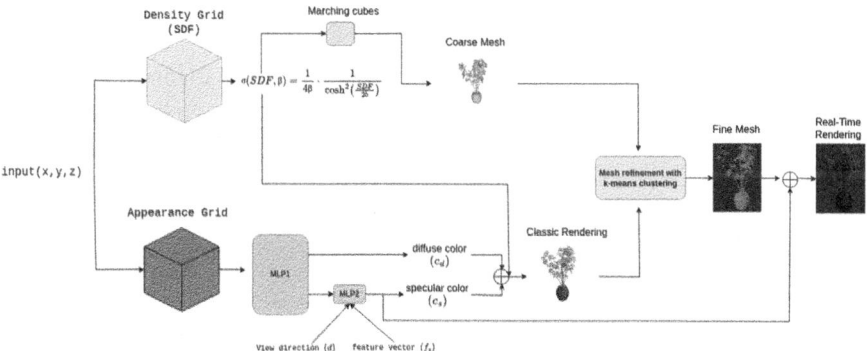

Fig. 1. Overview of the proposed method. Geometry is produced by a density grid, and view-dependent appearance is handled by a separate appearance grid.

Geometry Initialization. Geometry is modeled with a density grid and a shallow MLP:

$$SDF = \mathrm{MLP}(E^{\mathrm{geo}}(\mathbf{x})), \tag{1}$$

where E^{geo} is a multi-resolution feature grid and $\mathbf{x} \in \mathbb{R}^3$ denotes spatial coordinates.

The density function $\sigma(SDF, \beta)$ is designed to concentrate density near the zero-level set using a hyperbolic cosine:

$$\sigma(SDF, \beta) = \frac{1}{4\beta} \cdot \frac{1}{\cosh^2\left(\frac{SDF}{2\beta}\right)}. \tag{2}$$

$$\beta(x) = \mathrm{MLP}_\beta(x) \tag{3}$$

Unlike prior methods that employ a fixed β value, the proposed approach learns a spatially varying $\beta(\mathbf{x})$ through an MLP, enhancing flexibility for complex surfaces. The ablation study compares the model's performance using a fixed beta value versus a spatially varying beta.

Appearance Initialization. Appearance is decomposed into view-independent diffuse color $\mathbf{c}_d$ and view-dependent specular color $\mathbf{c}_s$:

$$\mathbf{c}_d, \mathbf{f}_s = \psi(\mathrm{MLP}_1(E^{\mathrm{app}}(\mathbf{x}))), \quad \mathbf{c}_s = \psi(\mathrm{MLP}_2(\mathbf{f}_s, \mathbf{d})), \tag{4}$$

where $\mathbf{f}_s$ encodes specular information, $\mathbf{d}$ is the view direction, and ψ denotes a sigmoid activation. The final color is:

$$\mathbf{c} = \mathbf{c}_d + \mathbf{c}_s. \tag{5}$$

This separation allows for efficient baking of diffuse textures and flexible handling of specularities using lightweight shaders, following [3]. As in NeRF2Mesh, lighting is pre-baked into textures to avoid complex light estimation.

3.2   Mesh Refinement and Optimization

After initial convergence, a coarse mesh is extracted via Marching Cubes. This mesh undergoes refinement to improve both geometry and appearance.

Differentiable rendering with nvdiffrast is employed to align the projected mesh with input images. The appearance model is reused to accelerate convergence. Vertex positions are optimized by learning per-vertex offsets, using gradients from a pixel-wise rendering loss.

To adapt mesh resolution, face refinement is performed heuristically: faces with high rendering errors are identified using k-means clustering. High-error faces are subdivided, while low-error regions are simplified, concentrating detail where needed. This adaptive process repeats during training.

For unbounded scenes, the environment is partitioned into regions of progressively lower resolution away from the center. Regularization terms are applied to ensure smooth surfaces and bounded vertex displacements.

3.3   Loss Function

Training minimizes a weighted combination of three objectives:

Photometric Rendering Loss. The rendering loss encourages consistency between rendered and ground-truth pixel colors:

$$\mathcal{L}_{\text{render}} = \frac{1}{N} \sum_{i=1}^{N} \|\hat{\mathbf{c}}_i - \mathbf{c}_i\|_2^2 . \tag{6}$$

Eikonal Loss. The Eikonal loss regularizes the SDF to behave like a true distance field:

$$\mathcal{L}_{\text{eik}} = (\|\nabla_{\mathbf{x}} \text{SDF}(\mathbf{x})\|_2 - 1)^2 . \tag{7}$$

Beta Regularization Loss. An L_1 penalty on $\beta(\mathbf{x})$ prevents unstable variations and encourages smooth transitions:

$$\mathcal{L}_\beta = \lambda_\beta \cdot |\beta(\mathbf{x})| . \tag{8}$$

Total Loss. The total training objective is:

$$\mathcal{L}_{\text{total}} = \mathcal{L}_{\text{render}} + \lambda_{\text{eik}} \mathcal{L}_{\text{eik}} + \lambda_\beta \mathcal{L}_\beta. \tag{9}$$

3.4   Evaluation

Datasets: The effectiveness and generalization capability of the proposed method are evaluated using two diverse datasets:

1. NeRF-Synthetic dataset [6]: A widely used benchmark introduced in the original NeRF, containing various synthetic scenes. Each scene has RGB images with known camera poses, ideal for testing novel view synthesis.
2. Mip-NeRF 360 dataset [1]: A collection of nine complex scenes (five outdoor and four indoor), each featuring a detailed central object or area surrounded by intricate backgrounds.

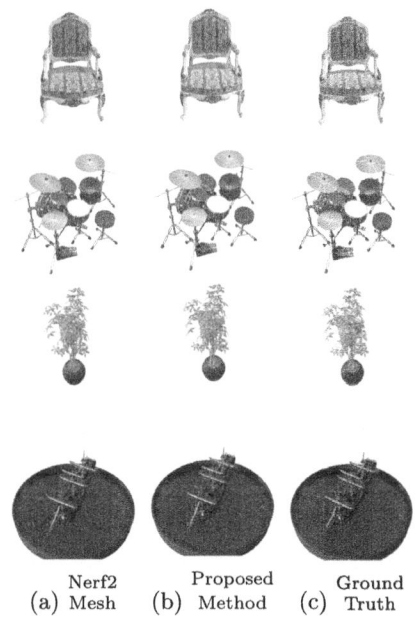

Nerf2   Proposed   Ground
(a) Mesh   (b) Method   (c) Truth

Fig. 2. Final output comparison on Nerf-Synthetic Dataset. The output is the combination of the fine mesh with the small MLP for illuminations.

Metrics: To evaluate geometric similarity between two 3D meshes, the Chamfer Distance (CD) is computed between surface points from each mesh. This metric quantifies the average distance between points on one surface and their nearest neighbors on the other, offering a robust measure of alignment and shape similarity. An observability grid of resolution 256 is constructed using rays cast from test views located within the mesh bounding box.

Normal Consistency measures how well the surface normals of a predicted mesh align with those of a reference mesh. For each predicted point, the closest reference point is found, and the cosine similarity between their normals is computed. Flipped normals are corrected before averaging. This metric captures the local surface orientation accuracy, which is critical for realistic shading, rendering, and simulation. Higher values indicate better geometric fidelity on a fine scale.

Peak Signal-to-Noise Ratio (PSNR) is used to evaluate the photometric accuracy of the rendered images by comparing them to ground-truth images.

$$\text{PSNR} = 20 \cdot \log_{10} \left(\frac{\text{MAX}}{\text{MSE}} \right) \tag{10}$$

NeRF-Synthetic Dataset: Qualitative comparison on the NeRF-Synthetic dataset is presented to assess the quality of the coarse mesh extracted during the initial stage of the pipeline (see Fig. 3), the fine mesh extracted in the last stage (see Fig. 4) and the final rendered image (see Fig. 2)

Figure 3 shows the coarse mesh extracted during the initial stage of the pipeline. The figure includes images from four different scenes in the dataset. Red squares highlight areas where fine geometric details are more accurately captured and the surface appears smoother. Blue squares indicate regions where the mesh surface is more complete and better filled, with fewer missing parts.

Figure 4 shows the fine mesh extracted in the final stage. Zoomed-in views are also provided, with red squares highlighting regions that exhibit smoother surfaces or improved mesh topology, characterized by fewer and more uniform faces.

Figure 4 demonstrates that the rendered output of the proposed method closely matches the ground truth, with minimal visible differences.

Across all scenes, the proposed method outperforms NeRF2Mesh by achieving lower Chamfer Distance (see Table 1), higher Normal Consistency (see Table 2), and higher PSNR (see Table 3). These results confirm its superiority in both geometric reconstruction and rendering quality.

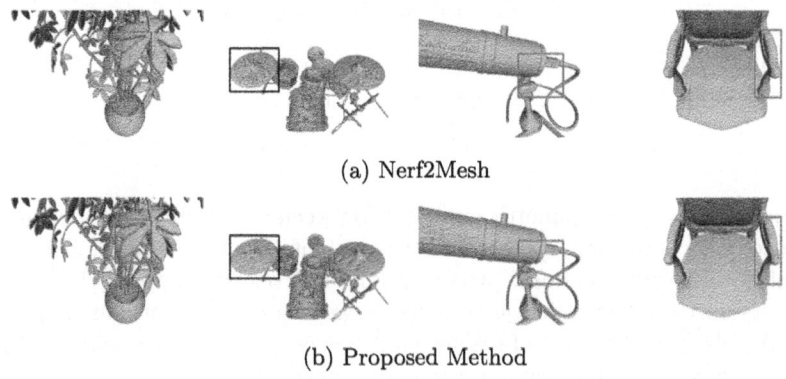

(a) Nerf2Mesh

(b) Proposed Method

Fig. 3. Comparison of generated coarse meshes on Nerf-Synthetic Dataset.

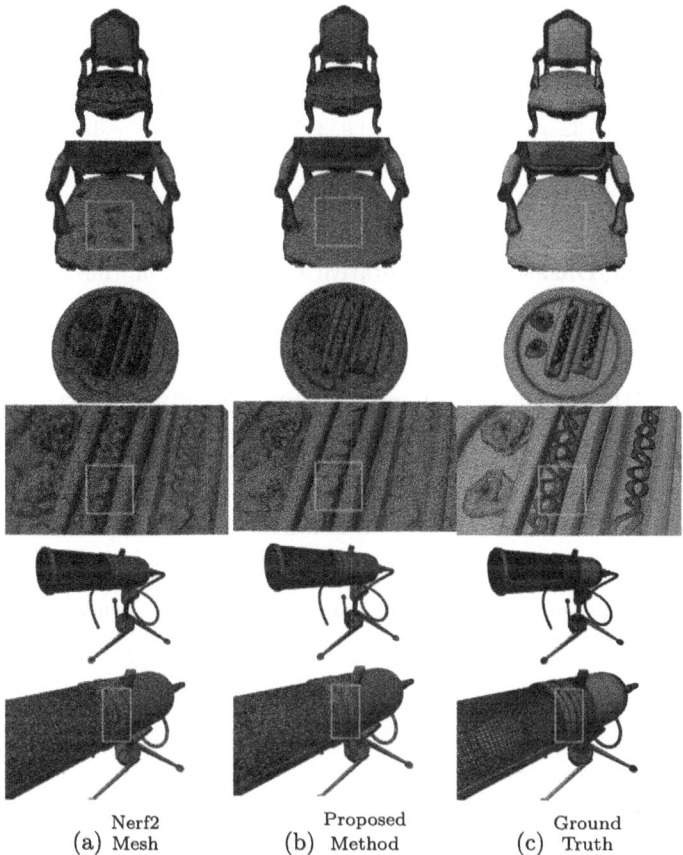

| Nerf2 | Proposed | Ground |
| (a)   Mesh | (b)   Method | (c)   Truth |

Fig. 4. Geometry comparison of the generated fine mesh on Nerf-Synthetic Dataset.

Table 1. Chamfer Distance (Unit is 10^{-3}) across NeRF synthetic dataset scenes. Bold numbers indicate best results per scene.

	Chair	Drums	Ficus	Hotdog	Materials	Mic	Ship
Nerf2Mesh [9]	**7.677**	0.6010	0.0439	70.68	**0.407**	0.2060	276.6
Proposed Method	7.825	**0.5761**	**0.0232**	**9.151**	0.468	**0.0680**	**51.56**

Table 2. Normal Consistency across NeRF synthetic dataset scenes. Bold numbers indicate best results per scene.

	Chair	Drums	Ficus	Hotdog	Materials	Mic	Ship
Nerf2Mesh [9]	**0.9077**	0.8316	0.8775	0.8243	**0.8781**	0.8636	0.6346
Proposed Method	0.9064	**0.8519**	**0.8819**	**0.8439**	0.8344	**0.8744**	**0.6387**

338   P. Frasiolas et al.

Table 3. PSNR across NeRF synthetic dataset scenes. Bold numbers indicate best results per scene.

	Chair	Drums	Ficus	Hotdog	Materials	Mic	Ship
Nerf2Mesh [9]	32.3917	25.0147	30.0653	**34.9732**	25.7856	31.9232	27.4562
Proposed Method	**32.4012**	**25.0215**	**30.4823**	34.9219	**25.9793**	**31.9340**	**27.5306**

Mip-Nerf 360 Dataset: A qualitative comparison on the Mip-NeRF 360 dataset is presented to assess the quality of the fine mesh extracted in the final stage (see Fig. 5) and the corresponding rendered images (see Fig. 6). Red squares indicate regions where the proposed method yields cleaner geometry, whereas Nerf2Mesh tends to generate a high number of mesh faces without significant geometric improvement. However, a limitation of the proposed method is its slightly lower performance in terms of PSNR (see Table 4)

Figure 6 demonstrates that the rendered output of the proposed method closely matches the ground truth, with minimal visible differences.

Table 4. PSNR across Mip-NeRF 360 dataset scenes. Bold numbers indicate best results per scene.

	Bicycle	Bonsai	Counter	Garden	Kitchen	Room	Stump
Nerf2Mesh [9]	**22.4764**	**26.1323**	**24.8213**	**22.6985**	**25.7852**	**26.882**	**23.1960**
Proposed Method	22.1009	25.1960	24.6931	22.6091	24.6916	24.3575	23.0895

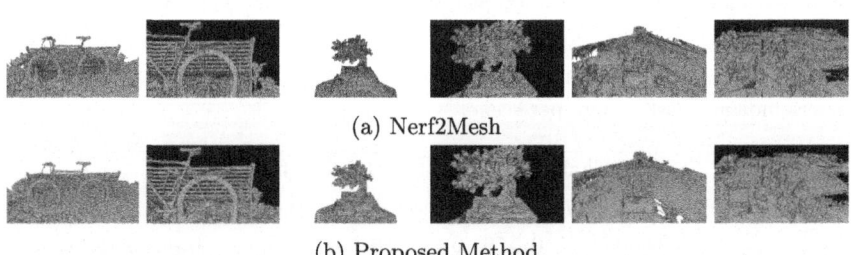

(a) Nerf2Mesh

(b) Proposed Method

Fig. 5. Geometry comparison of the generated fine mesh Mip-NeRF 360 scenes.

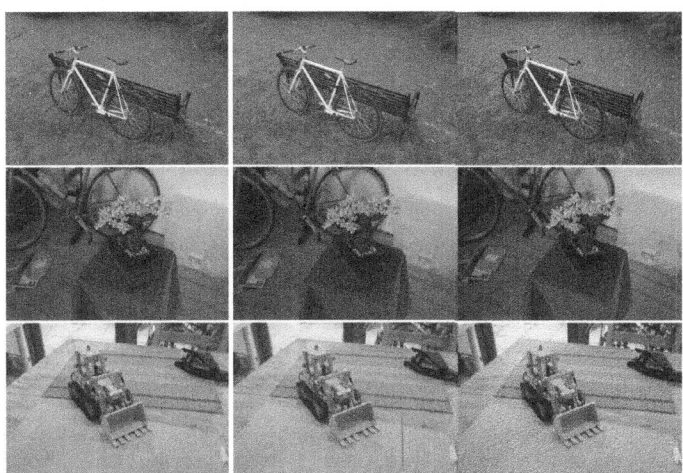

(a) Nerf2Mesh   (b) Proposed Method   (c) Ground Truth

Fig. 6. Final output comparison on Mip-Nerf 360. The output is the combination of the fine mesh with the small MLP for illuminations.

4   Conclusion

This paper presents a robust NeRF-based framework for extracting high-quality textured surface meshes using a signed distance function with spatially adaptive sharpness. A hyperbolic secant-based density formulation enables sharper surface localization and enhances mesh fidelity, particularly in areas with fine structural detail. The process begins with geometry and appearance initialized through a volumetric NeRF, followed by mesh refinement using differentiable rendering. While effective, the framework inherits common limitations of mesh-based pipelines, including baked-in lighting that hampers relighting capabilities and reduced accuracy in capturing complex view-dependent effects. Future work may focus on improving appearance modeling and adopting more expressive rendering techniques to further enhance realism and flexibility.

Acknowledgments. This   work   was   supported   by   the   European Union's Horizon Europe programme under grant number 101092875 "DIDYMOS-XR" (https://www.didymos-xr.eu).

References

1. Barron, J.T., Mildenhall, B., Verbin, D., Srinivasan, P.P., Hedman, P.: Mip-nerf 360: Unbounded anti-aliased neural radiance fields. In: Proceedings of the IEEE/CVF Conference on Computer Vision and Pattern Recognition, pp. 5470–5479 (2022)

2. Chen, W., et al.: Learning to predict 3d objects with an interpolation-based differentiable renderer. Adv. Neural Inf. Process. Syst. **32** (2019)

3. Chen, Z., Funkhouser, T., Hedman, P., Tagliasacchi, A.: Mobilenerf: Exploiting the polygon rasterization pipeline for efficient neural field rendering on mobile architectures. In: Proceedings of the IEEE/CVF Conference on Computer Vision and Pattern Recognition, pp. 16569–16578 (2023)

4. Hedman, P., Srinivasan, P.P., Mildenhall, B., Barron, J.T., Debevec, P.: Baking neural radiance fields for real-time view synthesis. In: Proceedings of the IEEE/CVF International Conference on Computer Vision, pp. 5875–5884 (2021)

5. Lorensen, W.E., Cline, H.E.: Marching cubes: A high resolution 3d surface construction algorithm. In: Seminal Graphics: Pioneering Efforts that Shaped the Field, pp. 347–353 (1998)

6. Mildenhall, B., Srinivasan, P.P., Tancik, M., Barron, J.T., Ramamoorthi, R., Ng, R.: Nerf: Representing scenes as neural radiance fields for view synthesis. Commun. ACM **65**, 99–106 (2021)

7. Müller, T., Evans, A., Schied, C., Keller, A.: Instant neural graphics primitives with a multiresolution hash encoding. ACM Trans. Graphics (TOG) **41**(4), 1–15 (2022)

8. Munkberg, J., et al.: Extracting triangular 3d models, materials, and lighting from images. In: Proceedings of the IEEE/CVF Conference on Computer Vision and Pattern Recognition, pp. 8280–8290 (2022)

9. Tang, J., et al.: Delicate textured mesh recovery from nerf via adaptive surface refinement. In: Proceedings of the IEEE/CVF International Conference on Computer Vision, pp. 17739–17749 (2023)

10. Wang, P., Liu, L., Liu, Y., Theobalt, C., Komura, T., Wang, W.: Neus: Learning neural implicit surfaces by volume rendering for multi-view reconstruction. arXiv preprint arXiv:2106.10689 (2021)

11. Yariv, L., et al.: Bakedsdf: Meshing neural sdfs for real-time view synthesis. In: ACM SIGGRAPH 2023 Conference Proceedings, pp. 1–9 (2023)

FaDeN: Fast Depth-Supervised NeRFs with RGB-D Cameras

Janne Mustaniemi$^{(\boxtimes)}$, Li Liu , and Janne Heikkilä

Center for Machine Vision and Signal Analysis (CMVS), University of Oulu, Oulu,
Finland
janne.mustaniemi@oulu.fi

Abstract. We address novel view synthesis from RGB-D images with
a limited number of input views. Existing methods often struggle under
such conditions and are sensitive to photometric inconsistencies and noisy
depth measurements. We propose FaDeN, a fast method that estimates
absolute scale, enhances sensor depth maps for supervision, corrects photometric variations, and synthesizes virtual views for improved scene coverage. FaDeN reduces training time by more than 200× compared to similar NeRF-based methods while producing novel views of comparable or
superior quality. Experiments on public and custom datasets demonstrate
improvements in novel view synthesis, enhancing both image quality and
geometric accuracy. In addition, we show that FaDeN improves the accuracy and completeness of 3D surface reconstructions.

Keywords: Novel view synthesis · 3D reconstruction · Depth map

1 Introduction

Neural Radiance Fields (NeRF) [5] and Gaussian Splatting [4] have significantly
advanced novel view synthesis and 3D scene reconstruction. However, most existing methods rely solely on color images, which poses challenges when the number
of input views is limited or the scene contains textureless regions. Additionally,
variations in exposure and white balance across input images introduce photometric inconsistencies that further degrade the quality of the synthesized images,
as shown in the top row of Fig. 3.

Several methods [2,7,9,12,13] integrate depth priors into NeRF optimization,
improving novel view synthesis in sparse view settings. These approaches utilize
depth cues from sparse SfM point clouds [2], monocular depth estimation (MDE)
models [9,12], or a combination of both [7,13]. Similarly, geometric priors such as
depth and normal maps have been leveraged to enhance Gaussian Splatting [11].

While recent methods have advanced novel view synthesis, several limitations
remain. NeRF-based approaches are typically slow, MDE models may struggle
to generalize across scenes, and most methods cannot recover the scene's absolute scale – important for tasks like 3D reconstruction. Variations in camera
settings during image capture are also often overlooked. Moreover, based on our

© The Author(s), under exclusive license to Springer Nature Switzerland AG 2026
M. Castrillón-Santana et al. (Eds.): CAIP 2025, LNCS 15621, pp. 341–352, 2026.
https://doi.org/10.1007/978-3-032-04968-1_29

experiments, the methods based on Gaussian Splatting are particularly sensitive to image degradations like motion blur.

Some methods, such as Yuan et al. [14], have explored leveraging RGB-D cameras for novel view synthesis. We also build on this idea, as RGB-D cameras offer several advantages over purely image-based approaches. They provide direct depth measurements, avoid the generalization issues of MDE models, and recover depth in textureless regions where image-based methods struggle. However, RGB-D cameras also introduce challenges, as their depth maps often contain noise, artifacts, and missing data.

Our method integrates RGB-D images from a consumer depth camera into NeRF training to enhance novel view synthesis. We address the scale ambiguity, enhance the sensor depth maps for supervision, correct photometric inconsistencies, and synthesize virtual views to improve scene coverage. Our approach reduces training time by more than 200× compared to similar methods while producing novel views of comparable or superior quality. Beyond view synthesis, we also demonstrate its effectiveness in improving the accuracy and completeness of 3D surface reconstructions.

2   Related Work

Neural Radiance Fields (NeRF) were first introduced by Mildenhall et a. [5] for novel view synthesis. Their impressive results have inspired many subsequent works, leading to improvements in rendering quality, computational efficiency, and scalability. Similarly, Gaussian Splatting [4] has emerged as an alternative approach for real-time rendering. This section reviews methods that incorporate depth information, such as those derived from structure from motion (SfM), monocular depth estimation (MDE), or RGB-D cameras.

DS-NeRF [2] utilizes sparse 3D points from SfM and integrates depth supervision into the NeRF training process. It introduces a loss function designed to align a ray's termination depth with the depth of the 3D point, incorporating reprojection error as a measure of uncertainty. The method has been demonstrated to enhance the quality of rendered images, reduce the number of required training images, and accelerate training by 2-3 times. While DS-NeRF primarily focuses on depth data from SfM, the approach can be adapted to other depth sources.

NerfingMVS [13] uses the SfM 3D points to finetune a monocular depth estimation (MDE) model for each specific scene. This allows NerfingMVS to generate refined depth priors, which are then employed to guide the sampling process within the NeRF framework. NerfingMVS mainly focuses on multi-view depth estimation, but the method is shown to improve the quality of rendered images.

DDP-NeRF [7] addresses the sparse-view setting using depth completion. Prior to the NeRF optimization, they train a model to estimate dense depth and uncertainty from the sparse SfM depth. These estimates are used to constrain the NeRF optimization and to guide the scene sampling. DDP-NeRF enables novel view synthesis on room-scale scenes using only 18-36 input images.

Yuan et al. [14] employ an RGB-D camera to generate a 3D mesh via Poisson surface reconstruction and render novel views from perturbed camera poses. These rendered images serve as pseudo ground truth during the NeRF pre-training phase, and the model is subsequently finetuned with a limited set of real captured images. While this approach leverages synthetic views, it does not use the depth measurements directly. Our method implements depth supervision with enhanced sensor depth maps and inpaints missing regions in the rendered color images, leading to improved image quality.

SCADE [12] utilizes a state-of-the-art MDE model to obtain depth priors. It predicts a multimodal distribution of depth estimates for each view to address the inherent ambiguities related to MDE. The method incorporates a space carving loss, which guides the NeRF representation to fuse multiple hypothesized depth maps into a consistent geometry.

DäRF [9] adapts a pre-trained MDE model to predict depth maps not only for the input views but also for unseen viewpoints. During training, it renders novel color images, which are then processed by the MDE model to generate pseudo ground truth depth maps. A depth loss is applied to regularize NeRF optimization by minimizing the difference between rendered depths and these pseudo depths. Additionally, DäRF proposes a patch-based fitting approach to address scale and shift ambiguities inherent in MDE.

DN-Splatter [11] incorporates depth and normal priors to regularize Gaussian Splatting optimization, improving scene geometry reconstruction. It introduces an edge-aware depth loss to mitigate inaccurate sensor depths near object boundaries. Additionally, normals predicted by a monocular network help align Gaussians with scene surfaces. These priors contribute to smoother reconstructions while also enhancing novel view synthesis.

3   Method

This section describes the components of our framework, as illustrated in Fig. 1. Given RGB-D images as input, the method produces a NeRF model that enables rendering of novel color images and depth maps.

3.1   Structure-From-Motion

We estimate camera poses and a sparse point cloud from color images using the structure-from-motion (SfM) software COLMAP [8]. A known limitation of image-based reconstruction is the inability to recover absolute scale, which complicates the use of sensor depth maps in NeRF optimization. We also experimented with estimating poses from depth maps using the iterative closest point (ICP) algorithm, but found that SfM-based poses were more accurate. In the following section, we describe how we estimate the unknown scale factor of the SfM reconstruction.

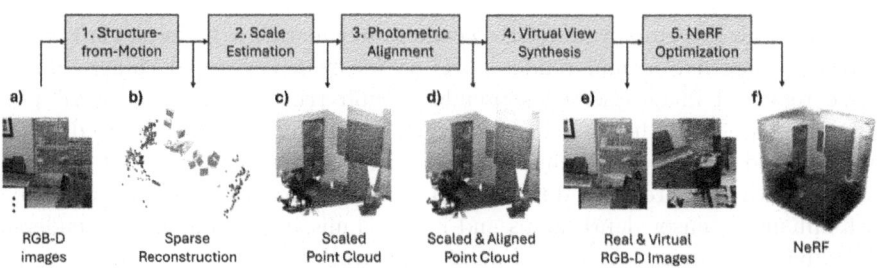

Fig. 1. Overview of the proposed framework. Camera poses and sparse point cloud are estimated from color images using SfM. The absolute scale of the reconstruction is recovered using depth maps. Color images are photometrically aligned to account for variable exposure and white balance settings. Virtual views and enhanced depth maps are synthesized from a mesh. Finally, the NeRF is optimized using real and virtual RGB-D images.

3.2   Scale Estimation

To resolve the unknown scale of the SfM-based camera poses, we estimate a global scale factor using depth measurements from the sensor. Let $d^i_{\text{sfm}}(p)$ denote the depth at the 2D point p in the i-th image as derived from SfM, and $d^i_{\text{sensor}}(p)$ the corresponding sensor depth. The scale factor s is estimated by minimizing:

$$s = \arg\min_s \sum_i \sum_{p \in \mathcal{V}_i} \left(s \cdot d^i_{\text{sfm}}(p) - d^i_{\text{sensor}}(p) \right)^2, \tag{1}$$

where $\mathcal{V}_i$ is the set of pixels with $d^i_{\text{sensor}}(p) > 0$. We apply RANSAC to robustly estimate s due to potential outliers in both depth sources. The resulting scale factor converts SfM poses to metric units. Sensor depth maps are then transformed into point clouds and aligned, as illustrated in Fig. 1c. The importance of scale estimation is validated in our experiments.

3.3   Photometric Alignment

Auto exposure and auto white balance often introduce variations in color and brightness across images, which can lead to artifacts in NeRF renderings (see the top row in Fig. 3). One possible strategy to address this is to learn per-image appearance embeddings, as supported, for example, in Instant-NGP [6]. However, we found that this generally leads to less sharp results. Instead, we address the issue with a photometric alignment method that enforces consistency across views.

For each image i, we estimate gain factors $\gamma^i = (\gamma^i_R, \gamma^i_G, \gamma^i_B)$ for the RGB channels. These gains are optimized using color correspondences obtained by projecting the 3D points (from the previous step) into each image. A point is considered visible if its projected depth agrees with the sensor depth.

Let $\mathbf{c}_k^i$ be the k-th color observed in image i. We minimize the following loss:

$$\underset{\{\gamma^i\}}{\operatorname{argmin}} \frac{1}{|\mathcal{P}|} \sum_{(k,i,j)\in\mathcal{P}} \mathcal{L}(\gamma^i \mathbf{c}_k^i - \gamma^j \mathbf{c}_k^j) + \lambda \sum_{c\in\{R,G,B\}} \left(\frac{1}{N}\sum_{i=1}^{N}\gamma_c^i - 1\right)^2, \quad (2)$$

where $\mathcal{P}$ is the set of color correspondences, $\mathcal{L}(\cdot)$ is the Huber loss, and N is the number of images. The regularization enforces that the average gain for each channel remains close to 1, preventing trivial solutions.

While this alignment promotes consistency across views, the overall white balance may still be incorrect if some input images are poorly balanced. Assuming most images are correctly balanced, we apply the following normalization:

$$\gamma_c^i \leftarrow \gamma_c^i \cdot \frac{1}{\operatorname{median}_i\left(\frac{3\,\gamma_c^i}{\sum_{c\in\{R,G,B\}}\gamma_c^i}\right)}, \quad (3)$$

which preserves the dominant white balance across the dataset. Figure 1(a, e) shows an image before and after alignment. This process significantly reduces artifacts in rendered images (Fig. 6).

3.4   Virtual View Synthesis

Novel view synthesis methods often struggle in sparse-view settings, where only a limited number of input images are available. While depth supervision helps alleviate this issue [2,7,9,12], we further improve performance by synthesizing additional virtual views. Our approach is inspired by Yuan et al. [14], but extends it by also synthesizing depth maps and inpainting missing regions in the virtual color images.

Given the photometrically aligned RGB-D images, we perform TSDF surface reconstruction using the Open3D library [15]. Since sensor depth maps often contain noise and artifacts, we generate enhanced depth maps from the reconstructed mesh via raycasting. Let $d_{\mathrm{mesh}}^i(p)$ be the enhanced depth at pixel p in image i. To filter out outliers (e.g., due to occlusions), we retain only values close to the original sensor depth $d_{\mathrm{sensor}}^i(p)$:

$$d_{\mathrm{fused}}^i(p) = \begin{cases} d_{\mathrm{mesh}}^i(p), & |d_{\mathrm{mesh}}^i(p) - d_{\mathrm{sensor}}^i(p)| \leq \beta \\ 0, & \text{otherwise} \end{cases} \quad (4)$$

We also synthesize novel virtual views by randomly perturbing the input poses, adding up to 30° of rotation and 1 m of translation, to improve scene coverage. The total number of views (real and virtual) is fixed to 120.

As shown in Fig. 2, virtual views may contain holes if the mesh is incomplete, which can occur due to missing depth data or unseen regions. To address this, we inpaint the virtual color images using LaMa [10]. Inpainting is especially beneficial when parts of the scene are entirely unseen, such as the partially missing door in Fig. 6.

Virtual RGB-D image   Without inpainting

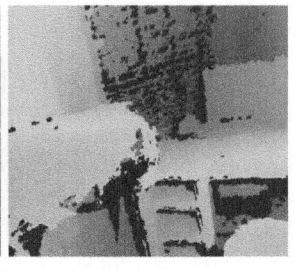

Fig. 2. An example of a virtual RGB-D image rendered from the mesh. The holes in the color image have been inpainted.

3.5   NeRF Training

We use Instant-NGP [6] for NeRF training, which is known for its fast training and high-quality renderings. The total loss combines color and depth supervision: $\mathcal{L}_{total} = \mathcal{L}_{color} + \lambda \mathcal{L}_{depth}$, where $\mathcal{L}_{color}$ is the Huber loss and $\mathcal{L}_{depth}$ is the logarithmic L1 loss. For depth supervision, we use the fused depths from Eq. 4 as targets, with λ fixed to 0.01 in all experiments.

Training is performed in two stages. We first pretrain the model for 5,000 iterations using both real and virtual views, followed by fine-tuning for 20,000 iterations using real views only. As Instant-NGP does not support changing the training set mid-training, we modify the pipeline to load all images at once and mask out unused views during the second stage. Training takes approximately 10 minutes on an NVIDIA GeForce RTX 2070 GPU.

4   Experiments

We evaluate our framework for novel view synthesis, assess its components through an ablation study, and demonstrate its use in 3D reconstruction.

4.1   Novel View Synthesis

We follow [7,9,12] and evaluate novel view synthesis on the ScanNet dataset [1]. Specifically, the methods are evaluated on three scenes, each containing 18âĂŞ20 images for training and 8 images for testing. Additionally, we captured a custom dataset using an Azure Kinect RGB-D camera, selecting 20 images for training and 8 for testing in each of two scenes. We compute PSNR, SSIM, and LPIPS to assess the quality of the synthesized color images. For depth maps, we report the absolute relative error (AbsRel) and RMSE.

Before computing color metrics, we match the mean and standard deviation of the predicted images to those of the ground truth. This step ensures a fair comparison, as differences in exposure and white balance between training and test images would otherwise distort the metrics. As a result, the reported metrics for all methods are generally higher than those presented in [7,9,12]. For depth evaluation, we apply scale and shift fitting between the predicted and ground-truth depth maps prior to computing the error metrics.

The evaluation focuses on recent approaches, including 3DGS [4], 2DGS [3], DN-Splatter [11], DDP-NeRF [7], SCADE [12], and DäRF [9]. Earlier methods such as NeRF, DS-NeRF, and NerfingMVS have already been outperformed [9, 12] and are therefore omitted. The method by Yuan et al. [14] is not included in the comparison, as its code release does not cover the preprocessing steps such as surface reconstruction and virtual view generation, which are essential for running the method on new datasets.

Table 1 shows the quantitative results on ScanNet [1]. Our method achieves superior performance while being approximately 200× faster than DDP-NeRF, SCADE, and DäRF. For example, training DäRF takes around 36 h, whereas our method requires only 10 minutes. Although Gaussian Splatting (GS) methods [3,4,11] are faster than these NeRF-based baselines, our method remains faster, requires significantly less memory, and achieves better quality. Notably, GS methods perform poorly on this dataset, which includes image degradations such as motion blur.

Qualitative comparisons in Fig. 3 show that our results are generally sharper than those of DDP-NeRF, SCADE, and DäRF. For example, the floor regions appear noticeably sharper. Moreover, the synthesized depth maps are of higher quality. DN-Splatter produces accurate depth maps and sharp images but the visual artifacts are quite noticeable. Our method avoids color artifacts near the door region in the first row, caused by variable exposure and white balance settings.

Table 2 and Fig. 4 present the results on our Azure Kinect dataset. We compare against 2DGS [3] instead of DäRF [9], as the implementation of DäRF is not fully available for datasets beyond ScanNet. The trends are similar to those on ScanNet, with our method outperforming others. The most noticeable differences appear in the top right corner of the image, where SCADE [12], 2DGS [3], and DN-Splatter [11] suffer from significant artifacts.

Table 1. Quantitative results on ScanNet [1]. Our method is around 200x faster than [7,9,12]. Depth metrics (AbsRel and RMSE) are computed only when available.

Method	PSNR ↑	SSIM ↑	LPIPS ↓	AbsRel ↓	RMSE ↓
3DGS [4]	18.24	0.582	0.407	-	-
2DGS [3]	18.93	0.634	0.377	0.130	0.354
DN-Splatter [11]	21.20	0.687	0.304	0.030	**0.120**
DDP-NeRF [7]	21.93	0.727	0.279	0.059	0.199
SCADE [12]	21.92	0.727	0.286	0.062	0.179
DäRF [9]	21.87	**0.738**	0.314	0.070	0.206
FaDeN	**22.95**	**0.738**	**0.258**	**0.028**	0.125

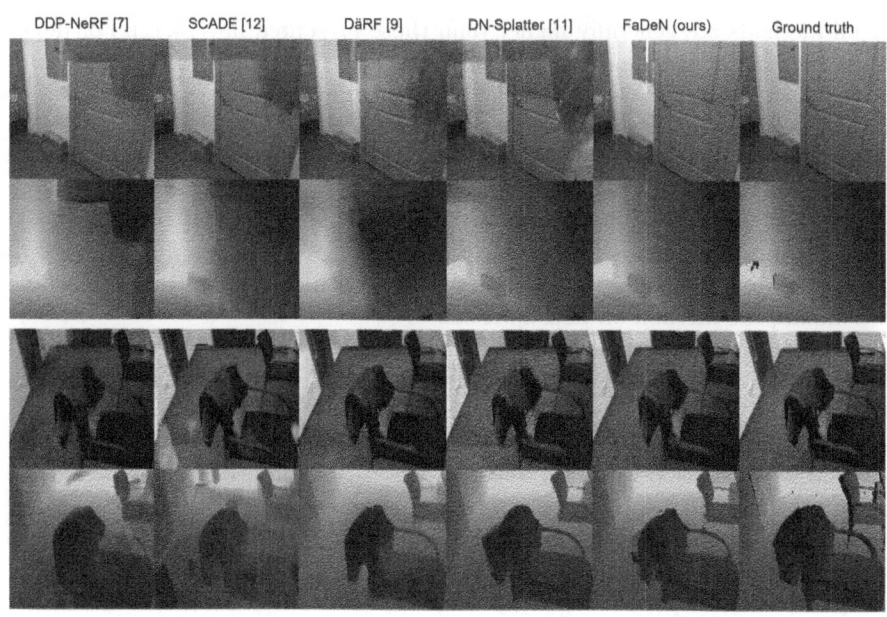

Fig. 3. Qualitative results on ScanNet [1].

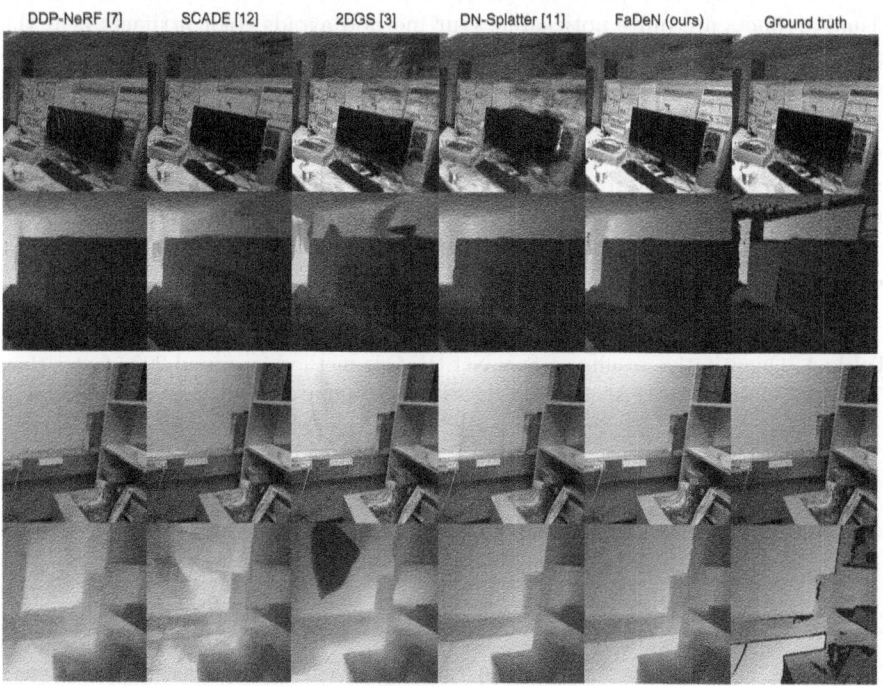

Fig. 4. Qualitative results on our Azure Kinect dataset.

Table 2. Quantitative results on Azure Kinect RGB-D.

Method	PSNR ↑	SSIM ↑	LPIPS ↓	AbsRel ↓	RMSE ↓
3DGS [4]	20.71	0.769	0.275	-	-
2DGS [3]	20.37	0.778	0.279	0.127	0.296
DN-Splatter [11]	21.18	0.790	0.257	0.026	**0.091**
DDP-NeRF [7]	22.67	0.790	0.293	0.045	0.123
SCADE [12]	23.29	0.801	0.275	0.074	0.202
FaDeN	**24.34**	**0.830**	**0.234**	**0.021**	0.101

4.2 Surface Reconstruction

In this experiment, we apply the proposed framework to RGB-D surface reconstruction using our RGB-D data. For each training view, our framework renders both a depth map and an ambient occlusion (AO) map. The AO map is used to filter out unreliable depth values. That is, if a pixel's AO value is below a threshold (0.1), the corresponding depth is discarded. The filtered depth maps are then fused into a triangle mesh using the scalable TSDF fusion method from the Open3D library [15].

Figure 5 (left) shows a mesh reconstructed directly from the original depth maps. Some fine details, such as the objects on the shelf, appear inaccurate and overly smooth. There are also some color artifacts and holes in the mesh. In contrast, the mesh reconstructed from the enhanced depth maps is noticeably more accurate and complete, with finer details preserved and fewer artifacts compared to the one computed directly from the sensor depth.

Fig. 5. Surface reconstruction using Azure Kinect RGB-D camera. The framework enhances sensor depth maps which leads to more accurate and complete reconstructions.

4.3   Ablation Study

We evaluate the components of the proposed framework in Table 3. Enabling depth loss without scale estimation significantly degrades performance. Photometric alignment further improves the results, as also shown in Fig. 6. It effectively removes blending artifacts (seams) caused by variable exposure and white balance. Table 3 also highlights the benefit of synthesizing virtual views and using them during training. Inpainting the holes in the virtual images is beneficial, as shown in Fig. 6, where it removes a black artifact in a region never observed in the input views. These components primarily improve the quality of the novel color images, while their effect on depth accuracy is minimal and in some cases slightly negative. This may be due to the inpainting step, which does not enforce multi-view consistency.

Fig. 6. Ablation study. Instant-NGP [6] is the baseline without depth supervision or any proposed components. We disable one component at a time: depth loss, photometric alignment, or virtual view synthesis. FaDeN represents the complete pipeline.

Table 3. Ablation study on the key components: depth loss (depth), scale estimation (scale), photometric alignment (photo), and virtual view synthesis (virtual).

Components				Color metrics			Depth metrics	
depth	scale	photo	virtual	PSNR ↑	SSIM ↑	LPIPS ↓	AbsRel ↓	RMSE ↓
-	-	-	-	17.68	0.604	0.430	0.225	0.571
✓	-	-	-	13.15	0.436	0.718	0.206	0.532
✓	✓	-	-	22.30	0.733	0.289	**0.026**	**0.116**
✓	✓	✓	-	22.57	0.723	0.280	0.038	0.147
✓	✓	✓	✓	**22.95**	**0.738**	**0.258**	0.028	0.125

5   Conclusion

We proposed FaDeN, a fast and effective framework for novel view synthesis using RGB-D images. The method enables high-quality synthesis from a limited number of input views. It handles key challenges such as scale ambiguity, photometric inconsistencies, noisy depth measurements, and incomplete scene coverage. FaDeN reduces training time by more than 200× compared to similar NeRF-based methods while achieving superior visual quality and geometric accuracy, also outperforming recent Gaussian Splatting methods. In addition to novel view synthesis, the method enables more accurate and complete 3D surface reconstruction.

Acknowledgment. This research was supported by the European Union through the EU Interreg Aurora project IMMERSE (20366448), and the Infotech Project FRAGES.

References

1. Dai, A., Chang, A.X., Savva, M., Halber, M., Funkhouser, T., Nießner, M.: Scan-Net: Richly-annotated 3D reconstructions of indoor scenes. In: IEEE Conference on Computer Vision and Pattern Recognition, pp. 5828–5839 (2017)
2. Deng, K., Liu, A., Zhu, J.Y., Ramanan, D.: Depth-supervised NeRF: Fewer views and faster training for free. In: Proceedings of the IEEE/CVF Conference on Computer Vision and Pattern Recognition, pp. 12882–12891 (2022)
3. Huang, B., Yu, Z., Chen, A., Geiger, A., Gao, S.: 2D Gaussian splatting for geometrically accurate radiance fields. arXiv preprint arXiv:2403.17888 (2024)
4. Kerbl, B., Kopanas, G., Leimkühler, T., Drettakis, G.: 3D Gaussian splatting for real-time radiance field rendering. ACM Trans. Graphics **42**(4) (2023)
5. Mildenhall, B., Srinivasan, P.P., Tancik, M., Barron, J.T., Ramamoorthi, R., Ng, R.: NeRF: representing scenes as neural radiance fields for view synthesis. Commun. ACM **65**(1), 99–106 (2021)
6. Müller, T., Evans, A., Schied, C., Keller, A.: Instant neural graphics primitives with a multiresolution hash encoding. ACM Trans. Graphics **41**(4) (2022)
7. Roessle, B., Barron, J.T., Mildenhall, B., Srinivasan, P.P., Nießner, M.: Dense depth priors for neural radiance fields from sparse input views. In: Proceedings of the IEEE/CVF Conference on Computer Vision and Pattern Recognition (2022)
8. Schönberger, J.L., Frahm, J.M.: Structure-from-motion revisited. In: Conference on Computer Vision and Pattern Recognition (CVPR) (2016)
9. Song, J., et al.: DäRF: Boosting radiance fields from sparse input views with monocular depth adaptation. Adv. Neural Inf. Process. Syst. **36** (2024)
10. Suvorov, R., et al.: Resolution-robust large mask inpainting with fourier convolutions. arXiv preprint arXiv:2109.07161 (2021)
11. Turkulainen, M., Ren, X., Melekhov, I., Seiskari, O., Rahtu, E., Kannala, J.: DN-Splatter: Depth and normal priors for gaussian splatting and meshing. In: IEEE/CVF Winter Conference on Applications of Computer Vision (2025)
12. Uy, M.A., Martin-Brualla, R., Guibas, L., Li, K.: SCADE: Nerfs from space carving with ambiguity-aware depth estimates. In: Proceedings of the IEEE/CVF Conference on Computer Vision and Pattern Recognition, pp. 16518–16527 (2023)

13. Wei, Y., Liu, S., Rao, Y., Zhao, W., Lu, J., Zhou, J.: NerfingMVS: Guided opti-
    mization of neural radiance fields for indoor multi-view stereo. In: IEEE/CVF
    International Conference on Computer Vision, pp. 5610–5619 (2021)
14. Yuan, Y.J., Lai, Y.K., Huang, Y.H., Kobbelt, L., Gao, L.: Neural radiance fields
    from sparse RGB-D images for high-quality view synthesis. IEEE Transa. Patt.
    Analy. Mach. Intell. (2022)
15. Zhou, Q.Y., Park, J., Koltun, V.: Open3D: A modern library for 3D data process-
    ing. arXiv preprint arXiv:1801.09847 (2018)

AnthroFormer3D: Automating 3D Body Measurement Extraction via Vision Transformers Using a Novel Dataset

Mohammad Baksh[1]([⊠]) [iD], Muhammad Mohsin Zafar[2] [iD],
and Kenneth E. Barner[1] [iD]

[1] Department of Electrical and Computing Engineering, University of Delaware,
Newark, DE 19716, USA
{mmbaksh,barner}@udel.edu
[2] MIQYAS Inc., Jeddah 23414, Saudi Arabia
mohsin@miqyas.net

Abstract. Anthropometric measurements are essential for various industries, including healthcare, sports science, and fashion. Traditionally, these measurements are obtained manually or using specialized 3D scanners, methods that are often time-consuming, costly, and inaccessible. Estimating 3D body measurements from 2D images presents a significant challenge due to the ill-posed nature of inferring 3D geometry from 2D projections and the scarcity of large-scale datasets with accurate 3D annotations. To address these challenges, we propose AnthroFormer3D, a deep learning framework that automates the estimation of anthropometric measurements directly from 2D images. Our model leverages Vision Transformers (ViTs) to capture global context and spatial relationships critical for modeling human body shape. Additionally, we introduce a large-scale synthetic dataset comprising over 1.2 million images from 150,000 virtual subjects, designed to represent broad variations in height, weight, body mass index, and natural human poses. A key aspect of our method is the integration of auxiliary height and weight information into the model, enabling it to exploit complex non-linear correlations between these attributes and body shape. We explore early and late fusion strategies for incorporating these features into the ViT architecture. Experimental evaluations show that our models achieve superior accuracy compared to baseline methods, particularly when employing gender-specific training. The key contributions of this work are threefold: (1) the development of a Vision Transformer-based model specifically designed for anthropometric measurement estimation, (2) the creation of a large-scale, publicly available synthetic dataset to support training and evaluation, and (3) the design of an effective framework that integrates auxiliary attributes, such as height and weight, to improve predictive performance. Collectively, these contributions advance automated anthropometry and enable practical applications.

Keywords: Anthropometry · 3D Body Measurements · Vision Transformer · Realistic Synthetic Dataset

M. Castrillón-Santana et al. (Eds.): CAIP 2025, LNCS 15621, pp. 353–363, 2026.
https://doi.org/10.1007/978-3-032-04968-1_30

1   Introduction

Anthropometric measurements, quantitative descriptors of human body dimensions, are integral to numerous domains, including healthcare, ergonomics, apparel design, and e-commerce. In clinical settings, these measurements facilitate the assessment of growth patterns, nutritional status, and the evaluation of physical rehabilitation outcomes [1]. In the fashion and e-commerce industries, precise body dimensions are critical for optimizing garment fit and enhancing the customer experience, contributing to reduced return rates and improved satisfaction [2].

Conventionally, anthropometric data are obtained through manual measurement or specialized 3D scanning technologies. While these methods yield high accuracy, they are often limited by practical constraints: manual techniques are susceptible to observer variability, and 3D scanning systems entail considerable cost, technical expertise, and infrastructure [3,4]. These challenges have motivated efforts to automate measurement processes using computer vision.

Early computational approaches relied on techniques such as edge detection and silhouette analysis but were hindered by variability in pose, lighting, and clothing and the need for manual intervention [5]. The emergence of deep learning (DL), particularly convolutional neural networks (CNNs), has markedly advanced the field by enabling automatic feature extraction from complex visual data [6]. However, CNN-based models typically require extensive labeled datasets for effective training, and the scarcity of large-scale, annotated anthropometric datasets remains a significant barrier [7].

To address these limitations, we introduce AnthroFormer3D, a novel framework that employs Vision Transformers (ViTs) for the automated estimation of anthropometric measurements from 2D images. ViTs offer distinct advantages over traditional CNNs by capturing global contextual information and long-range spatial dependencies, making them particularly well-suited for tasks involving human body geometry [8]. Our method incorporates multi-view image inputs, along with auxiliary information such as height and weight, to enhance the prediction of key 3D body measurements, including waist, hip, and chest circumferences.

A central contribution of this work is the creation of a comprehensive synthetic dataset consisting of 1.2 million images representing 150,000 unique virtual subjects. These subjects exhibit broad variation in body shape, pose, and environmental conditions, thereby providing a robust foundation for training and evaluating measurement models. This dataset addresses the critical shortage of accessible and diverse anthropometric data, supporting the development of more generalizable and scalable solutions.

Through integrating ViT architectures and a rigorously constructed dataset, AnthroFormer3D advances the field of automated anthropometry and offers promising implications for applications in healthcare, apparel design, and personalized digital experiences.

2 Related Works

Recent advances in anthropometric measurement estimation have focused on developing accurate and practical solutions, particularly with the integration of learning-based models and computer vision techniques. However, the lack of diverse and publicly accessible datasets remains a significant challenge, limiting the generalizability of models to real-world scenarios involving varied body types, clothing styles, and background conditions.

2.1 Datasets for Anthropometric Measurement

Progress in automated anthropometric measurement has been closely tied to the availability of reliable datasets. The CAESAR dataset [9] remains a foundational resource, offering high-resolution 3D scans and over 40 traditional anthropometric measurements from a demographically diverse population. However, its limited accessibility and relatively constrained diversity in body types and clothing styles restrict its utility for modern learning-based models.

To address these limitations, synthetic datasets have gained prominence. The SURREAL dataset [10], for instance, provides millions of rendered images of human models with variations in pose, shape, and environment, enabling learning of complex mappings between 2D and 3D representations. Despite their utility, many synthetic datasets lack the precision or scope required for detailed anthropometric analysis.

Parametric 3D body models have also facilitated dataset generation. The SCAPE model [11] pioneered data-driven shape and pose variation modeling, while the more recent SMPL model [12] introduced a gender-aware, skinned linear representation capable of disentangling shape and pose. SMPL's flexibility has made it a preferred choice for generating synthetic bodies and simulating diverse anthropometric characteristics.

Despite recent advances, a significant gap persists: no publicly available dataset offers large-scale, gender-specific, and richly annotated anthropometric measurements across a wide spectrum of body types, poses, and real-world scenarios. This limits the development of robust and generalizable models.

2.2 Methods for Anthropometric Measurement Estimation

Early approaches to anthropometric estimation relied on simple image analysis techniques, which proved inadequate in the face of real-world variability. More recent efforts have leveraged machine learning, combining image features with regression models to estimate parameters like weight and BMI [13–15]. While these studies demonstrate potential, they often treat anthropometric attributes in isolation and lack comprehensive measurement outputs.

Pose estimation, a critical component in body measurement, has evolved from CNN-based methods to Transformer-based architectures. ViTPose [16], for instance, demonstrated the effectiveness of Vision Transformers in capturing

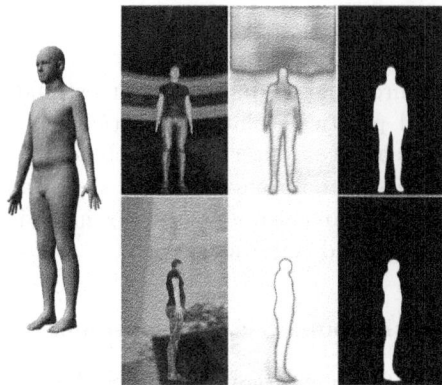

Fig. 1. The data generation pipeline: model generation from SMPL parameters, RGB rendering, confidence estimation, and binary mask extraction.

global spatial relationships for pose estimation, which is foundational for accurate anthropometric inference.

Several studies have explored the direct estimation of body measurements from images. For example, Yan and Kämäräinen [17] proposed predicting multiple circumferences using paired 2D views, while others [18,19] employed silhouette-based methods using SCAPE or SMPL models to infer 3D shape and dimensions. However, these methods often depend on specific camera setups or controlled environments, limiting their general applicability.

More recent work, such as BfSNet [20], integrates the SMPL model with deep networks to predict anthropometric measurements from images. While BfSNet achieved promising results, its reliance on private datasets and mixed-gender training reduced its adaptability and precision, particularly for gender-specific metrics.

In contrast, our proposed method leverages the strengths of ViT architectures and a novel synthetic dataset to address these limitations. By incorporating gender-specific models and integrating auxiliary information such as height and weight, our approach aims to deliver more accurate and generalizable anthropometric estimates.

3 Data Generation Process

Creating the open-source anthropometric measurement dataset involves several stages, combining real-world data, parametric modeling, and photorealistic rendering. This section details the methodology, as illustrated in Fig. 1.

3.1 Acquisition and Preparation of the CAESAR Dataset

The dataset generation begins with the CAESAR dataset [9], which provides high-resolution 3D body scans and detailed anthropometric measurements.

Each scan, consisting of approximately 32,900 vertices, was standardized to the SMPL model topology (6,890 vertices) using Non-Rigid Iterative Closest Point (NRICP) registration [21], ensuring mesh consistency while preserving anatomical accuracy.

3.2   Synthetic Human Model Generation with SMPL

The SMPL model was utilized to generate diverse human body shapes and poses, with shape parameters β defining individual variations. The vertex positions $v \in \mathbb{R}^{3V}$ are expressed as:

$$v = M\beta + \mu, \tag{1}$$

where $M \in \mathbb{R}^{3V \times |\beta|}$ are PCA-derived shape components, μ is the mean shape, and β are the shape coefficients. These coefficients were sampled from a multivariate Gaussian distribution:

$$\beta \sim \mathcal{N}(\mu_\beta, \Sigma_\beta), \tag{2}$$

where μ_β and Σ_β were estimated from the standardized CAESAR meshes. We synthesized 150,000 models using this distribution, equally split between male and female subjects.

3.3   Rendering Synthetic Images

Each synthetic model was rendered using Blender, applying textures from the SURREAL dataset to emulate realistic clothing and skin tones. Two fixed camera views (frontal and side) were used, producing 1.2 million images (600,000 per gender). Lighting, background, and camera parameters were randomized to simulate real-world variability. Each rendering included corresponding segmentation masks and confidence maps.

3.4   Generation of Anthropometric Measurements

Anthropometric measurements were derived using a landmark-based approach aligned with ISO 7250-1:2017 standards. Landmarks from CAESAR were mapped onto SMPL models to ensure anatomical fidelity. Measurements were calculated as Euclidean distances between selected landmark pairs:

$$d_{ij} = \|l_i - l_j\|, \tag{3}$$

where $l_i, l_j \in \mathbb{R}^3$ are the 3D coordinates of the landmarks. This method mirrors standard manual anthropometry, ensuring consistency and interpretability.

3.5   Weight-to-Volume Regression Model

To assign realistic weights to synthetic subjects, mesh volumes V_{mesh} were computed, and weights W were estimated via linear regression:

$$W = \alpha V_{\text{mesh}} + \beta, \tag{4}$$

with α and β derived from CAESAR data. This improved the accuracy and realism of the dataset's anthropometric attributes.

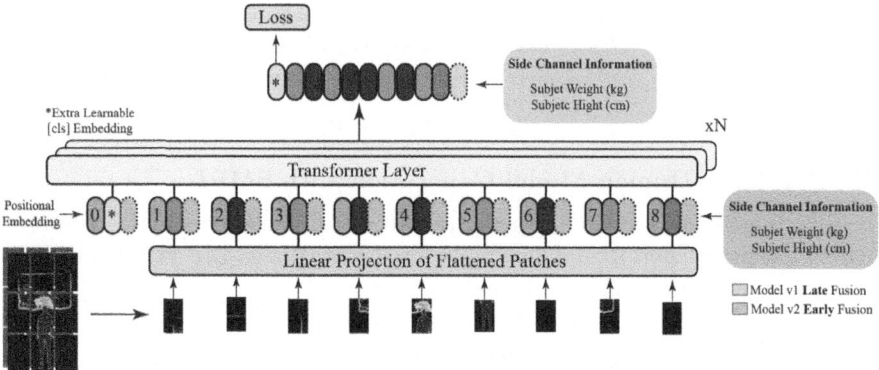

Fig. 2. Overview of the proposed ViT-based framework for anthropometric measurement estimation. Input segmentation masks are divided into patches and projected into embeddings, processed via transformer layers. Height and weight are fused either at the input (early fusion) or output (late fusion).

3.6 Dataset Composition and Accessibility

The final dataset includes diverse body shapes, standard A-poses, and comprehensive annotations, including SMPL parameters, anthropometric measurements, camera details, and segmentation masks. Combining real and synthetic data, it supports training robust, inclusive, and scalable learning-based models. The dataset, released as open-source[1], enables research across computer vision, healthcare, apparel design, and ergonomics. It incorporates gender-specific models and measurements, including female-specific dimensions such as under-bust circumference, ensuring greater precision and inclusivity in anthropometric estimation.

4   Methodology

The proposed AnthroFormer3D framework employs a Vision Transformer (ViT) to estimate anthropometric measurements from 2D segmentation masks, augmented with subject-specific attributes (height and weight). The model processes binary masks and confidence estimates and outputs precise anthropometric measurements. An architectural overview is shown in Fig. 2.

We explore two fusion strategies for integrating auxiliary inputs: Model v1 (Late Fusion) incorporates height and weight at the final regression stage, while Model v2 (Early Fusion) embeds them alongside the image patches at the input. These differing strategies led to distinct architectural designs optimized for each case, summarized in Table 1.

In Model v1, a deeper transformer stack (12 layers, 768 embedding dim) is used to learn high-level visual abstractions before merging auxiliary data. In

---

[1] Dataset is available under request

contrast, Model v2 leverages early conditioning through a lighter configuration (8 layers, 384 embedding dim), which is sufficient due to the early integration of structured input. This divergence is deliberate, reflecting each fusion method's computational and representational needs. Both models terminate in fully connected regression heads to predict anthropometric outcomes.

Table 1. Optimal hyperparameter configurations for Model v1 (Late Fusion) and Model v2 (Early Fusion).

Hyperparameter	Model v1 *(Late fusion)*	Model v2 *(Early fusion)*
Learning Rate	1×10^{-4}	5×10^{-4} (Male), 5×10^{-5} (Female)
Number of Epochs	20	20
Number of Attention Head.	8	8
Feedforward Network Dim.	3072	1536
Embedding Dimension	768	384
Number of Layers	12	8
Optimizer	AdamW	Adam (Male), RAdam (Female)
Loss Function	Huber (Male, Female)	Huber (Male), MSE (Female)

Model v1 is trained using AdamW and Huber loss for both genders. Model v2 employs Adam for males and RAdam for females, using Huber and Mean Squared Error (MSE) losses respectively. The loss functions are defined as:

$$L_{\text{Huber}}(y, \hat{y}) = \begin{cases} \frac{1}{2}(y - \hat{y})^2 & \text{if } |y - \hat{y}| \leq \delta \\ \delta|y - \hat{y}| - \frac{1}{2}\delta^2 & \text{otherwise} \end{cases} \quad (5)$$

while the MSE loss is given by:

$$L_{\text{MSE}} = \frac{1}{n} \sum_{i=1}^{n} (y_i - \hat{y}_i)^2. \quad (6)$$

For benchmarking, we reimplemented BfSNet [20] using its original compound loss and hyperparameters, enabling a direct comparison with our ViT-based models.

5   Experimental Results

The performance of the proposed models, Model v1 (*Late Fusion*) and Model v2 (*Early Fusion*), was evaluated against a re-implemented baseline, BfSNet [20], using male and female datasets. Quantitative results are presented in Tables 2 and 3, showing mean squared errors and standard deviations across key anthropometric measurements.

Table 2. Quantitative comparison of anthropometric measurement estimation methods for male datasets. Column 3 represents our implementation of BfSNet [20]. Columns 1 and 2 show the results of our proposed late and early fusion models, respectively.

Measurement	Model v1 *(Late fusion)*	Model v2 *(Early fusion)*	BfSNet [20] *(Our implem.)*
A. Head circ.	10.22 ± 7.67	**10.19 ± 7.61**	11.69 ± 8.84
B. Neck circ.	**11.68 ± 8.90**	12.74 ± 9.55	20.02 ± 14.42
C. Shoulder to cro.	**11.17 ± 8.39**	16.24 ± 12.27	18.64 ± 12.46
D. Chest circ.	**24.96 ± 19.34**	34.90 ± 26.02	49.07 ± 37.39
E. Waist circ.	**32.47 ± 24.85**	45.36 ± 34.03	58.14 ± 43.92
F. Pelvis circ.	**31.11 ± 24.00**	44.37 ± 32.57	69.18 ± 51.63
G. Wrist circ.	**5.60 ± 4.24**	5.90 ± 4.43	7.24 ± 5.49
H. Bicep circ.	**12.37 ± 9.31**	14.54 ± 11.00	34.97 ± 21.64
I. Forearm circ.	14.18 ± 4.53	**9.78 ± 7.35**	14.98 ± 11.01
J. Arm length	**10.71 ± 8.14**	21.17 ± 16.10	25.35 ± 13.46
K. Inside leg leng.	**12.19 ± 9.19**	32.19 ± 24.25	13.51 ± 10.08
L. Thigh circ.	19.15 ± 14.50	20.49 ± 15.42	**18.64 ± 12.46**
M. Calf circ.	**11.97 ± 9.07**	12.94 ± 9.79	17.07 ± 12.89
N. Ankle circ.	**7.71 ± 5.83**	8.34 ± 6.28	14.43 ± 9.26
O. Acromial heig.	**7.56 ± 5.70**	46.82 ± 35.59	19.57 ± 10.21
P. Shoulder bread.	**9.10 ± 6.85**	9.45 ± 7.07	12.21 ± 9.14
Mean error (mm)	**14.18 ± 4.53**	21.59 ± 7.63	25.82 ± 9.06

Table 3. Quantitative comparison of anthropometric measurement estimation methods for female datasets. Column 3 represents our implementation of BfSNet [20]. Columns 1 and 2 show the results of our proposed late and early fusion models, respectively.

Measurement	Model v1 *(Late fusion)*	Model v2 *(Early fusion)*	BfSNet [20] *(Our implem.)*
A. Head circ.	**10.77 ± 8.33**	13.18 ± 9.61	36.45 ± 20.22
B. Neck circ.	12.71 ± 9.91	**12.12 ± 8.55**	42.90 ± 26.73
C. Shoulder to cro.	**10.78 ± 8.21**	15.62 ± 10.75	57.94 ± 34.01
D. Chest circ.	39.37 ± 33.55	**26.73 ± 18.18**	104.86 ± 75.78
E. Under bust circ.	37.33 ± 31.47	**23.23 ± 15.73**	130.76 ± 81.31
F. Waist circ.	41.80 ± 36.11	**25.75 ± 17.12**	122.46 ± 85.31
G. Pelvis circ.	35.74 ± 32.52	**29.41 ± 19.65**	104.38 ± 75.93
H. Wrist circ.	**11.77 ± 9.57**	12.18 ± 9.68	25.95 ± 14.97
I. Bicep circ.	**13.58 ± 12.47**	14.13 ± 9.82	32.94 ± 24.84
J. Forearm circ.	**9.20 ± 7.42**	10.29 ± 7.32	35.48 ± 20.82
K. Arm length	**10.58 ± 7.96**	14.20 ± 9.14	61.29 ± 37.82
L. Inside leg leng.	**11.62 ± 8.75**	15.74 ± 9.32	63.72 ± 44.38
M. Thigh circ.	22.64 ± 19.12	**19.67 ± 13.11**	45.14 ± 34.31
N. Calf circ.	16.30 ± 12.68	**13.98 ± 9.50**	26.51 ± 19.83
O. Ankle circ.	**8.23 ± 6.24**	9.16 ± 6.56	26.63 ± 16.14
P. Acromial heig.	**6.85 ± 5.17**	26.43 ± 15.74	131.98 ± 88.63
Q. Shoulder bread.	**8.73 ± 6.59**	9.47 ± 6.70	48.18 ± 21.54
Mean error (mm)	18.12 ± 9.07	**17.13 ± 5.94**	64.56 ± 32.58

For the male dataset, Model v1 achieved the lowest overall mean error (14.18 ± 4.53 mm), outperforming both Model v2 (21.59 ± 7.63 mm) and BfSNet (25.82 ± 9.06 mm). Model v1 demonstrated superior accuracy in measurements such as neck, wrist, and acromial height, benefiting from ViT's capacity for long-range spatial modeling. BfSNet performed marginally better in thigh circumference, while Model v2 showed improved performance in head and forearm circumferences, indicating that early integration of height and weight may enhance specific feature extraction.

In contrast, Model v2 showed better results for female subjects, achieving the lowest mean error (17.13 ± 5.94 mm) compared to Model v1 (18.12 ± 9.07 mm) and BfSNet (64.56 ± 32.58 mm). Model v2 outperformed others in measurements such as chest, under-bust, and waist circumferences, where complex body geometry and soft tissue representation are critical. The gender-specific training of our models proved advantageous, particularly over BfSNet's mixed-gender approach, which resulted in significantly higher errors for female-specific metrics. However, further error analysis revealed that both models exhibit larger errors in upper body girths (chest and waist) compared to limb measurements, highlighting the challenge of representing soft tissue variation and posture-induced shape changes.

These findings highlight the strengths of ViT in anthropometric estimation. Model v1 excels for male subjects through effective late-stage feature integration, while Model v2 captures nuanced features necessary for accurate female measurements. Despite these strengths, both models faced challenges with certain upper body measurements, particularly chest and waist circumferences, likely due to limitations in synthetic data diversity and the inherent complexity of soft tissue modeling from 2D images.

It is important to contextualize these results in terms of dataset design. While BfSNet was initially trained on a private, large-scale dataset, we retrained its architecture on our 1.2 million-sample synthetic dataset to ensure a fair comparison under identical training conditions. Therefore, the performance gap observed in our results reflects not a limitation of BfSNet's original method but rather how effectively each architecture generalizes within our standardized experimental setting.

Additionally, BfSNet was designed as a unified model across genders, whereas our framework explicitly trains male and female models separately, allowing for more specialized learning. Furthermore, the ViT backbone in AnthroFormer3D provides stronger global context modeling than traditional CNN-based encoders, enhancing prediction accuracy for both localized and holistic measurements.

In summary, AnthroFormer3D demonstrates marked improvements in accuracy and generalization over existing methods under matched conditions. By combining transformer-based modeling, structured auxiliary input fusion, and gender-specific training, our approach offers a robust, scalable, and anatomically aware solution for automated anthropometric measurement estimation from 2D imagery.

Overall, AnthroFormer3D demonstrates substantial improvements over existing methods, offering more accurate, gender-specific, and generalizable solutions for automated anthropometric measurement.

6   Conclusions

This work presents AnthroFormer3D, a novel Vision Transformer-based framework for automated anthropometric measurement from 2D images. Leveraging a large-scale synthetic dataset and integrating auxiliary height and weight information, the proposed models demonstrate accurate and consistent estimation of key body measurements across diverse subjects.

A key contribution is the development of a comprehensive, publicly available dataset, addressing existing limitations in diversity and accessibility. The inclusion of gender-specific measurements, particularly female-specific metrics such as under-bust circumference, extends the applicability of automated anthropometry to previously underrepresented domains.

Experimental results indicate that the late fusion strategy is more effective for male subjects, while early fusion performs better for female measurements. The superiority of gender-specific models over mixed-gender baselines, such as BfSNet, underscores the importance of tailored approaches in anthropometric estimation.

While AnthroFormer3D significantly advances the field, challenges remain, particularly in estimating complex upper body measurements influenced by soft tissue and posture. Future work will focus on enhancing dataset diversity, refining model architectures, incorporating advanced loss functions and pose priors, and testing performance on real-world images to further improve accuracy and robustness.

References

1. Wells, J.C.K., Fewtrell, M.S.: Measuring body composition. Arch. Dis. Child. **91**(7), 612–617 (2006)
2. Rodríguez-Cano, A.M., Piña-Ramírez, O., Perichart-Perera, O.: Development and validation of anthropometric-based fat-mass prediction equations using air displacement plethysmography in Mexican infants. JHEALTH **4**(1), 1–10 (2024)
3. Hall, J.M., Lobo, M.A., Lieberman, J.S.: Comparison and validation of traditional and 3D scanning anthropometric methods to measure the hand. Proc. Hum. Factors Ergon. Soc. Ann. Meet. **64**, 101–105 (2020)
4. Rumbo-Rodríguez, L., Sánchez-SanSegundo, M., Ferrer-Cascales, R., García-D'Urso, N., Hurtado-Sánchez, J.A., Zaragoza-Martí, A.: Comparison of body scanner and manual anthropometric measurements of body shape: a systematic review. Int. J. Environ. Res. Public Health **18**(12), 6213 (2021)
5. Esparza, A.D., Esparza, A.A., Esparza, J.: Digital anthropometry: a systematic review on precision, reliability and accuracy of most popular existing technologies. Front. Digit. Health **4**, 1–15 (2022)

6. Wang, J., Zhang, T., Cheng, Y., Al-Nabhan, N.: Deep learning for object detection: a survey. Comput. Syst. Sci. Eng. **38**(2), 165–182 (2021)

7. Tuan, H.N.A., Dieu, P.D., Hai, N.D.X., Thinh, N.T., Vinh, L.G.: Anthropometric identification system using convolution neural network based on region proposal network. Journal Tàp Chí Y hôc Vit Nam **506**(1-2) (2021)

8. Jamil, S., Piran, M.J., Kwon, O.-J.: A comprehensive survey of transformers for computer vision. Drones **7**(5), 287 (2023)

9. Robinette, K.M., Daanen, H., Paquet, E.: The CAESAR project: a 3-D surface anthropometry survey. In: Proceedings of Second International Conference on 3-D Digital Imaging and Modeling, pp. 380–386 (1999)

10. Varol, G., et al.: Learning from synthetic humans. In: Proceedings of the IEEE Conference on Computer Vision and Pattern Recognition (CVPR) (2017)

11. Anguelov, D., Srinivasan, P., Koller, D., Thrun, S., Rodgers, J., Davis, J.: SCAPE: shape completion and animation for people. ACM Trans. Graph. **24**(3), 408–416 (2005)

12. Loper, M., Mahmood, N., Romero, J., Pons-Moll, G., Black, M.J.: SMPL: a skinned multi-person linear model. ACM Trans. Graph. (Proc. SIGGRAPH Asia) **34**(6), 248:1–248:16 (2015)

13. Jiang, M., Guo, G.: Body weight analysis from human body images. IEEE Trans. Inf. Forensics Secur. **14**, 2676–2688 (2023)

14. Barrón, C., Kakadiaris, I.A.: Estimating anthropometry and pose from a single uncalibrated image. Comput. Vis. Image Underst. **81**(3), 269–284 (2001)

15. Soneja, R., Prashanth, S., Aarthi, R.: Body weight estimation using 2D body image. Int. J. Adv. Comput. Sci. Appl. **12**(4) (2021)

16. Xu, Y., Zhang, J., Zhang, Q., Tao, D.: ViTPose: Simple Vision Transformer Baselines for Human Pose Estimation, arXiv preprint (2022)

17. Yan, S., Kämäräinen, J.-K.: Learning Anthropometry from Rendered Humans, arXiv preprint (2021)

18. Dibra, E., Jain, H., Oztireli, C., Ziegler, R., Gross, M.: Hs-nets: estimating human body shape from silhouettes with convolutional neural networks. In: 2016 Fourth International Conference on 3D Vision (3DV), pp. 108–117. IEEE (2016)

19. Dibra, E., Jain, H., Oztireli, C., Ziegler, R., Gross, M.: Human shape from silhouettes using generative HKS descriptors and cross-modal neural networks. In: 2017 IEEE Conference on Computer Vision and Pattern Recognition (CVPR), pp. 5504–5514. IEEE (2017)

20. Smith, B.M., Chari, V., Agrawal, A., Rehg, J.M., Sever, R.: Towards accurate 3D human body reconstruction from silhouettes. In: 2019 International Conference on 3D Vision (3DV), pp. 279–288. IEEE (2019)

21. Liang, L., et al.: Nonrigid iterative closest points for registration of 3D biomedical surfaces. Opt. Lasers Eng. **100**, 141–154 (2018)

Sphere-Depth: A Benchmark for Depth Estimation Methods with Varying Spherical Camera Orientations

Soulayma Gazzeh$^{(\boxtimes)}$ (ID), Giuseppe Mazzola (ID), Liliana Lo Presti (ID), and Marco La Cascia (ID)

Department of Engineering, University of Palermo, Palermo, Italy
soulayma.gazzeh@unipa.it

Abstract. Reliable depth estimation from spherical images is crucial for 360° vision in robotic navigation and immersive scene understanding. However, the onboard spherical camera can experience unintentional pose variations in real-world robotic platforms that, along with the geometric distortions inherent in equirectangular projections, significantly impact the effectiveness of depth estimation.

To study this issue, a novel public benchmark, called Sphere-Depth, is introduced to systematically evaluate the robustness of monocular depth estimation models from equirectangular images in a reproducible way. Camera pose perturbations are simulated and used to assess the performance of a popular perspective-based model, Depth Anything, and of spherical-aware models such as Depth Anywhere, ACDNet, Bifuse++, and SliceNet. Furthermore, to ensure meaningful evaluation across models, a depth calibration-based error protocol is proposed to convert predicted relative depth values into metric depth values using supervised learned scaling factors for each model.

Experiments show that even models explicitly designed to process spherical images exhibit substantial performance degradation when variations in the camera pose are observed with respect to the canonical pose. The full benchmark, evaluation protocol, and dataset splits are made publicly available at: https://github.com/sgazzeh/Sphere_depth.

Keywords: Depth calibration · pose variation · Sensitivity · Equirectangular images

1 Introduction

Depth estimation from spherical images has emerged as a key component to achieve immersive 360° scene understanding and autonomous navigation [1,14, 21]. Unlike conventional pinhole cameras, which provide a limited field of view and require multiple frames to capture the complete scene, spherical cameras provide continuous and full scene coverage. This capability is particularly valuable for robotic platforms operating in dynamic, unconstrained environments, such as autonomous vehicles [24], drones [10], and mobile robots.

© The Author(s), under exclusive license to Springer Nature Switzerland AG 2026
M. Castrillón-Santana et al. (Eds.): CAIP 2025, LNCS 15621, pp. 364–374, 2026.
https://doi.org/10.1007/978-3-032-04968-1_31

(a) Planar Image   **(b) Canonical Pose**   **(c) Changing Pose**

Fig. 1. (a) planar image obtained by cubic projection; (b) gravity-aligned image obtained with a canonical camera pose; (c) Equirectangular image with variations in camera pose.

The ability to accurately estimate depth values from spherical images is useful in tasks such as scene understanding, object detection [15], and motion planning [6], all of which are foundational to successful autonomous operation.

Current depth estimation models [2,16,19,20,22,25] operate under distinct assumptions about the camera geometry. *Perspective-based models* [9,12] are typically trained on planar images. Techniques such as cubemap projection [5] or planar conversion [18] are employed to flatten the spherical view into planar images (Fig. 1a) that can be processed by these models. This is needed because the equirectangular projection introduces geometric distortions in images that significantly alter spatial relationships in the scene and affect depth prediction accuracy, especially near the poles.

Recent *Spherical-aware models* extend perspective-based models to handle equirectangular images by adopting distortion-aware sampling [4] or tailored data augmentation techniques [21], by projecting features onto tangent planes to locally approximate perspective projection [8], by dividing the equirectangular image into vertical slices [17] or adopting specific padding techniques [25], by modifying the convolutional operation to consider the spherical geometry [7] or by using latitude-dependent [23] or rectangular-shaped kernels [26]. Additionally, methods [11,19] merge features extracted from cubemap and equirectangular projections, while models [2,13] process image patches, and use geometric priors to merge the patch-based depth maps into a final depth map.

The previous approaches mostly assume that spherical cameras acquire equirectangular images in a canonical pose, i.e., with the equatorial plane parallel to the floor plane. Images acquired under this assumption are called *gravity-aligned images* (Fig. 1b). However, this assumption may not hold in practice. For instance, spherical cameras mounted on mobile platforms can be subjected to uncontrollable pose perturbations (namely, camera orientation variations) due to the platform's motion. These perturbations significantly change the appearance of the acquired equirectangular images (Fig. 1c), and compromise the accuracy of the predicted depth values. The issue has received little attention in the literature, and prevents a full understanding of how well depth estimation models perform in practice.

In this paper, we introduce a novel benchmark designed to systematically evaluate the robustness of popular depth estimation models (especially spherical-aware models) under a range of camera pose variations in real settings. In contrast to existing benchmarks for 360° depth estimation, such as [3], our framework introduces structured pitch and roll rotations to simulate different levels of camera pose variations and measure errors in realistic settings by using known 3D point coordinates of landmarks displaced in the scene.

In this paper, our main contributions are:

- **A pose-aware benchmark** for monocular depth estimation from equirectangular images, designed to evaluate model generalization under camera pose variation with a focus on sensitivity to pitch and roll changes.
- **A depth calibration-based evaluation protocol** that introduces a learnable scaling factor λ to align predicted depths to the ground truth values, allowing fair comparison between models with different output scales.
- **A comparative study** of models trained on either perspective or spherical images, concluding that even the performance of models designed to process spherical images degrades significantly in the presence of camera pose perturbations.

Our results highlight that despite recent advances, current 360° depth estimation models still implicitly rely on cameras in a canonical pose and are not robust to spatial distortions caused by pose variations. This public benchmark provides a reproducible basis for future research on pose- and geometry-invariant depth estimation in omnidirectional settings.

2   Depth Estimation Models Selected for Benchmarking

To assess the state-of-the-art in depth estimation from equirectangular images, we selected a representative set of spherical-aware models based on their architectural diversity, input modalities, and strategies to handle spherical distortions, as summarized in Table 1. We also consider a popular perspective-based model, Depth Anything v2 [22], to evaluate its generalization capabilities to 360° imagery via cubemap projections.

All selected methods have an encoder-decoder architecture. In Depth Anything v2, the encoder is a Vision Transformer (ViT) and the decoder is a Dense Prediction Transformer (DPT) devoted to the regression of depth values. The model does not explicitly handle equirectangular projection (ERP).

BiFuse++ [19] is a two-branch encoder-decoder that geometrically aligns features extracted from the equirectangular image and features from the cube face images, while DepthAnywhere [21] exploits a transformer-based model within a teacher-student framework. The model processes equirectangular images. At training time, it uses Depth Anything as a teacher to independently process each of the cube faces and generate pseudo-labels. It also adopts data augmentation by randomly rotating the spherical images.

Other methods that process only equirectangular images are ACDNet [25] and SliceNet [17]. They both adopt a ResNet-based encoder. In ACDNet, the decoder exploits Adaptively Combined Dilated Convolutional layers (ACDConv)

Table 1. Theoretical comparison of selected depth estimation models.

Model	Full 360° Image?	Architecture	Handling Spherical Geometry
Depth Anything v2 [22]	✗	ViT + DPT	No ERP support; trained on perspective images
DepthAnywhere [21]	✓	Transformer-based teacher-student framework using [22] as teacher on unlabeled data	Depth Anything on unlabeled cube face images; random rotations of spherical images
BiFuse++ [19]	✓	Two-branch encoder + decoder (ResNet-34)	Process both equirectangular and cube face images; geometrical align features from different input
ACDNet [25]	✓	ResNet + ACDConv-based decoder	ACDConv and circular padding mitigate ERP distortions
SliceNet [17]	✓	ResNet-based encoder + multi-scale vertical feature slicing + ConvLSTM-based decoder	Leverages gravity-aligned slices without explicit projections

to obtain adaptive receptive fields. Furthermore, circular padding of the images is used to deal with the image circularity introduced by ERP. SliceNet includes multi-scale vertical feature slicing and a ConvLSTM-based decoder. The model directly processes gravity-aligned images.

2.1   Depth Value Extraction Strategies

The evaluated models differ in how they represent and extract depth information. They either learn relative disparities or generate direct metric depth maps. To maintain consistency, we adhere to the recommended protocols for each model.

Direct Depth Prediction: The models [17,25] generate metric depth maps directly from their respective architectures.

Disparity-Based Estimation: As done in these papers [19,21,22], the depth is estimated using intermediate disparity representations. The raw outputs $\hat{d}$ are converted to depth value d by the following inverse transformation:

$$d = \frac{1}{\alpha\sigma(\hat{d}) + \epsilon}$$

with $\sigma(\cdot)$ the sigmoid activation function, α a model-specific scaling factor, and ϵ a small constant added for numerical stability. All predicted values are clipped to the range $[0, 10]$ meters.

3   Dataset, Depth Calibration and Experimental Results

The dataset includes 8 high-resolution equirectangular images acquired indoors with two Garmin VIRB 360 cameras mounted in various locations and with different orientations. 37 landmarks located in the scene were chosen as reference points, and their 3D coordinates were measured with respect to a World coordinate system. For each image, the dataset includes manually selected pixel coordinates of visible landmarks, the 3D position of the camera, and its orientation (yaw, pitch, roll) relative to the chosen World coordinate system. Thus, it is possible to determine the distance (depth) of each landmark to the camera.

To address scale ambiguity in monocular depth estimation, we estimate a scaling factor λ that aligns predicted depth maps d with our ground-truth depth d_{gt}. To this purpose, we split the landmarks visible in the image in a training and a test set. The training landmarks are used to estimate λ by minimizing the mean squared error in the least squares sense. λ is defined as:

$$\lambda = \frac{\sum_i^M d_i d_{\text{gt},i}}{\sum_i^M d_i^2}$$

where i indexes the M available training landmarks. It is important to note that λ is computed using gravity-aligned images. Gravity-aligned images are computed by aligning the camera's reference axes with the 3D World coordinate system through a 3D rotation of the spherical image. For each image, we measured the mean squared error between the predicted scaled depth value λd and the true depth value d_{gt} over the test landmarks. In our experiments, we report the average error value ε over the processed images.

3.1   Depth Estimation from Equirectangular Images

We evaluate the robustness of spherical-aware depth estimation models on the 8 real images from our dataset and corresponding gravity-aligned images. Real images were split based on the degree of camera orientation variation into 4 *Small deformation* images (with pitch and roll angles in the range $\pm1°$) and 4 *High deformation* images (with pitch and roll angles exceeding $\pm15°$, up to $\pm30°$).

For each model, Table 2 presents the learned scaling depth factor λ, and the error ε measured in the above described scenarios. For all models except DepthAnywhere, the λ value is close to 1, indicating a good alignment of the ranges of the predicted depths with the true value range. DepthAnywhere consistently underestimates depth being $\lambda = 0.56$.

ACDNet performs well on gravity-aligned images, with a measured error of $\varepsilon = 0.21$. Its accuracy decreases as the camera orientation changes. However, on our dataset, ACDNet outperforms all other selected techniques. Indeed, DepthAnywhere, BiFuse++, and SliceNet struggle with large camera orientation variations while achieving comparable errors on gravity-aligned images. Among all the models, SliceNet struggles even with small camera pose changes.

3.2   Depth Estimation from Cube Face Images

In this experiment, we project a gravity-aligned 360° image into six cube faces and apply the *Depth Anything* model to generate depth predictions for each face.

Table 2. Depth error ε (meters) on gravity-aligned and real images

	λ	ε(G. Aligned)	ε(Small Def.)	ε(High Def.)
ACDNet [25]	1.11	**0.21**	**0.23**	**1.38**
DepthAnywhere [21]	0.56	0.26	0.34	2.32
Bifuse++ [19]	0.93	0.38	0.32	2.04
SliceNet [17]	0.92	0.41	0.65	2.09

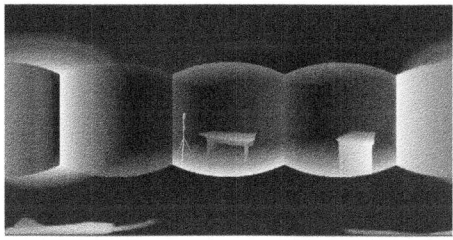

Fig. 2. Equirectangular depth map of the scene obtained by reprojecting six depth maps of the cube faces.

These predicted cubic depth maps are subsequently re-projected into an equirectangular format to reconstruct the spherical depth map of the scene (Fig. 2). As in the previous experiment, the error is calculated on the test landmarks. In this case, the estimated scale factor is $\lambda = 1.38$, which means that, in general, Depth Anything tends to underestimate depth values, probably because it cannot use contextual information about the relationships between the cube faces. The estimated error ε (G. Aligned) = 1.72 and is understandably high considering the erroneously estimated depth values along the face borders. These findings reveal the limitations of perspective models with spherical data, as independently processing cube faces fails to account for the scene's geometry.

3.3   Depth Estimation Under Spherical Camera Pose Variation

To further investigate the robustness of depth estimation models against camera orientation changes, we conducted experiments simulating pitch and roll variations. Controlled rotations were applied to 360° equirectangular images to mimic camera misalignment with the gravity vector. The pitch and roll angles were varied in two sets:

(1) High changes from −40° to +40° in 10° increments.
(2) Small changes from −2° to +2° in 0.5° increments.

This experimental study examines each model's sensitivity to both significant and subtle orientation changes encountered in real-world scenarios using a single image sample.

(a) ACDNet

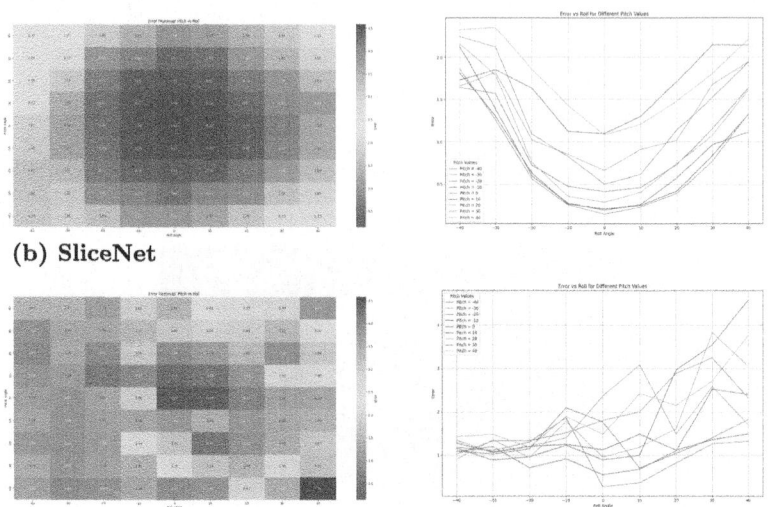

(b) SliceNet

Fig. 3. Error heatmaps (on the left) and error response curves (on the right) for ACD-Net and SliceNet under large pitch and roll angle variations.

Sensitivity to Large Camera Pose Variations. Due to the strong behavior alignment among the models' results, we present ACDNet's result as a representative example in Fig. 3(a). ACDNet, DepthAnyWhere, and BiFuse++ show consistent patterns, with optimal performance near the canonical camera orientation (pitch $= 0°$, roll $= 0°$) and increasing error at the extremes, indicating sensitivity to significant pose variations while managing moderate variation well. In contrast, SliceNet behaves differently (Fig. 3(b)), exhibiting asymmetric patterns with cooler regions in high-variation corners. It maintains consistent performance despite considerable changes in pitch and roll and displays a monotonic increase in error with roll at positive pitch values, deviating from the typical symmetric U-shape. This suggests that SliceNet might be less sensitive to variations, but could be less robust in certain angular configurations. For completeness, qualitative results for ACDNet under no and large camera pose variation (pitch $= -40°$, roll $= 40°$) are provided in Fig. 4.

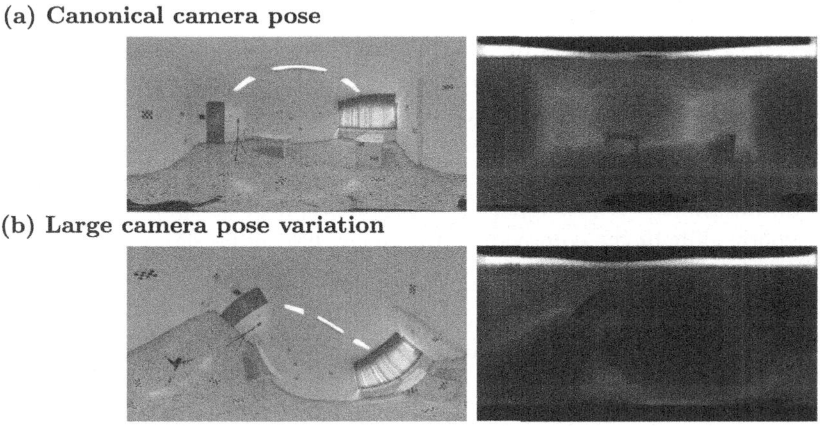

(a) Canonical camera pose

(b) Large camera pose variation

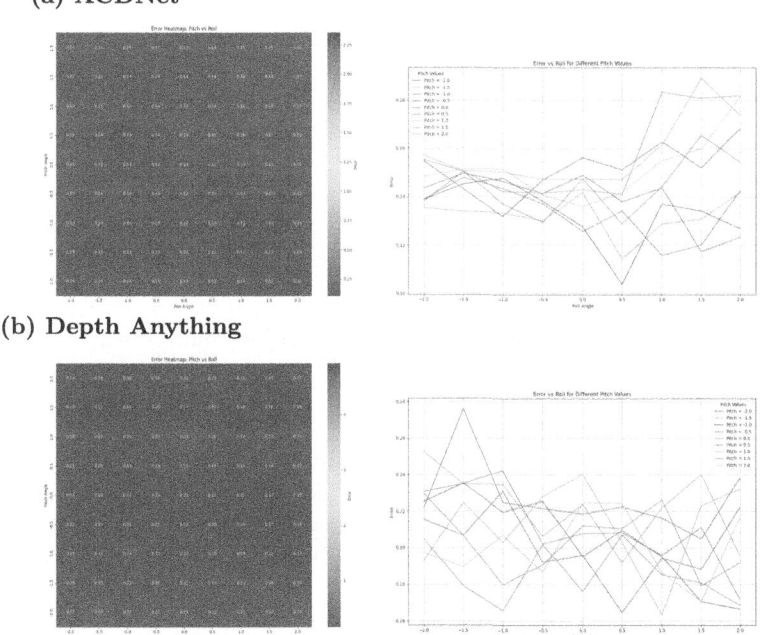

Fig. 4. Qualitative results for ACDNet under conditions of no camera pose variation and significant camera pose variation.

(a) ACDNet

(b) Depth Anything

Fig. 5. Error heatmaps (on the left) and error response curves (on the right) for ACD-Net and Depth Anything under small pitch and roll angle variations.

Sensitivity to Small Camera Pose Variations. To evaluate depth estimation stability under small camera orientation changes, we simulated $\pm 2°$ pitch and roll variations. Figure 5 shows the performance for ACDNet, which previously achieved the best results on gravity-aligned images with an error of $\varepsilon = 0.14$, as well as Depth Anything models. For both models, the error heatmap is mostly uniform and relatively flat, with no strong asymmetry between pitch and roll variations. As the orientation slightly deviates, the error increases slightly but remains low overall. This reflects a small but measurable degradation in performance.

Yaw variations were excluded from the experiment as they primarily cause a horizontal shift in the equirectangular image, having no significant impact on depth estimation.

3.4   Performance Across Depth Ranges

The LOWESS curves were computed for all data points per model to ensure a fair trend estimation, crucial for understanding depth-specific biases. This analysis indicates that while some models perform well for close-range depth recovery, others excel in distant scenes, which is vital for applications like navigation in large indoor spaces.

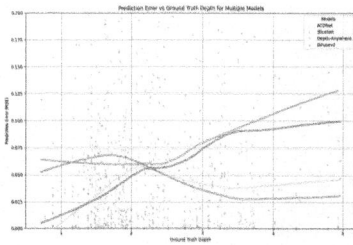

Fig. 6. LOWESS curves describing error trend across equirectangular models.

We analyzed prediction errors across varying ground truth depth values to assess the models' spatial performance consistency. The LOWESS curves in Fig. 6 reveal error trends relative to depth, showing which models are better for nearby or distant structures. ACDNet and Depth-Anywhere maintain stable errors across the depth spectrum, with Depth-Anywhere slightly better in far-range regions (depth > 3 m). SliceNet is effective for close-range (depth < 2 m) but struggles with long-range predictions. BiFuseV2 shows increasing error, particularly beyond 2.5 m, indicating less robustness in deep scenes.

4   Conclusion

We introduced a novel benchmark to evaluate the robustness of monocular depth estimation models on spherical imagery under realistic camera pose variations. By simulating deviations in pitch and roll, the sensitivity of models across a range of perturbation intensities is studied, capturing performance degradation trends from minor to significant orientation changes.

Our depth calibration-aware evaluation protocol utilizes a learnable scale factor to consistently compare models with varying output magnitudes. A multi-axis error analysis reveals that depth accuracy deteriorates with pose variation and at different depth ranges, with many models biased toward canonical viewpoints.

Our results show that even advanced spherical geometry-aware models struggle to generalize beyond their training data and are vulnerable to camera pose distortions. This benchmark creates a practical testing environment, highlighting current limitations and guiding the development of pose-invariant, distortion-aware depth estimation techniques for 360° vision systems.

Acknowledgments. This work was partially supported by European Union – "Next Generation EU" - PNRR Mission 4 "Istruzione e Ricerca" Component 2 "Dalla Ricerca all'Impresa" - Investment line 1.3 Project "Future Artificial Intelligence - FAIR" cod. PE0000013, CUP J53C22003010006 Title: "CAESAR: Self-conscious behaviour in Embodied AI Agents.

References

1. Ai, H., Wang, L.: Elite360d: towards efficient 360 depth estimation via semantic- and distance-aware bi-projection fusion. In: Proceedings of the IEEE/CVF Conference on Computer Vision and Pattern Recognition, pp. 9926–9935 (2024)
2. Ai, H, et al.: HRDFuse: monocular 360deg depth estimation by collaboratively learning holistic-with-regional depth distributions. In: Proceedings of the IEEE/CVF Conference on Computer Vision and Pattern Recognition, pp. 13273–13282 (2023)
3. Albanis, G., et al.: Pano3D: a holistic benchmark and a solid baseline for 360deg depth estimation. In: Proceedings of the IEEE/CVF Conference on Computer Vision and Pattern Recognition, pp. 3727–3737 (2021)
4. Athwale, A., et al.: Darswin: distortion aware radial swin transformer. In: Proceedings of the IEEE/CVF International Conference on Computer Vision, pp. 5929–5938 (2023)
5. Cheng, H.-T., et al.: Cube padding for weakly-supervised saliency prediction in 360 videos. In: Proceedings of the IEEE Conference on Computer Vision and Pattern Recognition, pp. 1420–1429 (2018)
6. Cho, E., et al.: Obstacle avoidance of a UAV using fast monocular depth estimation for a wide stereo camera. IEEE Trans. Ind. Electron. (2024)
7. Cohen, T.S., et al.: Spherical CNNs. arXiv preprint arXiv:1801.10130 (2018)
8. Eder, M., et al.: Tangent images for mitigating spherical distortion. In: Proceedings of the IEEE/CVF Conference on Computer Vision and Pattern Recognition, pp. 12426–12434 (2020)
9. Eigen, D., Puhrsch, C., Fergus, R.: Depth map prediction from a single image using a multi-scale deep network. In: Advances in Neural Information Processing Systems, vol. 27 (2014)
10. Jaisawal, P.K., et al.: Airfisheye dataset: a multi-model fisheye dataset for UAV applications. In: 2024 IEEE International Conference on Robotics and Automation (ICRA), pp. 11818–11824. IEEE (2024)

11. Jiang, H., et al.: Unifuse: unidirectional fusion for 360 panorama depth estimation. IEEE Robot. Autom. Lett. **6**(2), 1519–1526 (2021)
12. Laina, I., et al.: Deeper depth prediction with fully convolutional residual networks. In: 2016 Fourth International Conference on 3D Vision (3DV), pp. 239–248. IEEE (2016)
13. Li, Y., et al.: Omnifusion: 360 monocular depth estimation via geometry-aware fusion. In: Proceedings of the IEEE/CVF Conference on Computer Vision and Pattern Recognition, pp. 2801–2810 (2022)
14. Mohadikar, P., Duan, Y.: Omnidiffusion: reformulating 360 monocular depth estimation using semantic and surface normal conditioned diffusion. In: 2025 IEEE/CVF Winter Conference on Applications of Computer Vision (WACV), pp. 8068–8078. IEEE (2025)
15. Park, C., et al.: Odd-m3d: object-wise dense depth estimation for monocular 3D object detection. IEEE Trans. Consum. Electron. (2024)
16. Patni, S., Agarwal, A., Arora, C.: Ecodepth: effective conditioning of diffusion models for monocular depth estimation. In: Proceedings of the IEEE/CVF Conference on Computer Vision and Pattern Recognition, pp. 28285–28295 (2024)
17. Pintore, G., et al.: Slicenet: deep dense depth estimation from a single indoor panorama using a slice-based representation. In: Proceedings of the IEEE/CVF Conference on Computer Vision and Pattern Recognition, pp. 11536–11545 (2021)
18. Su, Y.-C., Jayaraman, D., Grauman, K.: Pano2Vid: automatic cinematography for watching 360° videos. In: Lai, S.-H., Lepetit, V., Nishino, K., Sato, Y. (eds.) ACCV 2016. LNCS, vol. 10114, pp. 154–171. Springer, Cham (2017). https://doi.org/10.1007/978-3-319-54190-7_10
19. Wang, F.-E., et al.: Bifuse++: self-supervised and efficient bi-projection fusion for 360 depth estimation. IEEE Trans. Pattern Anal. Mach. Intell. **45**(5), 5448–5460 (2022)
20. Wang, H., et al.: PDDepth: pose decoupled monocular depth estimation for roadside perception system. IEEE Trans. Circuits Syst. Video Technol. (2025)
21. Wang, N.-H.A., Liu, Y.-L.: Depth anywhere: enhancing 360 monocular depth estimation via perspective distillation and unlabeled data augmentation. In: Advances in Neural Information Processing Systems, vol. 37, pp. 127739–127764 (2024)
22. Yang, L., et al.: Depth anything v2. In: Advances in Neural Information Processing Systems, vol. 37, pp. 21875–21911 (2024)
23. Yoon, Y., et al.: Spheresr: 360deg image super-resolution with arbitrary projection via continuous spherical image representation. In: Proceedings of the IEEE/CVF Conference on Computer Vision and Pattern Recognition, pp. 5677–5686 (2022)
24. Zheng, J., et al.: Physical 3D adversarial attacks against monocular depth estimation in autonomous driving. In: Proceedings of the IEEE/CVF Conference on Computer Vision and Pattern Recognition, pp. 24452–24461 (2024)
25. Zhuang, C., et al.: ACDNet: adaptively combined dilated convolution for monocular panorama depth estimation. In: Proceedings of the AAAI Conference on Artificial Intelligence, vol. 36, pp. 3653–3661 (2022)
26. Zioulis, N., et al.: Omnidepth: dense depth estimation for indoors spherical panoramas. In: Proceedings of the European Conference on Computer Vision (ECCV), pp. 448–465 (2018)

Image Valuation in NeRF-Based 3D Reconstruction

Grigorios-Aris Cheimariotis$^{(\boxtimes)}$(ID), Antonis Karakottas(ID), Vangelis Chatzis(ID),
Angelos Kanlis(ID), and Dimitrios Zarpalas(ID)

Information Technologies Institute (ITI), Centre for Research and Technology Hellas
(CERTH), Thessaloniki, Greece
{acheimar,ankarako,chatzise,a.kanlis,zarpalas}@iti.gr

Abstract. Data valuation and monetization are becoming increasingly
important across domains such as eXtended Reality (XR) and digital
media. In the context of 3D scene reconstruction from a set of images
-whether casually or professionally captured - not all inputs contribute
equally to the final output. Neural Radiance Fields (NeRFs) enable photorealistic 3D reconstruction of scenes by optimizing a volumetric radiance field given a set of images. However, in-the-wild scenes often include
image captures of varying quality, occlusions, and transient objects,
resulting in uneven utility across inputs. In this paper we propose a
method to quantify the individual contribution of each image to NeRF-
based reconstructions of in-the-wild image sets. Contribution is assessed
through reconstruction quality metrics based on PSNR and MSE. We validate our approach by removing low-contributing images during training
and measuring the resulting impact on reconstruction fidelity.

Keywords: Data valuation · 3D reconstruction · Neural Radiance
Fields

1 Introduction

3D scene reconstruction has broad applications, including virtual tourism [5],
cultural heritage preservation [10], or healthcare [12]. Recent advances enable
photorealistic novel view synthesis from 2D image collections. Among the most
prominent methods are Neural Radiance Fields (NeRFs) [11] and 3D Gaussian
Splatting [7], both of which learn scene representations directly from images.
The Phototourism dataset [6] offers a compelling use case, enabling the reconstruction of explorable 3D scenes from in-the-wild photo collections. However, it
also introduces challenges such as variable image quality, transient objects, and
appearance inconsistencies. Techniques like NeRF-W [9] and HA-NeRF [2] have
demonstrated effectiveness in addressing these issues.

The original version of the chapter has been revised. The author name is corrected. A
correction to this chapter can be found at
https://doi.org/10.1007/978-3-032-04968-1_34

However, an image with transient objects, differences in appearance and/or low quality can negatively impact reconstruction quality and it is preferable to be excluded from training. In NeRF-W, images undergo a coarse pre-filtering step based on two heuristic criteria. Firstly, using Neural Image Assessment (NIMA) [15]—a model producing continuous scores (1 to 10) based on human-rated aesthetic judgments—images with a NIMA score below 3 are omitted. Secondly, images are filtered out where transient objects, detected by a DeepLab v3 [1] model, occupy more than 80% of the image's area. These methods result in a binary classification (keep/discard) applied prior to NeRF training, without explicit validation of an image's actual impact on the NeRF's reconstruction performance during training. Therefore, it cannot measure the potential impact of a training image in the reconstructed scene. Therefore, in NeRF-W, images are heuristically classified as either helpful or harmful; however, this categorization is not explicitly validated - images deemed harmful may still contribute positively to training, and vice versa.

This study aims to assign continuous contribution scores to individual images based on their impact on the final NeRF-based scene reconstruction. The motivation stems from two key data valuation objectives: (1) enabling fair compensation to different data providers in data marketplaces, and (2) improving the performance of data-driven systems. Reconstruction accuracy is evaluated using Peak Signal-to-Noise Ratio (PSNR) computed on a set of held-out test images for each scene.

Numerous studies have explored data valuation methods in the context of machine and deep learning, reflecting growing interest in the field [3, 4, 16]. In this work, we draw inspiration from Data Valuation using Reinforcement Learning (DVRL) [16], particularly its ability to estimate sample importance without retraining the model - an appealing property given the computational cost of NeRF training. Unlike, DVRL, which uses a learned policy to adjust training dynamics, our approach repurposes the reward signal directly as a continuous contribution score for each image in the NeRF pipeline.

Rather than proposing a new fundamental data valuation method, this work explores the applicability and adaptation of existing data valuation principles (specifically DVRL) to the unique demands of NeRF training and 3D rendering. Our main contributions lie in this adaptation, the comprehensive examination of various metrics, and the evaluation of their robustness and consistency. Additionally, we achieve the efficient extraction of explicit 'contribution' scores that are fast to obtain (especially compared to fundamental data valuation methods applied to NeRFs) and whose empirical utility is demonstrated by our results.

2   Related Work

2.1   Neural Radiance Fields in the Wild

Neural Radiance Fields (NeRFs) [11] have become a powerful approach for synthesizing photorealistic views of scenes from sparse input images by modelling a scene as a continuous 5D function parametrized by a neural network, NeRFs deliver impressive results in controlled environments. However, their performance often deteriorates in real-wold scenarios due to challenges such as transient occluders, variable lighting, weather conditions, and scene complexity. Several variants have been developed to address these issues. Ha-NeRF [2] improves scene representation through enhanced hierarchical sampling, while NeRF-W [9] introduces a mechanism to separate static and transient components, allowing more robust reconstructions from in-the-wild-photo collections. Additionally, NeRF On-the-go [14] leverages pre-trained DINOv2 [13] features to predict per-pixel uncertainty and remove transient objects from neighboring rays via cosine similarity in feature space.

2.2   Data Valuation

Traditional image data valuation methods include Data Shapley [3], which computes the marginal contribution of each through game-theoretic Shapley values but is computationally prohibitive for large datasets, influence functions, which estimate the effect of removing an image on a model's loss but rely on unstable Hessiangradient approximations and local convexity assumptions. These approaches suffer from high computational overhead, limited scalability, and static valuation that cannot adapt during training. In contrast, Data Valuation using Reinforcement Learning (DVRL) frames valuation as a sequential decision process: a neural data-value estimator learns sampling probabilities via policy gradients, guided by rewards from a small validation set. DVRL avoids costly second-order calculations, scales efficiently through multinomial sampling, and adaptively updates values online to reflect evolving model and data characteristics.

For Neural Radiance Fields (NeRFs), which rely on multi-view image consistency to reconstruct 3D scenes, the quality and relevance of each image play a crucial role in performance. In real-world scenes, transient objects such as people, vehicles, or lighting variations can degrade the reconstruction quality. Recent approaches, such as NeRF-W, incorporate image valuation mechanisms to mitigate this by leveraging a learned weighting strategy during training. Complementary techniques include the use of pre-trained Neural Image Assessment (NIMA) models to predict aesthetic or quality-related properties, helping identify and exclude images with low utility, particularly those containing significant transient content. This combination of learned valuation and perceptual quality assessment enables more robust training pipelines, especially for unconstrained datasets, such as Phototourism [6].

3   Method

The proposed valuation scheme is compatible with any NeRF variant that samples rays from a single image per training iteration. At each iteration, we evaluate the PSNR on a fixed test set and associate it with the image sampled at the moment I^n. For subsequent appearances of I^n, we compute the change in PSNR relative to its previous PSNR value. These PSNR deltas are then aggregated across iterations to quantify each image's overall influence on reconstruction

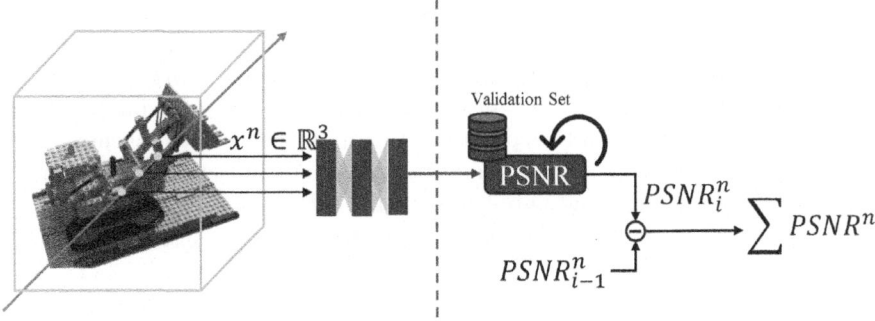

Fig. 1. Method overview: During NeRF training, we evaluate a small validation set after each training iteration i, recording the PSNR associated with image I^n. When image I^n is revisited, we compute the change in PSNR by subtracting the previous value from the current one. These PSNR differences are then aggregated to estimate each image's DV_{psnr}.

quality, which we refer to as DV_{psnr} (Fig. 1). In the following sections we first describe our training setup (Sect. 3.1). In Sect. 3.2 we describe our approach to identify fair metrics for valuating each image's contribution score. In Sect. 3.3 we present the metric utilized and, finally, in Sect. 4 we present our experimental setup and results.

3.1   Training Configuration

Our NeRF training set-up follows a conventional structure. Each training step samples rays from a single image. All training images are used once per epoch, and the order of image sampling is shuffled at the beginning of each epoch. This configuration is consistent with standard NeRF training practices and allows for systematic tracking of each image's influence on validation set performance. Impact is measured as the change in test $L1$ loss and PSNR. This design also offers two key advantages for image valuation: first, the per-image sampling enables attribution of performance changes to specific inputs; second, the randomized order across epochs helps mitigate ordering bias and supports robust aggregation of impact scores. Together, these factors make the training process suitable for analysing contribution of data in dynamic, real-world datasets.

3.2   Identifying Fair Contribution Scores

Since the model's parameters evolve throughout training, the impact of a given image depends on the model's current state. To ensure a fair comparison across images, we consider three complementary strategies: (a) Weighting the contribution of each image by training progress, for instance, assigning higher importance to updates made in later training stages. (b) Aggregating impact estimates across epochs, to account for randomness in image sampling order. (c) Reverting the model to its pre-update state before measuring the effect of each training image.

Option (a), while intuitively appealing, was not adopted in this study due to the lack of a principled framework to define the weighting function. Any choice of weight over training iterations, such as linear, exponential, or heuristic decay, would introduce arbitrary assumptions that are not grounded in theoretical or empirical validation. Moreover, weighting alone does not fully account for the interaction between the image and the model state, which is itself non-linearly dependent on training history.

Option (b) was selected as the main approach due to its practicality and empirical stability. It involves aggregating the impact of each image across all epochs, excluding the first epoch to avoid the outsized influence of early updates when the model is not trained. The resulting metrics - DV_{psnr} and DV_{loss} - are computed as the cumulative change in validation PSNR and L1 loss, respectively, each time an image is used. Although random image ordering across epochs does not fully eliminate model-state bias, we observed that the model quickly converges to a stable performance baseline (e.g., validation PSNR $\approx$ 14) after the first epoch. This suggests that, in later epochs, each image is processed under a comparably trained model, mitigating unfairness introduced by early training dynamics.

Option (c) was considered as a more controlled alternative, where the model would be reverted to a consistent state before each image-specific update. However, this approach was found to be computationally intensive and unstable in practice. In particular, measurements taken in the final epoch under this scheme exhibited high variance across runs with different random seeds, indicating that local updates can lead to unpredictable validation effects. This instability undermines the reliability of impact estimates and limits their utility for consistent image valuation.

To assess the reproducibility of image contribution scores, we identified and measured key source of variability - namely, the random ordering of training images and the random sampling of pixels within image. These are influenced by the random seed used during training. Correlation analyses across runs with different seeds revealed that reproducibility improves with more epochs and larger per-image sampling, but this trend held consistently only for one metric. Notably, the aggregated PSNR difference (DV_{psnr}) demonstrated a satisfactory correlation when 500 pixel-rays per image were used in each iteration.

3.3   Contribution Scores

At each training step, the L1 loss and PSNR were computed on a fixed validation set to calculate contribution scores for individual images. The L1 loss is a commonly used loss function in regression tasks and also drives NeRF training. It measures the absolute difference between the predicted values and the ground truth values.

$$\mathcal{L}_{L1} = \sum_{i=1}^{N} |I_i - \hat{I}_i| \tag{1}$$

where N is the number of data points, I_i is the original image, and $\hat{I}_i$ is the NeRF rendered one. L1 loss is considered more robust to outliers than Mean Squared Error (MSE), as it penalizes deviations linearly rather than quadratically, thereby limiting the influence of large errors.

PSNR is calculated using the MSE as follows:

$$\text{MSE} = \frac{1}{MN} \sum_{i=1}^{M} \sum_{j=1}^{N} \left(I_{i,j} - \hat{K}_{i,j} \right)^2 \tag{2}$$

where $I_{i,j}$ is the original image, $\hat{I}_{i,j}$ is the reconstructed image, and (M, N) are the dimensions of the image. The PSNR in decibels is then given by:

$$\text{PSNR} = 10 \cdot \log_{10} \left(\frac{L^2}{\text{MSE}} \right) \tag{3}$$

where L is the maximum possible pixel value of the image (e.g., 255 for 8-bit unsigned integer images).

PSNR Change and Aggregation for a Training Image. The instantaneous PSNR difference in our experiments was calculated with:

$$\Delta\text{PSNR}_i^{(t)} = \text{PSNR}(I_i; \theta^{(t+1)}) - \text{PSNR}(I_i; \theta^{(t)}) \tag{4}$$

where $\text{PSNR}^{(t)}$ is the calculated PSNR over the validation set at training step t. I_i is the i-th training image, $\theta^{(t)}$ are the NeRF MLP parameters before training on I_i, $\theta^{(t+1)}$ are the NeRF MLP parameters after training on I_i, and $\Delta\text{PSNR}_i^{(t)}$ is the change in PSNR due to training on I_i at step t. This quantifies the immediate effect of the training image I_i on test PSNR. The $\Delta\text{PSNR}_i^{(t)}$ values were aggregated for each image for all epochs, excluding the first. The outcome contribution score is referred to as DV_{psnr}.

4   Experiments

Dataset. We evaluated our method using 4 scenes from the PhotoTourism dataset [6], which contains internet-sourced photo collections of famous landmarks with significant viewpoint, transiency and appearance variations. These scene present challenges such as inconsistent lighting, occlusions from crowds, and diverse camera parameters, making them ideal benchmark for assessing robustness in unconstrained, in-the-wild conditions.

We focused on the Brandenburg Gate scene to evaluate the consistency of the proposed contribution score. This scene consists of 1363 images. NeRF-W [9] reserved 10 for testing, 763 for training, 96 for validation, and the rest are excluded from the process due to the automatic assessment which they proposed [15]. In addition, we also measured contribution scores for three more scenes (Sacre Coeur, Taj Mahal, Trevi Fountain) to assess the impact of training NeRF with images highly valued by our proposed method.

Reproducibility Evaluation with Correlation. A primary objective of this study was to develop contribution metrics that are minimally affected by the image loading sequence. As shown in Fig. 2, the DV_{psnr} scores exhibit strong consistency across runs, achieving a correlation coefficient of 0.8 for the Brandenburg scene. In each training step, PSNR was measured in a small validation set (10–14 images). For the Brandenburg scene, all images except the test images and validation images were valuated, totalling 1310 images. For reproducibility evaluation, for the other three scenes, a subset of 100 images was valuated twice with different seeds to measure consistency, and the correlation coefficients were 0.90 (Sacre Coeur), 0.87 (Trevi fountain), and 0.86 (Taj Mahal). However, to measure the impact of the training composition, all images except the test and validation images, were valuated once for all the scenes.

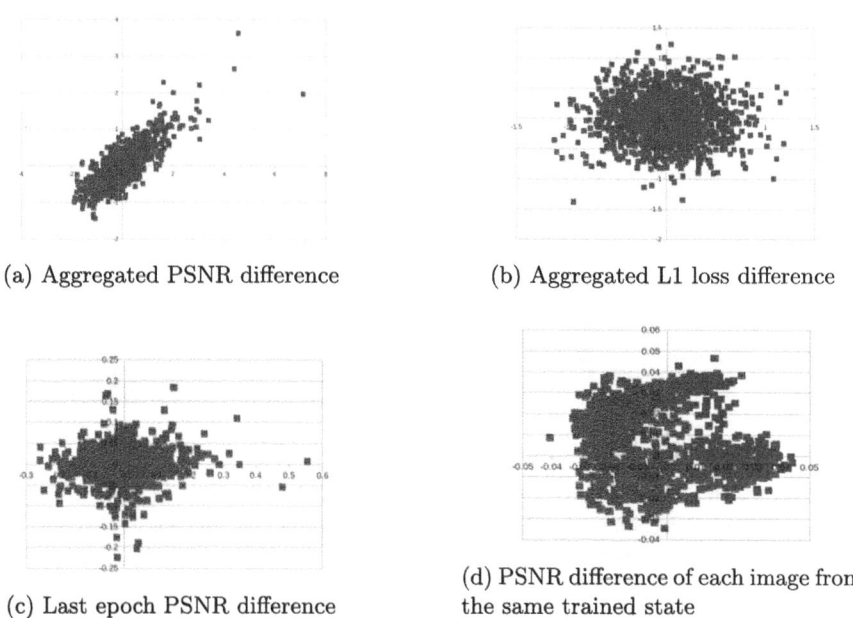

(a) Aggregated PSNR difference

(b) Aggregated L1 loss difference

(c) Last epoch PSNR difference

(d) PSNR difference of each image from the same trained state

Fig. 2. Contribution score correlations on the Brandenburg dataset. The x-axis shows contribution scores from one training run, and the y-axis from a different run with a different seed. Each (x, y) point compares scores for the same image.

Impact of Training Set Composition. Another primary objective was to assess the impact of using different training subsets. To expedite the computation of the contribution scores, PSNR was measured at each training iteration. For the Brandenburg scene, based on the calculated scores, 716 images were identified as having a positive impact on improving PSNR in the small validation set. To evaluate the effect of these selections, two different training sets were compared using 43 held-out test images, which were not part of the training or the

Table 1. PSNR achieved for the same pipeline with NeRF-W training set and with training set which was selected by the proposed data valuation.

PSNR	Brandenburg		Sacre Coeur		Taj Mahal		Trevi fountain	
	val	test	val	test	val	test	val	test
NeRF-W	19.27	**17.72**	16.94	15.62	17.93	16.22	17.51	17.33
DV_{psnr}	**19.96**	16.79	**17.45**	**16.23**	**18.33**	**16.55**	**17.65**	**17.50**

validation sets used for computing contribution scores. The training set selected by the proposed method achieved higher PSNR in a 10-image validation set but lower in a larger 43-image test set that included finer details, as demonstrated in Table 1.

Also, in Table 1, it is observed that, in the other 3 scenes, the training sets selected by our data valuation framework consistently result in higher PSNR on the test sets compared to the training set selected in NeRF-W. In these 3 scenes, validation and test sets are of almost equal size (Sacre Coeur 11:11, Taj Mahal 14:13, Trevi fountain 10:9) and present a more balanced and representative distribution of images.

(a) Test image rendered by NeRF trained with 716 images selected by DV_{psnr}.

(b) Test image rendered by NeRF trained with 763 images (NeRF-W selection).

(c) Test image rendered by NeRF trained with 1717 images selected by DV_{psnr}.

(d) Test image rendered by NeRF trained with 1716 images(NeRF-W selection).

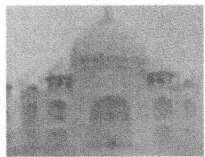

(e) Test image rendered by NeRF trained with 954 images selected by DV_{psnr}.

(f) Test image rendered by NeRF trained with 950 images(NeRF-W selection).

(g) Test image rendered by NeRF trained with 899 images selected by DV_{psnr}.

(h) Test image rendered by NeRF trained with 896 images(NeRF-W selection).

Fig. 3. Rendered images from NeRF trained on DV_{psnr} selections (a),(c),(e),(g) and on NeRF-W training sets (b),(d),(f),(h). PSNR = (a) 19.51, (b) 17.59, (c) 21.3, (d) 20.7,(e) 17.31, (f) 17.03, (g) 24.38, (h) 22.9.

In addition, for test views taken from more distant vantage points, the DV_{psnr}-selected training set yields better performance, as illustrated in Fig. 3(a–b). The model trained on the proposed dataset more accurately reconstructs the statue on top of the Brandenburg Gate and achieves a higher PSNR (19.51). In the example test images of the Trevi Fountain scene and Sacre Coeur scene (Fig. 3(c–f)), it is noticed that some details mainly on the bottom side and on the left side are sharper in the images rendered by the NeRF trained on DV_{psnr} selection. In the example test image of Taj Mahal (Fig. 3(g–h)), there are no obvious differences. However, the image rendered by the NeRF trained on DV_{psnr} selection achieved higher PSNR. Although the data valuation framework may not be fully optimal, it demonstrates a reasonable ability to discriminate between useful and less useful training images.

As shown in Fig. 4, the proposed data valuation framework identified an image included in NeRF-W's training set - although it contains numerous transient objects - as potentially harmful. In contrast, an image excluded from the NeRF-W training, validation, and test sets was rated highly valuable by our method. Each image presents both strengths and limitations: image (a) is brighter and may reveal more fine detail in unoccluded regions, while image (b) is darker with some less distinct areas but exhibits significantly fewer occlusions. These differences in valuation may also be influenced by the image downscaling used during analysis.

(a) Image included in the NeRF-W (b) Image excluded from NeRF-W training set, is valuated as "harmful" training-scored a high contribution for training NeRF by DV_{psnr} DV_{psnr} score.

Fig. 4. NeRF-W NIMA valuation and DV_{psnr} valuation.

5   Discussion

In this paper, we present a method for image valuation in NeRF-based 3D reconstruction. The proposed approach yields meaningful results, demonstrated both quantitatively and qualitatively, by showing that training on a subset of positively scored images can improve reconstruction performance in certain cases.

A primary focus of the study was the reproducibility of the contribution scores. We evaluated scores based on PSNR for their consistency across runs with different random seeds. Only aggregated PSNR-based scores (DV_{psnr}) exhibited a strong correlation between training runs.

Crucially, the practical value of the contribution scores was assessed by their effectiveness in guiding the selection of training sets. The DV_{psnr}-selected dataset slightly outperformed the NeRF-W [9] training set and achieved better results for distant viewpoints. This is due to image downscaling during valuation. It is arguable that high-performing NeRF would deal differently with full-resolution images. However, scores derived from downscaling were shown to be meaningful at double the valuation resolution, producing consistent PSNR improvements (Table 1). From this, we could hypothesize that data valuation scores may be transferable to other resolutions.

One limitation of this study is its reliance on two versions of an experimental NeRF pipeline, which do not reach state-of-the-art performance. For efficiency, we used NeRFs inspired by HA-NeRF [2] and NeRF-W [9], which reduce training time, but struggle with accuracy. However, the proposed valuation method is designed to be model-agnostic and could be integrated into any NeRF architecture. If the constraint of using only one image in each step of contribution scores' extraction is met, the method independently computes explicit image contribution scores.

Due to memory and time constraints, incorporating data valuation into high-end pipelines remains a challenge, as contribution scoring adds computational overhead. On an NVIDIA GeForce RTX 3060 (12 GB), a standard training step takes 0.185 s, with DV_{psnr}'s contribution scoring adding approximately 0.5 s per step. For a typical 50,000-iteration run, this is about 9.5 h of total training time. Compared to data valuation methods like Data Shapley, which demand multiple re-runs, our approach offers a considerably faster alternative. Although these timings are hardware-dependent and can be faster with more powerful resources, we believe this is a manageable cost for the continuous, explicit image valuation provided. Our future research aims to further optimize this by directly inferring contribution scores from a learnable model.

This work focused exclusively on the PhotoTourism dataset [6], given its rich variation in image quality and its frequent use in NeRF evaluation. Future work should explore the application of DV_{psnr} to other datasets, NeRF variants, as well as more recent 3D Gaussian Splatting techniques such as [8], to further validate its generalizability.

Acknowledgments. This work was supported by the European Union's Horizon Europe programme under grant number 101070250 <<XRECO: XR media eCOsystem>> (https://xreco.eu/). The computational resources were granted with the support of GRNET.

References

1. Chen, L., Papandreou, G., Schroff, F., Adam, H.: Rethinking atrous convolution for semantic segmentation (2017)
2. Chen, X., et al.: Hallucinated neural radiance fields in the wild. In: Proceedings of the IEEE/CVF Conference on Computer Vision and Pattern Recognition, pp. 12943–12952 (2022)
3. Ghorbani, A., Zou, J.: Data shapley: equitable valuation of data for machine learning. In: International Conference on Machine Learning, pp. 2242–2251. PMLR (2019)
4. Guo, H., Rajani, N.F., Hase, P., Bansal, M., Xiong, C.: Fastif: scalable influence functions for efficient model interpretation and debugging. arXiv preprint arXiv:2012.15781 (2020)
5. Helmy, M., et al.: Navigating the world with an intelligent tourist guide using generative AI. In: 2024 International Telecommunications Conference (ITC-Egypt), pp. 1–6. IEEE (2024)
6. Jin, Y., et al.: Image matching across wide baselines: from paper to practice. Int. J. Comput. Vision **129**(2), 517–547 (2021)
7. Kerbl, B., Kopanas, G., Leimkühler, T., Drettakis, G.: 3D gaussian splatting for real-time radiance field rendering. ACM Trans. Graph. **42**(4), 139-1 (2023)
8. Kulhanek, J., Peng, S., Kukelova, Z., Pollefeys, M., Sattler, T.: WildGaussians: 3D gaussian splatting in the wild. In: NeurIPS (2024)
9. Martin-Brualla, R., Radwan, N., Sajjadi, M.S., Barron, J.T., Dosovitskiy, A., Duckworth, D.: Nerf in the wild: neural radiance fields for unconstrained photo collections. In: Proceedings of the IEEE/CVF Conference on Computer Vision and Pattern Recognition, pp. 7210–7219 (2021)
10. Mazzacca, G., et al.: Nerf for heritage 3D reconstruction. Int. Arch. Photogram. Remote Sens. Spatial Inf. Sci. **48**(M-2-2023), 1051–1058 (2023)
11. Mildenhall, B., Srinivasan, P.P., Tancik, M., Barron, J.T., Ramamoorthi, R., Ng, R.: Nerf: representing scenes as neural radiance fields for view synthesis. Commun. ACM **65**(1), 99–106 (2021)
12. Molaei, A., et al.: Implicit neural representation in medical imaging: a comparative survey. In: Proceedings of the IEEE/CVF International Conference on Computer Vision, pp. 2381–2391 (2023)
13. Oquab, M., et al.: Dinov2: learning robust visual features without supervision. arXiv preprint arXiv:2304.07193 (2023)
14. Ren, W., Zhu, Z., Sun, B., Chen, J., Pollefeys, M., Peng, S.: Nerf on-the-go: exploiting uncertainty for distractor-free nerfs in the wild. In: Proceedings of the IEEE/CVF Conference on Computer Vision and Pattern Recognition, pp. 8931–8940 (2024)
15. Talebi, H., Milanfar, P.: Nima: neural image assessment. IEEE Trans. Image Process. **27**(8), 3998–4011 (2018)
16. Yoon, J., Arik, S., Pfister, T.: Data valuation using reinforcement learning. In: International Conference on Machine Learning, pp. 10842–10851. PMLR (2020)

VolE^{++}: A Text-Guided Point-Cloud Framework for Food 3D Reconstruction and Volume Estimation

Umair Haroon[1]([✉]) [iD], Ahmad AlMughrabi[1] [iD], Ricardo Marques[2] [iD], and Petia Radeva[1,3] [iD]

[1] Universitat de Barcelona, Barcelona, Spain
umairharoon@ub.edu
[2] Universitat Pompeu Fabra, Grup de Tecnologies Interactives (GTI), Barcelona, Spain
[3] Institut de Neurociències, Universitat de Barcelona, Barcelona, Spain

Abstract. Accurate food volume estimation is crucial for health monitoring, medical nutrition management, and food intake applications. Current 3D food volume estimation methods are too generic, missing the context of the estimated objects, and thus their performance is suboptimal. We present VolE^{++}, a framework designed to achieve food objects' 3D reconstruction and volume estimation. This approach enables users to specify a target food item through text input, allowing for precise segmentation of specific food objects in a real-world environment. Once segmented, the object is reconstructed using the VolE 3D reconstruction framework. This process uses Multi-View Stereo techniques to transform a point cloud into a refined mesh, ensuring high spatial fidelity for accurate 3D volume estimation. Extensive evaluations of the FoodKit and MetaFood3D datasets demonstrate the effectiveness of our method in isolating and reconstructing food items, with improvements across multiple datasets achieving a 0.2% MAPE, highlighting its superior performance in food volume estimation.

Keywords: Text-Guided Food Selection · 3D Reconstruction · Food Volume Estimation · Real-World Scale · Training-Free

1 Introduction

Accurate dietary assessment is crucial for understanding health trends and individual nutritional needs. While significant progress has been made in food segmentation and recognition through advancements in machine learning and computer vision, accurately estimating food volume remains a difficult challenge. This difficulty arises primarily due to the inherent difficulty in recovering accurate 3D data from 2D images, which is a fundamental requirement for reliable volume estimation. Additionally, in real-world scenarios, food items are often arranged in complex ways, making it challenging to isolate and analyse specific

M. Castrillón-Santana et al. (Eds.): CAIP 2025, LNCS 15621, pp. 386–397, 2026.
https://doi.org/10.1007/978-3-032-04968-1_33

objects. This situation emphasises the need for methods that enable users to guide the 3D reconstruction and volume estimation processes.

Traditional dietary assessment methods are often costly and prone to human error, leading to a growing interest in image-based solutions for automating food segmentation, recognition, and volume estimation [20]. Perhaps recent view synthesis methods like Neural Radiance Fields (NeRF) [16] and Gaussian splatting [14] allow recovering the 3D geometric information of a real-world scene from 2D images. Still, they require accurate camera poses from Structure from Motion (SfM) [19] tools, and cannot determine real-world scale without prior information [13]. Further difficulties arise from low-quality images and varied backgrounds, impacting reconstruction accuracy. Existing volume estimation techniques often rely on specialized and expensive hardware [8] or predefined models [24], struggling to accommodate diverse food shapes. While supervised learning can yield high accuracy, it often requires extensive annotated data and fixed environments, limiting practicality [9]. Many existing methods also lack user control for focusing on specific objects in complex scenes, highlighting the need for a framework that allows for guided 3D reconstruction and accurate volume estimation. To address these limitations, we introduce VolE^{++}, an advanced VolE framework designed for accurate volume estimation in challenging free-motion and multi-object environments, which also allows for text-guided control. Building on the original VolE pipeline for 3D reconstruction [12], VolE^{++} produces dense 3D point clouds using images and camera locations from the FoodKit dataset, captured with AR-enabled mobile devices. A key advancement is the integration of text-guided, decoupled video segmentation, which improves user control over identifying and analyzing individual food items. By utilizing Decoupled Video Segmentation (DEVA) [7], VolE^{++} enhances object tracking and refines segmentations. This user-controlled framework overcomes existing limitations, enabling text-guided 3D reconstruction for more consistent outcomes. Our key contributions include (Fig. 1):

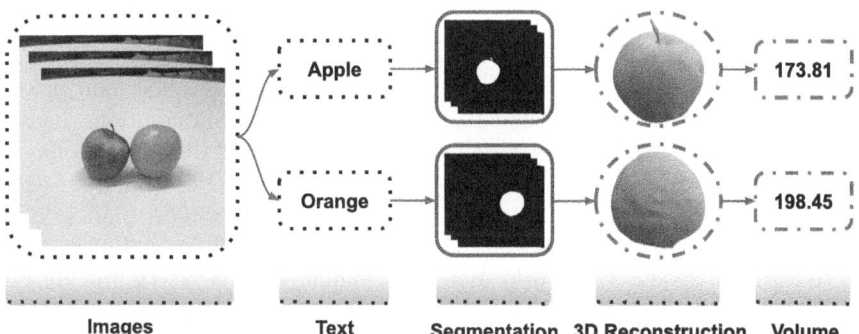

Images   **Text**   **Segmentation** **3D Reconstruction** **Volume**

Fig. 1. Visualization of the processing pipeline: Given 2D input images and a text prompt (e.g., "apple"/"orange"), our framework performs text-guided segmentation to isolate food items and reconstruct them in 3D for volume estimation.

– We propose an innovative approach that allows users to accurately identify and isolate specific food items within a scene using simple text prompts. This represents the first exploration of user-prompted guidance in food volume estimation, enabling text-guided segmentation of the desired food object for subsequent accurate volume measurement.
– Building upon the VolE pipeline, we leverage its advanced capability for generating scaled 3D reconstruction of semantically selected objects and estimating the selected object volume, ensuring spatial fidelity and consistency.
– We conduct extensive experiments using the challenging MTF dataset [13] and our FoodKit dataset. Our evaluations thoroughly assess the framework's performance regarding segmentation quality, 3D reconstruction accuracy, and overall volume estimation precision in real-world food scenarios.

The remainder of this paper is structured as follows: Sect. 2 offers a review of related work, while Sect. 3 describes the proposed method in detail. In Sect. 4, we present the experimental results, and Sect. 5 summarizes our contributions and outlines future research directions.

2   Related Work

Recent advancements in accurate food volume estimation have been driven by improvements in 3D reconstruction using visual data. Techniques like SfM and COLMAP [19] have laid the groundwork for this progress by reconstructing 3D scene geometry from multiple images. Innovations like NeRF [16] and related implicit neural representations have further transformed 3D vision, creating dense volumetric representations from sparse 2D images and often relying on SfM for camera pose estimations. Despite enhancements in speed and quality from methods like InstantNGP [17] and NeuS/2 [22,23], challenges like scale ambiguity and diverse food shapes and textures still hinder accurate volume estimation.

Advanced 3D reconstruction techniques for estimating food volume have become more popular, particularly within initiatives like the MetaFood CVPR challenge [13]. A leading approach, VolETA [2], combines SfM with neural surface reconstruction methods, such as NeuS/2 [23], to produce detailed food meshes and accurate volume measurements. Other significant contributions include ININ-VIAUN [13], which integrates deep learning with Multi-View Stereo for depth information, and FoodRiddle [13], which employs 3D Gaussian splatting to address challenges related to data scarcity and complex food shapes. However, many methods, including VolETA, rely on reference objects, limiting their effectiveness in diverse real-world scenarios. While some datasets provide pre-computed object masks, practical applications require on-the-fly mask generation. Models like FoodMem [1] enable near real-time food segmentation in videos, but integrating this segmentation into 3D reconstruction for text-guided control remains an active research area. Additionally, some methods are constrained by the need for specialized hardware [8] or fixed environments [21], limiting their broader applicability.

Recognizing the limitations of existing methods, the original VolE [12] framework was introduced as a novel solution for improving 3D reconstruction and volume estimation in food scenes without needing a reference object or depth sensors. Unlike previous approaches that often relied on reference objects [2], depth sensors, or specialized hardware [13], VolE [12] leverages standard mobile device capabilities through technologies like ARCore [10] and ARKit [3]. This allows for capturing real-world measurements using device location and IMU data during video recording. This innovative approach effectively addresses scale ambiguity and enhances adaptability to various food shapes by integrating advanced segmentation techniques with robust 3D reconstruction methods.

Building on the solid foundation of VolE [12], VolE^{++} addresses a significant gap in existing reconstruction pipelines: the lack of explicit user control over the segmentation process. While earlier methods often rely on automated segmentation, real-world food volume estimation frequently involves scenes with multiple food items, which necessitates selective segmentation for accurate volume calculations. To address this issue, VolE^{++} integrates DEVA [7], which enhances segmentation accuracy through a bi-directional propagation mechanism that refines masks over time, reducing inter-frame inconsistencies. This integration enables precise user-specified food segmentation using textual prompts, enhancing volume estimation accuracy. Thus, VolE^{++} offers improved control and accuracy for dietary assessment tools.

3   Our Proposal: VolE^{++}

Our VolE^{++} framework introduces a robust, text-guided approach to estimating food volumes by combining advanced 3D reconstruction with text-driven video segmentation. This section outlines our pipeline, focusing on parameter extraction, text-guided segmentation, point cloud masking, 3D mesh reconstruction, mesh refinement, and volume estimation.

3.1   Overview

Our VolE^{++} framework consists of three phases: (a) Parameters Extraction, (b) 3D Mesh Reconstruction, and (c) Volume Estimation. Initially, we process input image sequences and their associated camera 3D Coordinates to refine camera poses and, crucially, employ a text-guided segmentation module to isolate the desired food object. The segmented data and refined poses feed into the 3D Mesh Reconstruction phase, generating a detailed mesh of the selected food item. In the final stage of Volume Estimation, we calculate the volume of the reconstructed 3D mesh.

3.2   Parameters Extraction

Accurate 3D reconstruction fundamentally relies on precise camera parameters and robust object segmentation. Given a sequence of input images, $\mathcal{I} =$

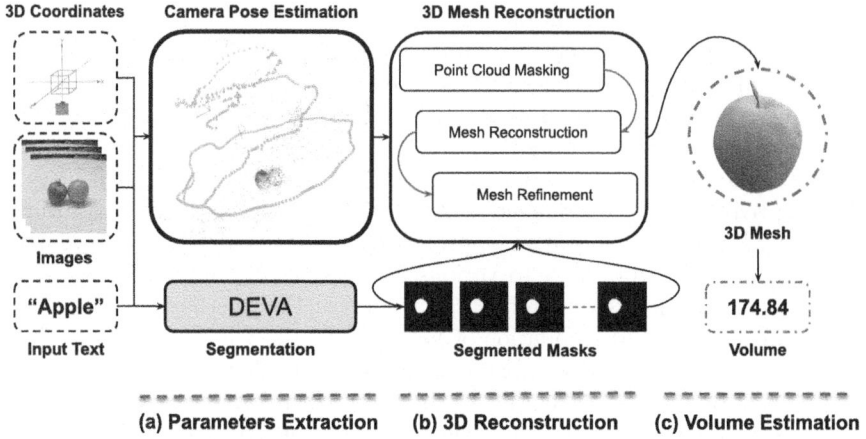

Fig. 2. Overview of the VolE^{++} Framework: (a) Parameter Extraction: Camera poses are refined with 3D coordinates via COLMAP, while DEVA generates segmentation masks for food objects based on prompts (e.g., "Apple"/"Orange"). (b) 3D Reconstruction: Refined poses and segmented images create a dense 3D point cloud, which is then converted into a refined mesh. (c) Volume Estimation: The volume of the food mesh is computed through tetrahedral decomposition.

$\{I_i | i = 1 \ldots N_I\}$, we first refine the camera intrinsics and extrinsics, denoted as $\mathcal{C} = \{C_i | i = 1 \ldots N_I\}$. Unlike previous approaches relying solely on image data, VolE^{++} leverages AR captured 3D coordinates data and COLMAP [19], a robust SfM pipeline, to enhance these initial camera poses. COLMAP refines poses by performing feature extraction (e.g., SIFT [15]), feature matching across views, and geometric verification to ensure spatial consistency. In parallel to pose estimation, the semantic isolation of specific food items is a key feature for an accurate volume estimation. Here, VolE^{++} leverages DEVA [7], a robust framework that excels in video object segmentation. DEVA's architecture combines an image segmentation model (trained for task-specific hypotheses at the frame level) with a universal temporal propagation model (developed with class-agnostic mask propagation datasets). This decoupled design allows DEVA to generalize effectively, even in scenarios with limited labeled training data. Crucially for VolE^{++}, we adapt DEVA to enable text-guided object segmentation via text prompts. The user provides a textual label, such as "apple" or "orange", to specify the target food object. This input guides DEVA to produce a set of precise segmentation masks $\mathcal{S} = S_1, S_2, \ldots, S_T$, where T is the total number of frames in the sequence. These masks are then applied to the input images, isolating the desired objects from the background. This crucial step ensures that only the relevant regions corresponding to the preselected objects contribute to the subsequent 3D reconstruction process, as illustrated in Fig. 2(a).

3.3   3D Mesh Reconstruction

We reconstruct the 3D food mesh using Point Cloud Masking with segmented images and refined camera poses. We utilize the point cloud from COLMAP and masks from DEVA, projecting each 3D point P_i onto the 2D masks. The camera poses are represented as $mathcalC$, and the point cloud as P. We compute the image coordinates with the intrinsic matrix K and retain 3D points within the masks, defining valid points as $M_j = \{P_i \mid p_{ij} \in \mathcal{S}_j\}$. M_j contains valid points P_i projected into the jth segmented image area $\mathcal{S}_j$. The final segmented point cloud, $\mathcal{P}$, is obtained by the intersection of the valid points across all images: $\mathcal{P} = \bigcap_{j=1}^{N_I} M_j$, focusing on the object of interest discarding background noise.

After point cloud masking, we perform Mesh Reconstruction using a Multi-View Stereo (MVS) approach [11]. The filtered point cloud $\mathcal{P}$ is transformed into an initial tetrahedral mesh $\mathcal{T}$ through Delaunay triangulation, represented as $\mathcal{T} = f_D(\mathcal{P})$ [6]. A graph-cut optimization then labels each tetrahedron as inside or outside the object, denoted as $\mathcal{L} = f_G(\mathcal{T})$. Finally, the marching cubes algorithm extracts the mesh surface, yielding an accurate representation of the object's geometry, $\mathcal{M} = \mathbf{M}(\mathcal{T}, \mathcal{L})$. The mesh refinement process enhances the quality of the reconstructed mesh, improving accuracy and surface representation while eliminating artifacts. The process involves several steps: mesh simplification reduces vertices for efficiency, followed by mesh smoothing using techniques like Laplacian or bilateral filtering to create a more uniform surface. Additional denoising removes remaining noise through filtering, and finally, mesh optimization, which includes vertex relaxation and edge flipping, refines triangle quality. This produces a clean and precise food mesh, denoted as $\hat{\mathcal{M}}$, ensuring optimal results as shown in Fig. 2(b).

3.4   Volume Estimation

After refining the 3D mesh of the food item, the final step is accurately determining its volume, which relies on AR-generated 3D coordinates scaled to real-world dimensions. To calculate the volume of our closed triangular mesh $\hat{\mathcal{M}}$ with N faces, we employ the divergence theorem [18], connecting volume and surface integrals. We compute the internal volume by summing the signed volumes of tiny tetrahedra. Each tetrahedron is formed by the three vertices of a mesh triangle v_1^k, v_2^k, v_3^k, and a common origin point as its fourth vertex. The formula used is: $V = \frac{1}{6}\sum_{k=1}^{N}(v_1^k \cdot (v_2^k \times v_3^k))$, where the scalar triple product $(v_2^k \times v_3^k)$ gives the signed volume of the parallelepiped formed by vectors, with a factor of 1/6 representing the volume of the tetrahedron [5]. By summing the signed tetrahedron volumes for all triangle faces, we can efficiently calculate the total volume of the 3D food object in a single pass.

4   Experimental Results

This section evaluates VolE^{++} performance in object volume estimation and 3D reconstruction. We conducted experiments on two datasets to assess its accuracy, robustness, and text-guided control. $VolE^{++}$ is compared with SOA methods, including those evaluated on the MTF [13] and FoodKit [12] dataset. Our analysis includes quantitative metrics and qualitative visual assessments to provide a comprehensive evaluation of $VolE^{++}$ effectiveness in real-world scenarios.

4.1   Implementation Settings

All $VolE^{++}$ experiments were conducted on a system with an NVIDIA GeForce RTX 3090 GPU (24 GB). We configured our 3D reconstruction pipeline for the FoodKit and MTF datasets with a point cloud masking "max-resolution" of 512. Mesh reconstruction used a "close-holes" setting of 50 and a "smooth" factor of 5 to achieve accurate surface regularization.

4.2   Evaluation Protocol

To evaluate the accuracy of our framework for volume estimation, we use the Mean Average Percentage Error (MAPE) metric. MAPE is calculated as follows: $MAPE = \frac{1}{n} \sum_{i=1}^{n} \left| \frac{V_{est,i} - V_{gt,i}}{V_{gt,i}} \right| \times 100\%$, where $V_{est,i}$ represents the estimated volume, $V_{gt,i}$ denotes the ground truth volume, and n is the total number of objects. For 3D reconstruction evaluation, we use the Chamfer distance [4]. This metric measures the average closest distance between points in the reconstructed model and the real ground truth model, and vice versa. This 2-way assessment provides a reliable assessment for assessing 3D reconstruction quality.

4.3   Datasets

We evaluate $VolE^{++}$ on 2 datasets: the FoodKit [12] and the MTF dataset [13].

Foodkit Dataset. [12] is essential for evaluating food volume estimation frameworks in real-world conditions. It overcomes the limitations of existing datasets by providing diverse food items in free-motion scenarios, allowing accurate volume estimation with standard smartphone cameras. The dataset includes 21 food items with ground truth measurements for volume and mass, serving as a benchmark for estimation techniques. Data was collected using a combination of augmented reality (AR) and traditional 3D reconstruction, ensuring precise real-world alignment. Ground truth validation used the water displacement method (± 5 mL error margin) and digital scale mass measurements. FoodKit includes video scenes, image sets, ARKit-estimated 3D coordinates, image masks, and associated metadata, making it a valuable resource for advancing food volume estimation techniques.

MTF Dataset. [13] is a key benchmark for food volume estimation, consisting of 20 food scenes categorized into three difficulty tiers: "easy" with eight scenes (around 200 images each), "medium" with 7 scenes (approximately 30 images each), and "hard" with single-image scenes. Each image includes food masks and depth. For our evaluation with VolE^{++}, we focused on the easy and medium scenes, as our framework needs multiple input images for 3D reconstruction. We used the reference board from the MTF dataset to ensure accurate scaling of the reconstructed scenes to their original physical dimensions.

4.4 Comparative Analysis

We evaluated VolE^{++} against various state-of-the-art methods for volume estimation and 3D reconstruction. This comparative analysis focuses on relevant, recent, and reproducible techniques, enabling a clear assessment of VolE^{++}'s strengths and weaknesses.

Qualitative Results. Visual comparison of 3D reconstructions shows that VolE^{++} consistently outperforms other methods in detail and geometric accuracy. Compared to VolE [12] on the FoodKit dataset, we observe comparable or slightly improved performance. The framework effectively handles variations in shape, size, and surface texture, demonstrating adaptability to real-world food characteristics, as shown in Fig. 3. A comparison with VolETA [2] and VolE on the MTF dataset further highlights VolE^{++}'s superior ability to capture delicate geometries for more accurate 3D representations, as shown in Fig. 4.

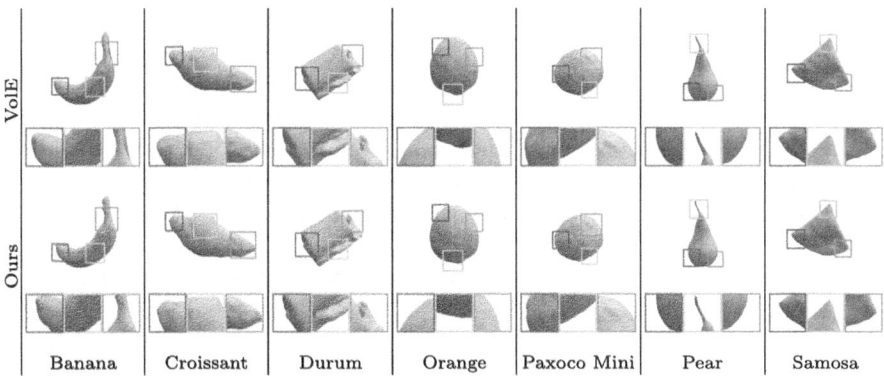

Fig. 3. Qualitative comparison to VolE [12] framework on the FoodKit.

Quantitative Results. We evaluated volume estimation using MAPE and 3D reconstruction accuracy with CD. The results indicate that VolE^{++} performs competitively, achieving impressive MAPE values on both FoodKit and MTF datasets, often surpassing existing methods. An analysis of the FoodKit dataset demonstrates its high accuracy in estimating food volumes compared to the VolE

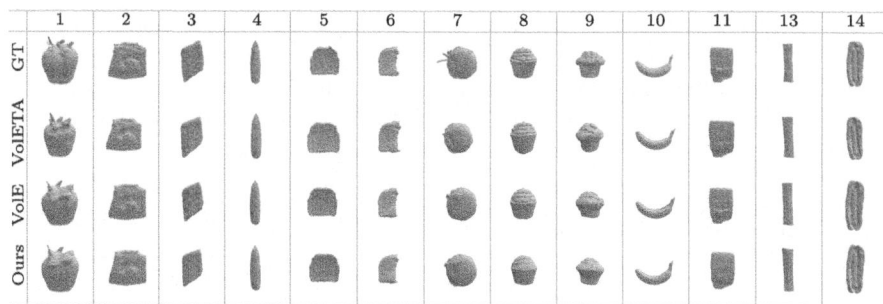

Fig. 4. Our framework 3D reconstruction Visual Results on MTF dataset in comparison with VolETA [2] and ground truth (GT).

[12] framework, as shown in Table 1. VolE^{++} also shows strong performance against VolETA [2], ININ [13], FoodR. [13], and VolE [12] consistently yielding superior or comparable volume estimation (lower MAPE) and 3D reconstruction quality (lower CD) as shown in Table 2. Overall, VolE^{++} is a robust contender in food volume estimation and 3D reconstruction.

4.5 Discussions

The experimental results demonstrate that VolE^{++} effectively estimates object volumes in complex scenarios, handling intricate geometries and diverse textures even in unbounded scenes. Its innovative use of AR-capable devices for initial spatial understanding, particularly showcased with the FoodKit dataset, enhances accuracy. Coupled with a novel text-guided segmentation method using DEVA, VolE^{++} allows for precise identification of target food items, streamlining reconstruction and improving reliability. It significantly outperforms SOA methods in volume estimation accuracy, as indicated by lower MAPE and CD. These results were confirmed across multiple datasets, highlighting VolE^{++}'s versatility in accurate volume estimation, especially in dietary assessments.

4.6 Limitations

The VolE^{++} framework demonstrates promising results; however, it still faces several limitations. One significant issue is the high computational demand associated with camera pose estimation and Multi-View stereo reconstruction, which can hinder real-time processing speeds. While our framework represents a substantial advancement toward practical applications, achieving true real-time performance, particularly with more complex datasets, remains a challenge.

Table 1. Comparison of volume estimation metrics between VoIE [12] and ours for various food items, including estimated volumes, absolute errors, and accuracies.

Items	Images	GT (±5)	Estimated Volume (5x)				Absolute Error (5x)				Accuracy ↑	
			Mean		Std Dev. ↓		Mean ↓		Std Dev. ↓			
			VoIE	Ours	VoIE	Ours	VoIE	Ours	VoIE	Ours	VoIE	Ours
Apple	1005	175	176.68	173.81	0.93	**0.76**	0.96	**0.68**	0.53	**0.43**	99.04	**99.32**
Orange	1001	200	201.06	198.45	3.45	**0.99**	1.35	**0.78**	1.02	**0.49**	98.65	**99.22**
Aguacate	1078	85	83.11	83.31	**1.11**	0.77	2.23	**1.99**	1.30	**0.90**	97.77	**98.01**
Lemon	887	140	134.74	134.14	3.19	**0.98**	1.76	**0.75**	1.33	**0.58**	98.24	**99.25**
Donut	780	245	242.24	242.64	2.57	**1.32**	1.13	**0.96**	1.04	**0.54**	98.87	**99.04**
Durum	1006	200	200.79	198.19	**0.47**	1.20	**0.40**	0.90	**0.24**	0.60	**99.60**	99.10
Pear	849	170	168.13	168.53	**0.71**	0.89	1.10	**0.86**	**0.42**	0.52	98.90	**99.14**
Choc. Cake	781	195	195.38	193.18	1.21	**0.90**	**0.38**	0.93	0.50	**0.46**	**99.62**	99.07
Choc. Croissant	1122	275	274.99	271.99	3.69	**1.26**	**0.95**	1.10	0.82	**0.46**	**99.05**	98.90
Samosa	848	145	144.10	143.10	2.76	**1.11**	1.53	**1.31**	1.10	**0.76**	98.47	**98.69**
Apple Pie	1201	135	135.52	133.32	**1.02**	1.19	**0.51**	1.24	**0.66**	0.88	**99.49**	98.76
Choc. Bomb	1111	200	197.65	197.05	4.39	**1.84**	2.08	**1.47**	1.06	**0.92**	97.92	**98.53**
Empanadilla	926	95	94.86	93.66	**1.12**	1.40	**0.89**	1.73	**0.65**	0.97	**99.11**	98.27
Falafel	929	48	47.58	46.98	2.49	**0.35**	3.96	**2.12**	2.87	**0.72**	96.04	**97.88**
French Bread	1139	163	162.49	161.49	1.54	**1.03**	**0.78**	0.93	**0.50**	0.63	**99.22**	99.07
Paxoco Mini	911	150	148.08	148.68	**1.62**	2.05	1.40	**1.26**	**0.89**	0.92	98.60	**98.74**
Napolitanas	1071	233	232.79	231.99	1.28	**0.71**	**0.41**	0.43	**0.32**	**0.31**	**99.59**	99.57
Capsicum	881	320	318.64	318.44	2.54	**0.74**	0.76	**0.49**	0.35	**0.23**	99.24	**99.51**
Choc. Panettone	1209	293	290.79	290.99	1.74	**1.19**	0.75	**0.69**	0.59	**0.41**	99.25	**99.31**
Banana	1156	150	153.03	152.63	1.74	**0.80**	2.02	**1.76**	1.16	**0.53**	97.98	**98.24**
Yellow Cane	715	350	350.07	349.47	1.45	**0.81**	0.30	**0.23**	0.24	**0.13**	99.70	**99.77**
Mean		**188.90**	188.23	**187.24**	1.95	**1.06**	1.22	**1.08**	0.84	**0.59**	98.78	**98.92**

Table 2. Comparison of volume estimation and 3D reconstruction methods on the MTF dataset, showing predicted volumes, percentage errors, and Chamfer Distance for VolETA [2], ININ [13], FoodR. [13], VoIE [12], and ours method.

ID	Scene Name	Predicted Volume					GT	Error Percentage ↓					Chamfer Distance ↓				
		VolETA	ININ	FoodR.	VoIE	Our		VolETA	ININ	FoodR.	VoIE	Our	VolETA	ININ	FoodR.	VoIE	Our
1	Strawberry	40.06	37.65	44.51	37.47	37.57	38.53	3.97	**2.28**	15.52	2.74	2.49	0.0016	0.0020	**0.0011**	0.0028	0.0021
2	Cinnamon bun	216.90	325.44	321.26	275.38	276.57	280.36	22.64	16.08	14.59	1.78	**1.35**	0.0071	0.0036	0.0031	**0.0022**	0.0023
3	Pork rib	278.86	473.40	336.11	268.93	264.84	249.65	11.70	89.63	34.63	7.72	**6.08**	0.0137	**0.0049**	0.0053	0.0068	0.0063
4	Corn	279.02	294.32	347.54	277.56	276.66	295.13	5.46	**0.27**	17.76	5.95	6.26	0.0020	0.0038	**0.0015**	0.0046	0.0043
5	French toast	395.76	353.66	389.28	394.04	390.54	392.58	0.81	9.91	0.84	**0.37**	0.52	0.0137	**0.0020**	0.0040	0.0021	0.0028
6	Sandwich	205.17	237.88	197.82	215.21	216.13	218.31	6.02	8.96	9.39	1.42	**1.06**	0.0067	0.0038	**0.0025**	0.0039	0.0040
7	Burger	372.93	361.49	412.52	370.69	366.80	368.77	1.13	1.97	11.86	**0.52**	0.54	0.0047	0.0048	**0.0025**	0.0036	0.0039
8	Cake	186.62	172.32	181.21	176.43	171.56	173.13	7.79	**0.47**	4.67	1.91	0.91	0.0030	0.0019	**0.0010**	0.0012	0.0017
9	Blueberry muffin	224.08	253.01	233.79	233.95	230.98	232.74	3.72	8.71	**0.45**	0.52	0.75	0.0039	0.0029	0.0033	0.0029	**0.0027**
10	Banana	153.76	157.58	160.06	159.20	155.60	163.23	5.80	3.46	**1.94**	2.47	4.59	0.0027	0.0034	**0.0019**	0.0118	0.0081
11	Salmon	80.40	76.46	86.00	82.75	83.98	85.18	5.61	10.24	**0.96**	2.85	1.41	0.0034	**0.0015**	0.0015	0.0021	0.0022
13	Burrito	363.99	246.60	334.70	297.00	298.79	308.28	18.07	20.01	8.57	3.63	**3.08**	0.0052	**0.0026**	0.0041	0.0055	0.0052
14	Hotdog	535.44	495.10	517.75	541.58	544.68	589.82	9.22	16.06	12.22	8.18	**7.66**	0.0043	0.0044	0.0046	0.0082	0.0090
	MAPE ↓	7.84	14.47	10.26	3.08	**2.82**	**S.D. ↓**	6.36	23.47	9.48	2.63	**2.61**					
	CD (Sum) ↓												0.0720	0.0416	**0.0364**	0.0576	0.0547
	CD (Mean) ↓												0.0055	0.0032	**0.0028**	0.0044	0.0042

5   Conclusions

This paper introduces an advanced framework designed to enhance food volume estimation through a text-guided 3D reconstruction pipeline. Users can isolate target food items using text prompts, leveraging DEVA for accurate segmentation even in complex scenes. The framework also includes camera pose refinement to generate real-world scaled 3D meshes, eliminating the need for physical reference objects. Validation using the MTF and our FoodKit datasets demonstrates that VolE++ outperforms existing methods, effectively handling a diverse range of food shapes and textures. However, it does encounter challenges related to computational intensity, which affects real-time performance on edge devices, as well as occasional segmentation errors. Future work will focus on optimising the computational pipeline for mobile use and enhancing the robustness of segmentation, further establishing VolE++ as a valuable tool for dietary assessment.

Acknowledgment. This work was partially funded by the EU project MUSAE (No. 01070421), 2021-SGR-01094 (AGAUR), Icrea Academia'2022 (Generalitat de Catalunya), Robo STEAM (2022-1-BG01-KA220-VET000089434, Erasmus+ EU), DeepSense (ACE053/22/000029, ACCIÓ), CERCA Programme/Generalitat de Catalunya, and Grants PID2022141566NB-I00 (IDEATE), PDC2022-133642-I00 (DeepFoodVol), and CNS2022-135480 (A-BMC) funded by MICIU/AEI/10.13039/501100011033, by FEDER (UE), and by European Union NextGenerationEU/PRTR. A. AlMughrabi acknowledges the support of FPI Becas, MICINN, Spain. U. Haroon acknowledges the support of FI-SDUR Becas, MICINN, Spain.

References

1. AlMughrabi, A., Galán, A., Marques, R., Radeva, P.: Foodmem: near real-time and precise food video segmentation. arXiv preprint arXiv:2407.12121 (2024)
2. AlMughrabi, A., Haroon, U., Marques, R., Radeva, P.: Voleta: one-and few-shot food volume estimation. arXiv preprint arXiv:2407.01717 (2024)
3. Apple Developer: Arkit (2024). https://developer.apple.com/augmented-reality/arkit/. Accessed 10 Dec 2024
4. Barrow, H., Tenenbaum, J., Bolles, R., Wolf, H.: Parametric correspondence and chamfer matching: two new techniques for image matching. In: Proceedings of Image Understanding Workshop, pp. 21–27. Science Applications (1977)
5. Botsch, M., Kobbelt, L., Pauly, M., Alliez, P., Lévy, B.: Polygon Mesh Processing. CRC Press (2010)
6. Cernea, D.: Openmvs: open multiple view stereovision (2015). https://github.com/cdcseacave/openMVS/
7. Cheng, H.K., Oh, S.W., Price, B., Schwing, A., Lee, J.Y.: Tracking anything with decoupled video segmentation. In: CVPR, pp. 1316–1326 (2023)
8. Dehais, J., Anthimopoulos, M., Shevchik, S., Mougiakakou, S.: Two-view 3D reconstruction for food volume estimation. IEEE Trans. Multimedia **19**(5) (2016)
9. Ferdinand Christ, P., Schlecht, S., Ettlinger, e.: Diabetes60-inferring bread units from food images using fully convolutional neural networks. In: CVPR (2017)

10. Google Developers: Arcore overview (2024). https://developers.google.com/ar/develop/. Accessed 10 Dec 2024

11. Haroon, U., AlMughrabi, A., Marques, R., Radeva, P.: Mvsboost: an efficient point cloud-based 3D reconstruction. arXiv preprint arXiv:2406.13515 (2024)

12. Haroon, U., AlMughrabi, A., Zoumpekas, T., Marques, R., Radeva, P.: Vole: a point-cloud framework for food 3D reconstruction and volume estimation. arXiv:2505.10205 (2025)

13. He, J., et al.: Metafood CVPR 2024 challenge on physically informed 3D food reconstruction: methods and results. arXiv:2407.09285 (2024)

14. Kerbl, B., Kopanas, G., Leimkühler, T., Drettakis, G.: 3D gaussian splatting for real-time radiance field rendering. ACM Trans. Graph. **42**(4), 139-1 (2023)

15. Lowe, D.G.: Distinctive image features from scale-invariant keypoints. Int. J. Comput. Vision **60**, 91–110 (2004)

16. Mildenhall, B., Srinivasan, P.P., et al.: Nerf: representing scenes as neural radiance fields for view synthesis. Commun. ACM **65**(1), 99–106 (2021)

17. Müller, T., Evans, A., Schied, C., Keller, A.: Instant neural graphics primitives with a multiresolution hash encoding. ACM Trans. Graph. **41**(4) (2022)

18. O'Rourke, J.: Computational Geometry in C. Cambridge University Press (1998)

19. Schonberger, J.L., Frahm, J.M.: Structure-from-motion revisited. In: CVPR, pp. 4104–4113 (2016)

20. Tahir, G.A., Loo, C.K.: A comprehensive survey of image-based food recognition and volume estimation methods for dietary assessment. In: Healthcare, vol. 9 (2021)

21. Thames, Q., Karpur, A., Norris, W., Xia, F., et.al.: Nutrition5k: towards automatic nutritional understanding of generic food. In: CVPR, pp. 8903–8911 (2021)

22. Wang, P., Liu, L., Liu, Y., et al.: Neus: learning neural implicit surfaces by volume rendering for multi-view reconstruction. arXiv:2106.10689 (2021)

23. Wang, Y., Han, Q., et al.: Neus2: fast learning of neural implicit surfaces for multi-view reconstruction. In: CVPR, pp. 3295–3306 (2023)

24. Xu, C., He, Y., Khannan, N., Parra, A., Boushey, C., Delp, E.: Image-based food volume estimation. In: Proceedings of MADIMA, pp. 75–80 (2013)

Correction to: Computer Analysis of Images and Patterns

Modesto Castrillón-Santana⬤, Carlos M. Travieso-González⬤,
Oscar Deniz Suarez⬤, David Freire-Obregón⬤, Daniel Hernández-Sosa⬤,
Javier Lorenzo-Navarro⬤, and Oliverio J. Santana⬤

Correction to:
M. Castrillón-Santana et al. (Eds.): *Computer Analysis of Images and Patterns*, **LNCS 15621, https://doi.org/10.1007/978-3-032-04968-1**

In the originally published version of chapter 28 and 32, one of the author's name was incorrect. It was Grigorios and not Grigogios. This has been corrected

---

The updated version of these chapters can be found at
https://doi.org/10.1007/978-3-032-04968-1_28
https://doi.org/10.1007/978-3-032-04968-1_32

Author Index

M. Castrillón-Santana et al. (Eds.): CAIP 2025, LNCS 15621, pp. 399–401, 2026.
https://doi.org/10.1007/978-3-032-04968-1

The manufacturer's authorised representative in the EU is Springer
Nature Customer Service Centre GmbH, Europaplatz 3, 69115 Heidelberg,
Germany. If you have any concerns regarding our products, please
contact ProductSafety@springernature.com

Printed and bound by CPI Group (UK) Ltd, Croydon, CR0 4YY
28/04/2026
02098524-0006